Understanding Climate Change through Gender Relations

T0341312

This book explains how gender, as a power relationship, influences climate change related strategies, and explores the additional pressures that climate change brings to uneven gender relations. It considers the ways in which men and women experience the impacts of these in different economic contexts. The chapters dismantle gender inequality and injustice through a critical appraisal of vulnerability and relative privilege within genders. Part I addresses conceptual frameworks and international themes concerning climate change and gender, and explores emerging ideas concerning the reification of gender relations in climate change policy. Part II offers a wide range of case studies from the Global North and the Global South to illustrate and explain the limitations to gender-blind climate change strategies.

This book will be of interest to students, scholars, practitioners and policymakers interested in climate change, environmental science, geography, politics and gender studies.

Susan Buckingham works as an independent researcher, writer and consultant on gender-environment issues. She has recently published a four-volume anthology on gender and environment and has edited five other books on environmental issues. She is currently working on the second edition of *Gender and Environment* (2000), which has been a key text in this area in Europe, North America, Australia and New Zealand. She is currently the gender consultant to the EU research programme 'URBAN-WASTE'. Susan is also a yoga practitioner and teacher and writes on yoga in research.

Virginie Le Masson is a Research Fellow at the Overseas Development Institute (ODI), UK. Her research interests combine social inclusion, disaster risk reduction and climate change mitigation and adaptation. Her research also looks at the sustainable development of mountain communities. Before joining ODI, Virginie worked with the French Red Cross disaster risk management programme in the Indian Ocean, and with a study abroad programme on climate change and the politics of food, water and energy.

Routledge Studies in Hazards, Disaster Risk and Climate Change
Series Editor: Ilan Kelman,
*Reader in Risk, Resilience and Global Health at the Institute
for Risk and Disaster Reduction (IRDR) and the Institute for
Global Health (IGH), University College London (UCL).*

This series provides a forum for original and vibrant research. It offers contributions from each of these communities as well as innovative titles that examine the links between hazards, disasters and climate change, to bring these schools of thought closer together. This series promotes interdisciplinary scholarly work that is empirically and theoretically informed, with titles reflecting the wealth of research being undertaken in these diverse and exciting fields.

Published:

Cultures and Disasters
Understanding Cultural Framings in Disaster Risk Reduction
Edited by Fred Krüger, Greg Bankoff, Terry Cannon, Benedikt Orlowski and E. Lisa F. Schipper

Recovery from Disasters
Ian Davis and David Alexander

Men, Masculinities and Disaster
Edited by Elaine Enarson and Bob Pease

Unravelling the Fukushima Disaster
Edited by Mitsuo Yamakawa and Daisaku Yamamoto

Rebuilding Fukushima
Edited by Mitsuo Yamakawa and Daisaku Yamamoto

Climate Hazard Crises in Asian Societies and Environments
Edited by Troy Sternberg

The Institutionalisation of Disaster Risk Reduction
South Africa and Neoliberal Governmentality
Gideon van Riet

Understanding Climate Change through Gender Relations
Edited by Susan Buckingham and Virginie Le Masson

Climate Change and Urban Settlements
Mahendra Sethi

Understanding Climate Change through Gender Relations

Edited by Susan Buckingham and
Virginie Le Masson

Routledge
Taylor & Francis Group

LONDON AND NEW YORK

First published 2017 by Routledge

2 Park Square, Milton Park, Abingdon, Oxfordshire OX14 4RN
52 Vanderbilt Avenue, New York, NY 10017

Routledge is an imprint of the Taylor & Francis Group, an informa business

First issued in paperback 2018

British Library Cataloguing in Publication Data
A catalogue record for this book is available from the British Library

Library of Congress Cataloguing in Publication Data
Names: Buckingham, Susan, 1953– editor. | Masson, Virginie Le, editor.
Title: Understanding climate change through gender relations /
edited by Susan Buckingham and Virginie Le Masson.
Description: Abingdon, Oxon; New York, NY : Routledge, 2017. |
Series: Routledge studies in hazards, disaster risk, and climate change |
Includes bibliographical references and index.
Identifiers: LCCN 2016048350 | ISBN 9781138957671 (hbk) |
ISBN 9781315661605 (ebk)
Subjects: LCSH: Climatic changes–Social aspects. | Women and the
environment. | Women–Social conditions.
Classification: LCC QC903.U47 2017 | DDC 304.2/5–dc23
LC record available at https://lccn.loc.gov/2016048350

ISBN: 978-1-138-95767-1 (hbk)
ISBN: 978-0-367-21888-1 (pbk)

Typeset in Times New Roman
by Out of House Publishing

From Virginie, à Mamic, whose everyday practice of saving water, food and energy, influenced by her experience of the war, are both an inspiration and a reminder of our privileges.

From Susan, to Charlotte, Jack and James: representing the future generation, with hope that they can enjoy and contribute to a future in which all women and men are equally valued, and in which nature is respected.

Contents

Figures

Tables

Contributors

Gotelind Alber is an independent researcher and advisor on sustainable energy and climate change policy with a special focus on gender issues, climate justice and multi-level governance. She is co-founder and board member of GenderCC, and has served several years as focal point of the Women and Gender UNFCCC observer constituency. She holds an advanced university degree in physics and has almost 30 years of working experience in research, policy and management, among others as managing director of the Climate Alliance of European Cities. Recent projects include the preparation of a guidebook on the integration of the gender dimensions in urban climate policy; research on the role of cooperatives for mitigation at the local level; development of a learning platform on gender and low-carbon development; project and programme evaluation in the area of sustainable energy; advice on multi-level governance and climate policy; and capacity-building and training in the fields of gender, climate change and sustainable energy.

Beth A. Bee is currently an Assistant Professor at East Carolina University, USA. Her work explores the theoretical and empirical intersections between climate change, feminist theory, and rural livelihoods in Mexico. Her previous work investigated how the production of knowledge and gendered relations of power shape adaptation and decision-making adverse economic and ecological changes, such as El Niño induced droughts, increased male migration, and neoliberal agricultural policies. Currently, she is examining the multi-scalar politics of, and expertise in, forestry projects that comprise Mexico's Reducing Emissions, Deforestation, and Degradation (REDD+) early-action activities. Beth holds a PhD in Geography and Women's Studies from Pennsylvania State University.

Susan Buckingham has worked in UK universities for twenty-five years, latterly as a Professor in the Centre for Human Geography at Brunel University, and now works independently as a researcher, writer and consultant. She has recently published a four volume anthology on gender and environment and has edited five other books on environmental issues. She is currently working on the second edition of *Gender and Environment* (2000), which has been a key text in this area in Europe, North America, Australia and New Zealand. Susan has been the convenor of the 'Gender, Energy and Climate Change' programme in the EU COST network 'genderSTE' (2012–2016), including advising the EU on gender sensitivity in its research programme Horizon 2020. She is currently the gender consultant to the EU research programme 'URBAN-WASTE'. Susan is also a yoga practitioner and teacher and writes on yoga in research.

Kate Cahoon completed a Master of Arts (Political Science) at the University of Potsdam, Germany, and now works as a project coordinator at GenderCC – Women for Climate Justice. For the past five years, she has regularly attended the international climate negotiations and participated actively in the work and advocacy of the Women and Gender Constituency, one of the nine official observer groups within the UNFCCC. Her master's thesis dealt with the use of human rights language and narratives in the context of the UNFCCC, yet she has also co-authored a number of articles on urban climate policy and gender and is interested in feminist theory, intersectionality and social movements.

Doris Damyanovic is Assistant Professor of Landscape Planning at the Institute of Landscape Planning, Department of Landscape, Spatial and Infrastructure Sciences, University of Natural Resources and Life Sciences (BOKU), Vienna. She studied landscape planning at BOKU Vienna and at the University of Wageningen. Her research focuses on landscape planning and gender issues in urban and rural areas.

Candela de la Sota is a chemical engineer with an MSc in Technology for Human Development and Cooperation. She is researcher at the Innovation and Technology for Development Centre at the Technical University of Madrid (itdUPM), mainly working in energy, environment and human development. She is PhD Candidate in Environmental Engineering and her thesis deals with the study of the potential mitigation of climate change and contribution to sustainable human development of improved stoves in poor countries. She also has experience in evaluation and supporting the development of intervention strategies in renewable energies, improved stoves and cleaner fuels.

Christian Dymén holds a PhD degree in urban and regional studies and a MSc degree in engineering. He has developed an expertise within governance and planning of sustainable urban and regional development with a special emphasis on gender issues. He has led several projects with focus on climate change, planning and gender in a European and Nordic context. Especially noteworthy is the project NORDLEAD – Making Nordic Cities Successful Climate Leaders (funded by the Nordic Council of Ministers), where he was leading the work of Nordregio in developing a framework to analyse govern-ance structures, triggering factors, success factors, challenges, attitudes, needs and gaps in relation to local climate change responses. Between 2008 and 2013 he participated as a researcher in the project 'Another Climate: Gendered Structures of Climate Change Response in Selected Swedish Municipalities,' funded by the Swedish Research Council (Vetenskapsrådet), The Swedish National Space Board (Rymdstyrelsen) and the Swedish Research Council (FORMAS). He also has considerable experience from projects within the European ESPON programme, such as the projects ESPON TIPSE (Territorial Dimension of Poverty and Social Exclusion in Europe) and ESPON DeTeC (Detecting Territorial Potentials and Challenges). In 2010–2012, he participated as a Nordic expert in an OECD project on renewable energy and rural development. He is a member of the Council of the Baltic Sea States Expert Group on Sustainable Development as well as Swedish member in the European network EU COST genderSTE network. He has previously worked as a planning practitioner and project manager at the municipal and regional level in Stockholm.

Luz Fernández holds a Bachelor of Environmental Sciences, an MSc in Environmental Engineering and Management and an MA in Environment and Development. She was awarded a PhD in Environmental Engineering at the Technical University of Madrid (UPM) in 2014 with a thesis focused on the pro-poor co-benefits of the carbon markets. She has more than ten years of professional experience as a project manager, consult-ant and researcher working with socio-environmental and climate change projects at different levels and with different donors in Europe and in Latin American. Currently, she is consultant in climate finance tracking and reporting at the Inter-American Development Bank.

Britta Fuchs studied Landscape Planning and Landscape Management at the University of Natural Resources and Life Sciences in Vienna (BOKU Vienna). After finishing her PhD she worked in the project management of the restoration of a historic park in London, UK. Since 2009 she has been affiliated with the Institute of Landscape Planning at the BOKU Vienna. Her research and teaching covers landscape and open space planning with a recent focus on climate change and climate change adaptation.

Noémi Gonda holds a PhD from the Department of Environmental Sciences and Policy of Central European University, Budapest, Hungary. Her PhD research was a feminist ethnography focused on the politics of gen-der and climate change adaptation in rural Nicaragua. She has a Master of Science in tropical agriculture and development from Montpellier SupAgro (France) and two Masters of Engineering in agricultural tech-niques and tropical agriculture from Bordeaux Sciences Agro (France) and Montpellier SupAgro respectively. She has extensive professional experience in rural development and natural resources management mainly in Central America where she has been working between 2002 and 2010 with smallholder farmers, indigenous groups and international organisations. She has contributed to research on small-scale farming, land rights, gender equality and rural women's empowerment. Her recent work on gender and climate change has been published in the journal *Gender, Technology and Development* and by the UNDP in Nicaragua.

Alex Haynes is CEO of Whittlesea Community Connections, a community services organisation in Melbourne, Australia. She holds a Bachelor of Architecture, Graduate Diploma Environmental Studies, and is completing her PhD (due 2017) through the Department of Social Work at Monash University. Her research includes an Oxfam–Monash partnership project in Bangladesh 'Gendered impacts of climate variability and slow onset events' (2011–2014), 'Standing Together Against Violence' in Solomon Islands, Melanesia (2014) and she is currently researching with refugees and asylum seekers in Melbourne to investigate employment barriers and social connection.

Martin Hultman is an associate professor and works as lecturer at Linköping University, Sweden. As a scholar and citizen he is involved in local environmental politics in forms of ecotourism, ridesharing, rights of nature and public seminars, also being coordinator of Environmental PostHumanities and SweMineTech research networks. He has edited special issues about environment in the leading gender journal TGV in Sweden as well as on ecopreneurship in *Small Business Journal*. Publications include 'The Making of an Environmental Hero: A History of Ecomodern Masculinity, Fuel Cells and Arnold Schwarzenegger' (2013), and with Jonas Anshelm in 2014, *Discourses of Global Climate Change* and 'A Green Fatwā? Climate change as a threat to the masculinity of industrial modernity'. His current research revolves around issues such as posthumanities ethics, ecofeminism, environmental utopias and ecopreneurship. At the moment he is writing a book together with Paul Pulé regarding Ecological Masculinities.

Annica Kronsell is a Professor and teaches International Relations and Gender, Feminist theory and Environmental and Climate Politics at the Department of Political Science at Lund University, Sweden. Recent publications include: *Rethinking the Green State. Environmental Governance Towards Environmental and Sustainability Transitions* (2015) with Karin Bäckstrand; 'The (In)visibility of Gender in Scandinavian Climate Policymaking', *International Feminist Journal of Politics*, with Gunnhildur Magnusdottir (2014); and with Anna Kaijser 'Climate Change through the lens of intersectionality', *Environmental Politics* (2013).

Richard Langlais is a PhD, and the founder and chairman of AREL Scientific AB, a provider of research and language consultation services, based in Lund, Sweden. In over thirty years of research, he has worked with integrating numerous fields into an overarching approach to sustainable development that includes gender studies and environmental philosophy, security studies, human ecology and climate change, from the circumpolar Arctic and North America, to Europe and East Asia.

Virginie Le Masson is a research officer working at the Overseas Development Institute (ODI) in London. With a PhD in human geography, her research interests combine social inclusion and gender aspects of environmental issues including disaster risk reduction and climate change mitigation and adaptation. She recently worked on the integration of gender within emergency projects in the water and sanitation sector in the Central African Republic. Before joining ODI, Virginie collaborated with local NGOs in the Philippines on participatory mapping projects. She also worked with the French Red Cross disaster risk management programme in Reunion Island, Indian Ocean. Overall, her approach to research focuses on qualitative and participatory methods of data collection with a strong emphasis on concepts of social vulnerability and capacities.

Julio Lumbreras is a chemical and environmental engineer. He received his PhD on Air Quality from the Technical University of Madrid (UPM) in 2003. Currently, he is Associate Professor at UPM, and an MC/MPA candidate at Harvard Kennedy School and the Spanish representative at the Task Force on Integrated Assessment Modelling within the UN Convention on Long-Range Transboundary Air Pollution (CLRTAP). He is co-author of more than 50 papers and book chapters and has participated in many international workshops, conferences, and research works on urban air quality, climate change and sustainable human development, including clean cookstoves projects.

Sherilyn MacGregor received her PhD from York University, Canada, and is now Reader in Environmental Politics, jointly appointed to the Sustainable Consumption Institute and the Politics Department at the University of Manchester, UK. Her research examines the intersections of gender politics and environmental politics, with a special focus on the theoretical and policy connections between gender equality, ecological sustainability, and democratic citizenship. She has written extensively on gender and environment topics and is editor of the journal *Environmental Politics* and the *Routledge Handbook of Gender and Environment* (forthcoming).

Javier Mazorra is a chemical engineer with an MSc in Technology for Human Development and Cooperation. He is a research fellow and member of the technical team of the Innovation and Technology for Development Centre (itdUPM), and a PhD Candidate, in Chemical Engineering at the Technical University of Madrid (UPM). His main research area is the assessment of impacts and effects of technological projects on developing countries, with special focus on energy access through improved cookstoves and its linkages with climate change and gender.

Angela Moriggi is a Researcher at the Natural Resources Institute LUKE, Finland, and PhD Fellow of the EU Marie Curie ITN programme SUSPLACE. Her research is aimed at revealing and enhancing the role of 'green care' interventions in fostering sustainable place-shaping. Previously, as a Research Fellow for Ca' Foscari University of Venice (2013–2016), her work focused mainly on the gender dimension of climate change policies and on public participation in environmental decision-making in the People's Republic of China. She has been Project Manager of two EU FP-7 Marie Curie IRSES projects GLOCOM (Global Partners in Contaminated Land Management) and EPSEI (Evaluating Policies for Sustainable Energy Investments), respectively for Ca' Foscari University of Venice and for the Euro Mediterranean Center for Climate Change (CMCC). She has spent extended periods of time in China, most recently as Marie Curie IRSES Fellow at the Chinese Research Academy of Environmental Sciences (CRAES) and Beijing Normal University (BNU). She holds an MA in International and Diplomatic Sciences and a BA in Oriental History, Cultures and Civilization from Bologna University.

Karen Morrow is a Professor in Environmental Law at Swansea University, UK. She is the founding co-editor of the *Journal of Human Rights and the Environment* and has published widely on public participation in environmental law with a particular emphasis on gender. Her publications include 'Perspectives on Environmental Law and the Law Relating to Sustainability: A Continuing Role for Ecofeminism?', in A. Philippopoulos-Mihalopoulos (ed.), *Law and Ecology: New Environmental Foundations* (2011); 'Ecofeminism and the environment: international law and climate change', in M. Davies and V. Munro (eds), *The Ashgate Companion to Feminist Legal Theory* (2013); 'Procedural Human Rights in International Law', in A. Grear and L. Kotze (eds), *Research Handbook on Human Rights and the Environment* (2015); and 'Climate Change, Major Groups and the Importance of a Seat at the Table: Women and the UNFCC Negotiations', in J. Penca and C. de Andrade (eds), *The Dominance of Climate Change in Environmental Law: Taking Stock for Rio +20* (2012): 26–36 reprinted in Susan Buckingham (ed.), *Gender and the Environment* (2015).

Patricia E. (Ellie) Perkins is Professor in the Faculty of Environmental Studies, York University, Canada, where she teaches and advises students in the areas of ecological economics, community organising, and critical interdisciplinary research design. Her research focuses on feminist ecological economics, climate justice, and participatory commons governance. She directed international research projects on community-based environmental and watershed education in Brazil and Canada (2002–2008) and on climate justice and equity in watershed management with partners in Mozambique, South Africa and Kenya (2010–2012) and is the editor of *Water and Climate Change in Africa: Challenges and Community Initiatives* *in Durban, Maputo and Nairobi* (2013). Previously, she taught economics at Eduardo Mondlane University in Maputo, Mozambique, and served as an environmental policy advisor with the Ontario government. She holds a PhD in economics from the University of Toronto.

Florian Reinwald is researcher at the Institute of Landscape Planning, Department of Landscape, Spatial and Infrastructure Sciences, BOKU Vienna. He studied Landscape Planning at the BOKU Vienna and TU Munich/Weihenstephan. The focus of his research work lies in settlement and community development, planning instruments and planning processes, village renewal and participation processes including new technologies. He has a special research interest in gender and justice regarding participation, spatial and urban planning and development as well as in the context of climate change adaption and mitigation strategies.

Ulrike Röhr is an engineer and sociologist by background and has been working on gender issues in planning, sustainable development, environment, and especially in energy and climate policy for about thirty years. Recently she has focused on promoting the mainstreaming of gender into climate policy on local and national levels. She has been involved in gendering the UNFCCC process since the very beginning and is co-founder of the global network GenderCC – Women for Climate Justice. In Germany, she is the head of genanet – focal point gender, environment, sustainability which is aiming to support gender mainstreaming in environmental policy by providing information, hosting a network of gender and environment experts, advising environment organisations and ministries, and carrying out research.

Lidewij Tummers has been a part-time tutor and researcher since 2006 in the Chair of Spatial Planning and Strategy Faculty of Architecture TU Delft, where she is currently writing a dissertation on Co-housing and Energy Transition. Central research themes are: the position of citizens in planning systems, spatial criteria for inclusive design, decentralised renewable energy networks, participatory design processes and a gendered perspective of spatial planning. She has published a number of papers and chapters on Gender and Planning as well as on co-housing, and is founding member of international networks of planning experts such as the European network of co-housing researchers and the European network Gender, Diversity and Urban Sustainability (GDUS). International collaboration includes guest professorship at Leibniz University Hannover (2009–2010, 2013); Le Studium Research fellow at Université François Rabelais in Tours (2011–2012) and gender-STE short stay grant at Brunel University London (2014). Tummers acts Europe-wide at consultancies on gender-aware planning for municipal and regional Departments for Urban Development. In her hometown of Rotterdam, she is an independent technical consultant for grass-roots housing coops and renewable energy initiatives.

Karin Weber studied Landscape Planning at the University of Natural Resources and Life Sciences (BOKU), Vienna. In her master's thesis she focused on gender aspects of natural disasters and its linkage to local planning instruments. Currently she is working on a project researching capacity-building in DRR for people with migration backgrounds in Austria.

Acknowledgements

The concept for this book emerged from the COST Action Network (TN1201) 'genderSTE' which was established, among other things, to embed an understanding of gender relations across a number of subject areas important to the European Union research agenda. Of the areas covered it is energy and climate change that are the focus of this book. (The other areas were transport and cities.) Our first thanks and acknowledgement, therefore, goes to Professor Ines Sanchez de Madariaga, as the instigator and director of the network, and Professor Marion Roberts, who was Susan's co-chair of the working group in which energy and climate change sat. Most of the chapters that we have included first saw the light of day as conference presentations at the 2014 genderSTE conference in Rome or at the session, which shared the title of this book, at the 2014 Royal Geographical Society/Institute of British Geographers Conference in London. Once the structure of the book was confirmed, we commissioned two further chapters, versions of which had been presented at a 'Disciplining Gender' session, convened by Dr Seema Arora-Jonsson, and part of the 'Undisciplined Environment', International Conference of European Network of Political Ecology in Stockholm in March, 2016. We would like to thank all the participants at these conference sessions for their constructive discussions, and the contributing authors for sharing their research findings, for responding to our comments patiently, and for delivering their chapters on time.

The consolidation of this book has also been a great opportunity to include the work of early career scholars and practitioners from different contexts and combine their findings with the perspectives of established experts in the domain of gender and the environment. We hope that this approach reflects how much we value and have enjoyed the collaboration between researchers coming from diverse points of view who share the same dedication to promote equality.

1 Introduction

Susan Buckingham and
Virginie Le Masson

Since the Kyoto Protocol was adopted in 1997, requiring wealthy countries to reduce their greenhouse gas (GHG) emissions to an average of five per cent against 1990 levels, only a handful of these so-called Annex 1 countries have met or exceeded the target.[1] The EU reduced its emission of GHG by a mere 0.89 per cent, despite its much vaunted environmental credentials, although this masks the achievements of some member states which have reduced GHG emissions by up to 13 per cent.[2] Meanwhile, the USA reduced its GHG emissions by 3.12 per cent and Australia by 0.01 per cent, while Russia increased its emissions by 0.53 per cent, Japan by 3.27 per cent and New Zealand by 3.47 per cent (World Resources Institute Interactive, 2016). This failure on the part of wealthy countries to meet their reduction target, agreed on the basis that they had already benefited from exploiting their own carbon-based resources to achieve global economic advantage, has failed to compensate for rising GHG emissions in fast-growing economies (such as China, Brazil, India, Indonesia, Malaysia). This does not augur well for the less than ambitious targets agreed for the Paris Agreement in December 2015 whereby, as Javier Mazorra and colleagues discuss in more detail, parties agreed to

> 'i) hold the increase in the global average temperature to well below 2°C while pursuing efforts to limit it to 1.5°C above pre-industrial levels; ii) increase the ability to adapt to climate change, foster climate resilience and lower greenhouse gas (GHG) emissions, without threatening food production and iii) make finance flows consistent with the previous statements.
>
> (UNFCCC, 2015: 2)

The agreement also recognises the need to support developing countries in their mitigation efforts as well as to meet the costs of adaptation (UNFCCC, 2015), however, the terms of implementation and financial support allocated to adaptation is still subject to debate between parties. At the time of writing, 61 out of 197 Parties to the Convention have ratified the agreement, accounting for 47.79 per cent of global GHG emissions.[3] The longer the implementation of the policy framework will take, the closer to the threshold of 1.5 degree warming the world will get. The consequences of a global increase of temperatures for ecosystems and human societies are as severe as they are unfair. The impacts of climate change already affect regions and people who have contributed the least to environmental degradation and carbon emissions (Garvey, 2010).

Human induced climate change is a contentious enough topic to research when approaching it with a general social development perspective. To increase the understanding of the production and consequences of climate change with attention to (in)equality, inclusion, marginalisation and rights, inevitably leads to an examination of pathways of unsustainable development. Increasing the evidence base of environmental degradation associated with industrialisation and economic growth, supports people who are primarily affected by changing climate patterns to advocate for climate change mitigation and adaptation and also questions current models of development that exploit and pollute the environment. As a consequence, social research on climate change might not serve the agenda of stakeholders who control dominant and carbon intensive ways of producing and trading energy, food and industrialised goods. To integrate a gender analysis to this research appears to have alienated these stakeholders still further, while organisations who might support social reform are sometimes so immured in what we might arguably call patriarchal systems of power that they are unable to see the need for a critical gender analysis of climate change (Buckingham and Kulcur, 2017).

Two major United Nations conferences in the 1990s – on sustainable development (UNCED in 1992) and on women (the Beijing conference in 1995) – agreed that it was critical for the sake of both that henceforth environmental degradation and women's inequality be considered in a coordinated way. However, international climate change negotiations and the agreements that have emerged from these remain largely untouched by gender concerns. It has fallen to women to push for gender considerations to inform climate change decision-making, and despite continuous pressure from women's groups, it was only in 2011 that the annual international climate change meeting – COP (Conference of the Parties) – agreed to do so. However, perhaps because it was women's groups that had been pressing for this inclusion, and perhaps because of the low level of comprehension of what constitutes gender among decision makers, there is a distinct air of additionality in this broadening of consideration. For example, the COP now includes a 'Women's Day' as part of its annual conference, which may excuse participants from considering women – and gender – in the rest of the conference. If gender, and gender equality, is to be a meaningful policy objective, it must be recognised that it comprises relations between women and men, and between and among different groups of women and men, not to mention between different conceptualisations of masculinity and femininity, which can each be practiced by either, and both, women and men.

The persistent failure to consider gender in all its complexity and subtlety, and the failure of the prevailing social and economic systems to achieve even the modest GHG reductions agreed under the Kyoto Protocol, has led us to conclude that there is room for a book which specifically explains how gender relations produce anthropogenic climate change, and how the world will fail to address climate change and its impacts until it understands and decides to address gender inequality. This inequality, as our chapter authors will argue, is an outcome of distributed power relations which are systematically gendered. All the contributors to this volume agree that understanding gender inequality also requires a simultaneous understanding of social and economic divisions based on class, ethnicity, age, disability, religion, sexuality, parenthood, among others, and how these divisions intersect to compound particular disadvantages and inequalities between and within social groups. There are some 'basics' underpinning intersectional gender relations that we will explain in the introduction which the reader who is just beginning to engage with gender will hopefully find helpful in getting the most out of the individual chapters, and which those more familiar with gender debates will probably skim or skip. We also review the literature and action which has pioneered thinking about gender in climate change. This has tended to focus either on women as victims and/or in the 'front line' of concern about climate change, or on women as decision makers who lack representation in discussions and policymaking about climate change, from the village council to the United Nations. From this, the reader will realise that the chapter authors are pushing this consideration further. The argument of the book, then, adds to the increasing macro-political-economy discussions towards de-growth and post-growth which are beginning to bring attention to the failure of prevailing political-economic structures and processes to address climate change as both a current and future threat (see e.g. Tim Jackson (2009) and Joseph Stiglitz (2010)). Further, it insists that reimagining existing social structures and processes start with the *sine qua non* of full and deep gender equality.

This book has two aims. First, to explore why there has been so little focus on understanding gender relations in relation to climate change causes, effects, mitigations and adaptations, and the implications of not adopting gender sensitivity in all these aspects. Second, to examine topical and regional case studies which illustrate and explain the limitations of climate change strategies emerging from a lack of sensitivity to gender relations. These case studies cover a range of geographical settings and scales, both in the Global North and the Global South and in both urban and rural areas in order to diversify the evidence base. The linkages between gender and climate change can be illustrated in many other ways than the typical experience of rural women walking long distance to fetch water in the Sahel. Differences in production and consumption patterns between cities and rural areas and between people according to social categories (class, age, ethnicity and gender) influence their income, their ways of working and travelling, their consumption choices over food and energy and their power over decision-making at every level (Hemmati, 2000). Where possible, examples in this volume suggest what a gendered approach to climate change-related strategies can offer whether in Sweden, Austria, China or Mexico. And in Ellie Perkins' chapter, she discusses the possibility of sharing information between the Global North and Global South through innovative student projects which reveal common understandings and similarities, as well as differences. This chapter and also Lidewij Tummers' chapter on co-housing suggest small-scale ways in which gender inequality and climate change can be addressed in synergetic and mutually supporting ways. Reflecting these two aims, the book is organised in two parts. The first part addresses conceptual frameworks and global and/or international themes concerning climate change and gender, while the second presents the cases studies from different global regions experiencing relative wealth and poverty.

Gender relations

Although 'gender' implies a relationship between male and female, feminism and masculinism, researchers and writers on gender and environmental issues, including climate change, have been overwhelmingly women. Writers from Carolyn Merchant (1980, 1996) to Joane Nagel (2015) have consistently argued that a particular form of masculinity is responsible for the parallel and related dominance of nature and women, but a more nuanced and plural consideration of masculinities and femininities has only recently been explored in depth. For example, Martin Hultman suggests in this volume that three types of masculinities have different impacts on the environment: industrial masculinity (that which is characterised as the historical dominant masculinity), ecological masculinity (in which men enjoy a sensitive relationship with nature, caring for its long-term sustainability), and a hybrid 'eco-modern' masculinity in which the proponent adopts an ecological modernity that provides superficial solutions but no long-term fundamental shifts. The example of eco-modern man which Hultman provides is Arnold Schwarzenegger who, as Governor of California, moderated his 'Terminator' SUV driving image with the championing of hydrogen power for motor vehicles (see also Hultman, 2013).

Nagel (2017) writes about how a particular understanding of nature, coupled with a desire to control it, leads to the development and use of science to solve problems which previous iterations of understanding and control have created. Climate change research in the US, Nagel explains, has emerged from a military research programme, and includes attempts to control climate through, for example, the geoengineering of weather systems. Angela Moriggi in Chapter 10 of this volume also describes how the national debate on climate change in China often fails to consider the 'human face' of climate change and how social aspects, and localised, small-scale manifestations of the issue, remain largely marginalised in favour of grand technological solutions and financial mechanisms to address climate change. The hyper-masculinity of such science overwhelms the increasing (though still relatively small) numbers of women being recruited so that its culture continues to be reproduced. However, more gender balance in quantity alone, without changes in culture, may not necessarily deliver more gender sensitive planning, as earlier discussion on 'critical mass' has suggested they might (see Childs and Krook (2006) for a review). Dymén and Langlais in this volume, note that in the Swedish context the recruitment of both male and female planners from an education system which is imbued with masculinist values appears to ensure their continuation (see also Magnusdottir and Kronsell, 2015). This clearly suggests that something more than an intrinsic 'maleness'/'femaleness' is at work in human/environment relations. The male/female gender binary is disrupted, however, by understanding our identities as multiple (Sandilands, 1999), privileging care work as productive (Merchant, 1996; and Alber, Cahoon and Röhr, in this volume), and adopting subaltern voices as alternative and more powerful 'standpoints' from which to appraise the related domination of women and nature (Shiva, 1989).

An attention to alternative standpoints draws attention to how different characteristics intersect to place different groups into more or less disadvantaged relations with nature, and hence climate change. Kimberlé Crenshaw (1989) is credited with being the first to write about the concept of intersectionality to explain that inequalities that beleaguer women and people of colour are not independent or mutually exclusive. While race and sex 'readily intersect' in the lives of women of colour, she argues, feminist and anti-racist movements tend to ignore that there are particular inequalities that affect women of colour, because of the intersection of their gender and ethnicity. Pertinent to her analysis is how she problematises demands for change as being effective only when the changes reflect the logic of the institutions they are challenging and that ' "[d]emands for change that do not reflect ... dominant ideology ... will probably be ineffective" Crenshaw, *supra* note 3, at 1367.' However, Crenshaw argues that despite these obstacles, using an 'intersectional sensibility ... should be a central theoretical and political objective of both antiracism and feminism' in order to confront the dominant ideology (Crenshaw, 1991:1243).

Gender analyses need to be more intersectional, in order to recognise that there are different ways in which different women and men are affected by and relate to climate change, mediated by power structures which they experience differently. In some cases this will involve power relations in which privileged women are complicit in the domination of poor women. And while women of colour and first nation/indigenous women are often campaigners for local environmental justice, this involvement is not well represented in the literature (Buckingham and Kulcur, 2009). Exceptions to this include work by Julie Sze (2005) on African American women's protests about health threatening traffic pollution in New York City, Dorceta Taylor on gender and minority ethnic bias in environmental governance in the USA (2014) and Cynthia Hamilton's account of minority women's protest against a solid waste incinerator in their

Los Angeles neighbourhood (1990). In a rare example of intersectional environmental justice analysis, Giovanna Di Chiro explicitly examines the 'intersectional, coalition politics forged by activists in US environmental justice and women's rights organisations' (2008: 276).

Using intersectionality as a tool, as Kaijser and Kronsell (2014) have done for analysing climate change, is, they argue, a way of revealing relationships which may previously have been regarded as natural, therefore paving a way for action to redress inequalities. In one of the few policy applications of intersectionality, the City of Vienna, which has led the way in planning and designing an urban infrastructure which is sensitive to women's and carers' needs (Irschik and Kail, 2013), is using the concept of 'gender+' in guiding socially and environmentally sustainable design and planning which facilitates care work, and the needs of children, older people, and those who are disabled (Damyanovic *et al.*, 2013).

Following the work of Donna Haraway (1991), divisions between male and female, human and animal, nature and technical, can no longer be seen as inevitably dualistic. Increasing attention to queer theory has also enriched our understanding of multiple gendered identities which are not necessarily easily categorisable, nor what they necessarily superficially appear to be (see Niamh Moore's description of grandmotherly protests against deforestation in Canada as examples of clowning and drag, 2017). Wendy Harcourt thinks about 'queer ecology [as] blurring the borders between humans and others (Harcourt *et al.*, 2015: 296); Tara Tabassi (ibid.: 291) wonders whether it can 'counter the powerful discourse of the reproductive binary, and bodies outside dualism such as the intersex or gender-non-conforming body'; and Greta Gaard advocates it as 'theorizing the ecological articulations of a diversity of genders and sexualities [as] a more strategic way to explore material dimensions of animal and ecological health' (2014: 231). Shiloh Kruper, with reference to 'Nuclia Waste's' drag-activist engagement with a nuclear site in Colorado, uses queer ecology to envisage humans as 'boundary creatures, neither fully human nor fully civilised' (2012: 317). Queer ecology, then, explores gender–nature relations more radically than does intersectionality, proposing the dissolution of a gender binary as a way of also unsettling the binary between human and non-human nature.

Gender and climate change policy

Given that Agenda 21, agreed at the same time as the UNFCCC, had a chapter explicitly requiring the inclusion of women and gender considerations in sustainable development planning, it is notable that it was not until 2011, almost 20 years after the United Nations Framework Convention on Climate Change (UNFCCC) was agreed in 1992, that gender considerations when appointing to international leadership positions or negotiating teams, let alone a systematic assessment of gender in the formation of climate change policies at the national level (Buckingham, 2010), were required. Karen Morrow, in Chapter 3, charts the complex nature of international climate change policy agreement, which involves the engagement of a range of stakeholders. However, she points to the lack of 'more obviously socially oriented/people centred aspects of participation into climate change governance' (p. 33). Despite this, Morrow, an environmental lawyer, argues that the UNFCCC's failure to engage with gender is 'difficult to justify' (ibid.). The Paris Agreement provides only timid recognition of human rights obligations, gender equality, the rights of indigenous peoples, migrants, children, persons with disabilities and people in vulnerable situations. Parties have agreed that adaptation and capacity-building should be gender-responsive which reflects the work of many international and civil society organisations (GGCA, WEDO, WECAN, Genre en Action, etc.) who have pushed for climate-related interventions to be more inclusive and integrate gender considerations.

However, 'women' as a binary category has dominated calls for climate change to become more gender sensitive. When this is as a call for more women to be involved in decision-making at different scales, it either refers to women's absence in different decision-making fora (WEDO, 2010) or calls attention to a specific 'feminine' quality that women can bring to decision-making, through their experiences (usually as carers of people and/or nature; e.g. UNDP, 2009). Early calls for greater gender sensitivity tended to portray women as helpless victims (Oxfam in Masika, 2002; and see Sherilyn MacGregor's discussion of feminist climate justice activists in Chapter 2 of this volume).

Even if countries say they have gender equality, that decision-making is 'balanced' numerically, and that countries are judged to be gender sensitive – it doesn't mean that they necessarily are. In this volume, Annica Kronsell (Chapter 7), and Christian Dymén and Richard Langlais (Chapter 15) discuss the continued masculinisation of climate change policy in Swedish municipalities, even when there are proportions of women policymakers that exceed what has been argued to be a 'critical mass', as referred to above. Virginie Le Masson (Chapter 13) interviews NGOs supporting Himalayan communities affected by climate change and finds that some women NGO leaders do not identify a need to consider gender. The main reason to explain this observation is due to the general belief among local communities and many practitioners that men and women already enjoy equality of rights and status, despite evidence that gender inequalities are persistent and that men and women hold different perspectives regarding climate change. Beth Bee, Chapter 12 in this volume, demonstrates how a country – Mexico – which has been nominated as the most gender friendly in REDD+[4] initiatives, also faces contradictory outcomes of promoting gender equality. The emphasis of Mexico's national REDD+ strategy (ENAREDD+) to increase women's access to credit and loans, still ignores their reproductive labour and potential contributions to decision-making within the programme. This framing ignores women's existing responsibilities and risks perpetuating the already uneven division of labour and workload between men and women. Bee argues that targeting women for credit-related projects 'is often based on essentialist assumptions about women's ability to empower their community, or notions of female solidarity that construct women as economically rational and homogenous groups, which does little to address the underlying cause of gender inequities' (pp 200–01). We need to be sure, then, what we mean by gender equality – and balance. For example Dorceta Taylor's research into the gender composition of environmental organisations in the United States shows that US environmental NGOs are promoting white women in order to tick the 'diversity' box, which is negatively impacting on non-white staff – especially women (2014). While contributing authors to this volume point out the lack of attention to gender in climate change assessments and climate change response, in Chapter 11, Noémi Gonda goes further to argue that climate change politics that already integrate gender concerns are not adequately informed by feminist research (with the exception of Lambrou and Piana's FAO report of 2006). In other words, while there exist national policies and NGO interventions that do pay attention to gender considerations, as Gonda shows, the objectives they pursue and the processes they use may not serve feminist purposes. Drawing on the case of Nicaragua, her analysis highlights how women have been given a primary place in climate change discourses and projects (through the discursive construction of nature as 'our own Mother'), while the mainstream masculinist and science-oriented discourse continues to underlie the way climate change adaptation interventions are conceived.

The repeated reference to the anecdotal, for example the widely cited fatal experiences of women and girls in the 1991 Bangladesh floods (Denton, 2002), tends to consolidate a version of women as helpless victims which contributors to this volume explore in more nuanced ways. In Chapter 9, Alex Haynes provides insights of how climate change and gender relations intersect and relate to everyday life in different rural districts in Bangladesh. Her findings stress that the recognition of power dynamics, particularly in households, and harmful social norms that maintain inequalities is crucial for the effectiveness of development and climate-related interventions. Britta Fuchs and colleagues (Chapter 16) also stress how the vulnerabilities of women of different age and class living in the Austrian Alps, lie not only in their unequal participation in risk-planning processes but also in the lack of time available to many women for getting involved in decision-making processes at the local level. The time spent on the daily commute or the time spent on the care for children and care-dependent relatives are important factors explaining differential vulnerabilities and capacities between social groups. This resonates with Virginie Le Masson's research findings in Ladakh that, once women are freed from the need to collect biomass for heating and cooking, through NGO supported new building and energy projects, they can be introduced to 'productive' work – that is, cash earning, although it remains to be seen whether this makes a difference to gendered power relations.

One of the reasons for this over-reliance of well worn and dramatic examples is the lack of a systematic collection of gendered data which can then be 'scaled up', as Mazorra and colleagues' chapter argues with respect to energy sources. People's different needs and priorities must be made visible via research that collects separate data for men and women (and for people of different ethnic origin, age or other relevant social differentiation). As well as the collection of gender disaggregated data, the ways in which this data is collected also needs to be gender sensitive. The method of separating men and women in Ladakhi communities to talk to a male and female (respectively) researcher enabled discussions to excavate the initial assertion that men and women were equal, and reveal substantial inequalities in workload and economic independence (Le Masson, 2014 and Chapter 13 in this volume). The purpose of MacGregor's chapter is to move the discussion past a simplistic emphasis on victimisation and material impacts by offering arguments for why gender analysis aids in the understanding of climate change. Development projects which support people's, especially women's, ability to earn an income, do not necessarily achieve their intended objectives. Haynes' experiences are that, while some of these projects she has observed in Bangladesh have opened up new opportunities for women to participate in the workforce and take on new roles,

> the power structures that maintain inequalities are dynamic and can adapt to re-establish control and capitalise on any new or redistributed resources [...] Public policy and programs that present as gender neutral or misunderstand the gender dynamics between men and women in rural communities risk being less effective and more disturbingly, have the potential to perpetuate or exacerbate existing gender inequality.
>
> (Haynes, pp 152–3, this volume)

A gender analysis is crucial to question unfair labour markets, denial of rights, and unequal access to resources, which all contribute to vulnerability to climate change and unsustainable development pathways.

Using a gender perspective, then, continues to challenge the status quo because it questions the allocation of power over natural resources, over economic opportunities and over decision-making processes (see Vainio-Matila, 2001). Gender studies have extensively shown that people's social identities, including their gender, influence who owns and controls power both within and outside the home, and that this privilege has been historically allocated to men. This translates into state governing bodies being male dominated, including the design of policy frameworks and the allocation of financial resources.

The collection of essays in this book is committed to dismantling such gender inequality and injustice, but does so through a critical appraisal of vulnerability, and relative privilege within genders. It also seeks to go beyond a categorisation of women as domestic decision makers and consumers which, while empowering them to act – review the Women's Environmental Network website which provides a good example of campaigns designed to empower women – is also in danger of reifying the caring role most often undertaken by women. As Sherilyn MacGregor and Gotelind Alber and colleagues argue in this volume, the value and practice of caring need to be redefined. One way in which this might take place is by changing the way in which most families in the Global North live. Here, Lidewij Tummers looks at the potential for co-housing to make a contribution to reducing greenhouse gas emissions through collective energy efficiency, and also to gender equality, by enabling the possibility for greater sharing of caring and household (reproductive) tasks.

We therefore find it more useful to argue that future research must move beyond focusing on gender differences, to studying the effects of the interplay between various aspects of gender and power. A deeper understanding of unequal manifestations of power and related influence on production and consumption patterns will inform how to transform lifestyles to become environmentally sound and less carbon-intensive.

Notes

1 http://cait.wri.org/historical/Country%20GHG%20Emissions?indicator[]=Total%20 GHG%20Emissions%20Excluding%20Land-Use%20Change%20and%20 Forestry&indicator[]=Total%20GHG%20Emissions%20Including%20Land-Use%20 Change%20and%20Forestry&indicator[]=Total%20GHG%20Emissions%20 Excluding%20Land-Use%20Change%20and%20Forestry%20PercentPer cent%20 Change&indicator[]=Total%20CO2%20(including%20Land-Use%20Change%20 and%20Forestry)%20PercentPer cent%20Change&year[]=2012&sortIdx=NaN&sort Dir=desc&chartType=geo
2 Sweden had reduced its GHG emissions by 5.51 per cent; Luxembourg by 5.99 per cent; Hungary by 6.14 percent; Greece by 5.32 per cent; Finland by 8.60 per cent; Denmark by 8.73 per cent; Cyprus by 13.63 per cent; and Croatia by 11.42 per cent. (http://unfccc.int/kyoto_protocol/items/2830.php),
3 http://unfccc.int/paris_agreement/items/9444.php
4 The United Nations Collaborative Programme on Reducing Emissions from Deforestation and Forest Degradation in Developing Countries, launched in 2008.

References

Buckingham, S. (2010) 'Call in the women', *Nature*, 468/ 25 November: 502.
Buckingham, S. and Kulcur, R. (2009) 'Gendered geographies of environmental injustice', *Antipode*, 41(4): 659–683.
Buckingham, S. and Kulcur, R. (2017) '"It's not just the numbers": challenging masculinist working practices in climate change decision-making in UK government and environmental non-governmental organizations', in Marjorie Griffin Cohen (ed.), *Gender and Climate Change in Rich Countries: Work, Public Policy and Action*. London: Routledge.
Childs, S. and Krook, M.L. (2006) 'Should feminists give up on critical mass? A contingent yes', *Politics & Gender*, 2(4): 522–530.
Crenshaw, K. (1989) 'Demarginalizing the intersection of race and sex: a black feminist critique of antidiscrimination doctrine, feminist theory, and antiracist politics', *University of Chicago Legal Forum*: 139–167.

Crenshaw, K. (1991) 'Mapping the margins: intersectionality, identity politics, and violence against women of color', *Stanford Law Review*, 43(6): 1241–1299.

Damyanovic, D., Reinwald, F. and Weikmann, A. (2013) *Manual for Gender Mainstreaming in Urban Planning*. Vienna: Department of Urban Development and Planning, City of Vienna.

Denton, F. (2002) 'Climate change vulnerability, impacts, and adaptation: why does gender matter?' *Gender and Development*, 10(2): 10–20.

Di Chiro, G. (2008) 'Living environmentalisms: coalition politics, social reproduction, and environmental justice', *Environmental Politics*, 17(2): 276–298.

Gaard, G. (2014) 'Towards new ecomasculinities, ecogenders, and ecosexualities', in C.J, Adams and L. Gruen (eds), *Ecofeminism. Feminist Intersections with Other Animals and The Earth*. London: Bloomsbury, 225–240.

Garvey, J. (2010) *Ethique des changements climatiques*. Paris: Editions Yago.

Hamilton, C. (1990) 'Women, home and community: the struggle in an urban environment', in I. Diamond and G.F. Orenstein (eds), *Reweaving the World: The Emergence of Ecofeminism*. San Franciscio: Sierra Club Books, 215–222.

Haraway, D. (1991) *Simians, Cyborgs and Women: The Reinvention of Nature*. London: Routledge.

Harcourt, W., Knox, S. and Tabassi, T. (2015) 'World-wise otherwise stories for our endtimes: conversations on queer ecologies', in W. Harcourt and I.L. Nelson (eds), *Practising Feminist Political Ecologies*. London: Zed Books.

Hemmati, M. (2000) 'Gender-specific patterns of poverty and (over-)consumption in developing and developed countries', in *Society, Behaviour, and Climate Change Mitigation*. Dordrecht: Kluwer Academic Publishers, 169–189.

Hultman, M. (2013) 'The making of an environmental hero: a history of ecomodern masculinity, fuel cells and Arnold Schwarzenegger', *Environmental Humanities*, 2: 79–99.

Irschik, E. and Kail, E. (2013) 'Vienna: progress towards a fair shared city', in I. Sanchez de Madariaga and M. Roberts (eds), *Fair Shared Cities: The Impact of Gender Planning in Europe*. Aldershot: Ashgate, 297–324.

Jackson, T. (2009) *Prosperity Without Growth: Economics for a Finite Planet*. London: Routledge (Earthscan).

Kaijser, A. and Kronsell, A. (2014) 'Climate change through the lens of intersectionality', *Environmental Politics*, 23(3): 417–433.

Kruper, S.R. (2012) 'Transnatural ethics: revisiting the nuclear cleanup of Rocky Flats, CO, through the queer ecology of nuclia waste', *Cultural Geographies*, 19(3): 303–327.

Lambrou, Y. and Piana, G. (2006) *Gender: The Missing Component of the Response to Climate Change*. Rome: Food and Agriculture Organization of the United Nations (FAO).

Le Masson, V. (2014) 'Seeing both sides: ethical dilemmas of conducting gender-sensitive fieldwork', in J. Lunn (ed.), *Fieldwork in the Global South. Ethical Challenges and Dilemmas*. London: Routledge.

Magnusdottir, G.L. and Kronsell, A. (2015) 'The (in)visibility of women in Scandinavian climate policy making', *International Feminist Journal of Politics*, 17(2): 308–326.

Masika, R. (ed.) (2002) *Gender, Development, and Climate Change*. Oxford: Oxfam.

Merchant, C. (1980) *The Death of Nature: Women, Ecology and the Scientific Revolution*. New York: Harper Collins.

Merchant, C. (1996) *Earthcare: Women and the Environment*. London: Routledge.

Moore, N. (2017) 'Mothers, grandmothers, and other queers in eco/feminist theory and activism', in T. Marsden (ed.), *Handbook The Sage Handbook of Nature*. London: Sage.

Nagel, J. (2015) *Gender and Climate Change: Impacts, Science, Policy*. London: Routledge.

Nagel, J. (2017) 'Men at work: scientific and technical solutions to the "problem" of nature', in T. Marsden (ed.), *Handbook The Sage Handbook of Nature*. London: Sage.

Sandilands, C. (1999) *The Good-Natured Feminist. Ecofeminism and the Quest for Democracy*. Minneapolis: University of Minnesota Press.

Shiva, V. (1989) *Staying Alive: Women, Ecology and Development*. London: Zed Books.

Stiglitz, J. (2010) *The Stiglitz Report: Reforming the International Monetary and Financial Systems in the Wake of the Global Crisis*. New York: The New Press.

Sze, J. (2005) 'Race and power: an introduction to environmental justice energy activism', in D.N. Pellow and R.J. Brulle (eds), *Power, Justice and the Environment*. Cambridge, MA: MIT Press.

Taylor, D. (2014) *The State of Diversity in Environmental Organisations: Mainstream NGOs, Foundations, Government Agencies.* Green 2.0. Online. Available online at http://orgs.law.harvard.edu/els/files/2014/02/FullReport_Green2.0_FINALReducedSize.pdf (accessed 3 April 2016).

UNDP. (2009) Resource Guide on Gender and Climate Change. Available online at www.undp.org/content/undp/en/home/librarypage/womens-empowerment/resource-guide-on-gender-and-climate-change.html (accessed 8 January 2017).

UNFCCC. (2015) *Adoption of the Paris Agreement*, FCCC/CP/2015/L9 United Nations Framework Convention on Climate Change. Available online at https://unfccc.int/files/meetings/paris_nov_2015/application/pdf/paris_agreement_english_.pdf (accessed 8 January 2017).

Vainio-Mattila, A. (2001) 'Navigating Gender. A framework and a tool for participatory development', *Ministry for Foreign Affairs,* Department for International Development Cooperation, Helsinki, Finland.

WEDO. (2010) *Open Letter to United Nations Secretary-General Ban Ki-moon*, 9 March 2010.

Women's Environmental Network. (2016) Available online at www.wen.org.uk/all-resources/ (accessed 12 September 2016).

World Resources Institute Interactive. (2016) Available online at http://cait.wri.org/historical/Country%20GHG%20Emissions?indicator[]=Total%20GHG%20Emissions%20Excluding%20Land-Use%20Change%20and%20Forestry&indicator[]=Total%20GHG%20Emissions%20Including%20Land-Use%20Change%20and%20Forestry&indicator[]=Total%20GHG%20Emissions%20Excluding%20Land-Use%20Change%20and%20Forestry%20Per cent%20Change&indicator[]=Total%20CO2%20(including%20Land-Use%20Change%20and%20Forestry)%20Per cent%20Change&year[]=2012&sortIdx=NaN&sortDir=desc&chartType=geo (accessed 25 September 2016).

Part I
Structures

2 Moving beyond impacts

More answers to the 'gender and climate change' question

Sherilyn MacGregor

AMY GOODMAN: *What does gender equality have to do with climate change?*
MARY ROBINSON: *Oh, there are huge gender impacts of climate, huge impacts. If you undermine poor livelihoods, who has to pick up the pieces? Who has to put food on the table? Who has to go further in drought for firewood? ... Those who are trying to adapt and be resilient, the vast majority of farmers in the developed world, are women. So, it is usually important that the gender dimensions are recognized, both the fact that women are more vulnerable, because even in natural disasters they are ... more likely to die than men – 14 times, because they care about their children and try to hug them, they can't run very fast in long clothes – a whole variety of reasons.*

(Democracy Now! Interview,
10 December 2015)[1]

Introduction

When pressed to explain what gender has to do with climate change, it is tempting to point to impacts and vulnerabilities. It seems reasonable to answer that women are hurt by climate change because they lack resources and power. Gender inequality is itself harmful to women and the global ecological crisis is making it worse. It is through this logic that recognising women's plight, and taking action to reduce their suffering, have become matters of climate justice. Feminist climate justice activists such as Mary Robinson and members of the Women and Gender Constituency (WGC) of the United Nations Framework Convention on Climate Change (UNFCCC) have worked tirelessly to make this case. And while it is difficult to refute it, I will argue that better answers are necessary. Better answers are necessary because the exclusion of gender equality from climate policy, and the persistent blindness to gender in scholarly work on climate change, suggest that truly effective arguments have been lacking thus far. More persuasive arguments are also possible because there are over thirty years of feminist environmental research and theorising on which to draw.

The process by which human societies have changed the earth's climate has been shaped by gender relations in complex ways. Climate change is perceived, experienced, and accommodated differently by diverse men and women in all parts of the world. There is empirical evidence to support the claim that women make up a significant proportion of those who have been, and will be, hurt by climate change, but it is factually and strategically questionable to claim, as noted ecofeminist Vandana Shiva does, that 'women are the worst victims of ecological destruction' (quoted in Shabbir, 2015). The purpose of this chapter is to move the discussion past this simplistic emphasis on victimisation and material impacts by offering arguments that draw upon different kinds of evidence. I make use of a number of feminist theoretical tools for understanding climate change as a political object. The challenge is to make these tools accessible so that they can help to change the nature of the debate.

For the sake of brevity, I do not include an opening section that explains climate change or its effects; neither do I offer an overview of feminist environmental thinking in its many guises. Assuming readers either already possess – or can independently acquire – the necessary intellectual footing, the chapter is organised around three arguments supported by reasons and evidence for why gender analysis aids in the understanding of climate change. These arguments roughly map onto the spheres of i) ideology, ii) politics, and iii) visions of the future. The conclusion is that feminist environmental theory can and should be considered a rich toolbox for building a robust response to the gender and climate question.

Answering the 'gender and climate change' question: three (more) arguments

The meta-argument that has driven feminist environmental research since the 1980s is that it will be impossible to understand the causes and impacts of human exploitation of the planet, or to identify reasonable paths towards a less unsustainable society, without an analysis of how gendered identities, roles, and power relations shape human life. A fundamental premise of feminist environmental scholarship is that gender is an axis of power that cannot be subsumed within general theories of social inequality or subalternity. It may be one of many intersecting categories, but its specificity as a way of marking out and separating the human population into two hegemonic identity patterns that are socially and culturally normalised and policed – masculinity and femininity – deserves serious and sustained scholarly analysis. One can apply the same argument to the analysis of race and class as other specific axes of social inequality and power. And many do: in the environmental social sciences, analysis of class, race and place has been firmly established in the body of literature identified by the term 'environmental justice' (cf. Walker, 2012). In the more recently emerging 'climate justice' literature the same is also true: the groups most discussed in the context of climate justice are poor people in the Global South. Unless they are provided by feminist researchers, investigations that foreground gender tend to be left out of volumes purporting to cast a wide net over the topic of global climate change.[2] As Gaard observes, 'the focus on race, class and environment has backgrounded gender, sexuality and species' (2011: 37).

In a review of the gender gap in environmental geography, Reed and Christie (2009) surmise that gender analyses may be absent because gender is not considered relevant to environmental research or that it is considered a 'women's issue' for women scholars to address. They stop short of apportioning blame or delving into why a discipline that is geared to understanding people and place could fail so spectacularly to look under its own nose. In response, they simply offer a set of opportunities for filling the gap with existing research. The opportunities and research to which they refer together provide support for the useful statement that:

> We need constantly to ask how gender shapes the human experiences of environmental change, conflict and management, and how it influences the commitments of women and men participating in these enterprises.
>
> (Reed and Christie, 2009: 253)

Such constant questioning is a well-established feminist method in many academic disciplines. But it is the content of the responses to one particular question that is at issue here. There are useful answers and counterproductive answers. I argue that answers to the 'why is gender relevant to climate change' question that simply list the negative impacts on women *and go no further* are probably counterproductive. To be convincing, and to do justice to the complexity of the issues involved, more needs to be said. Getting beyond impacts seems essential if opportunities to understand climate change more fully, more politically, are to be realised. The goal should be to advance a larger set of arguments based on sound reason, credible evidence and concrete examples that resonate with even the most resistant of listeners. What follows is my attempt to do just that.

Argument 1: Climate change discourse is gendered

Using quantitative, gender-disaggregated measures of climate causes and effects (e.g. who is more to blame, who is more hurt) can help to understand the problem, and plays an important role in making the case for the inclusion of gender as a category of analysis in climate change research. The problem, however, is that this strategy has resulted in a 'gender-equals-women' problem and gives the false impression that climate change is a gender issue primarily (or only) because it hurts women more than men. As Arora-Jonsson (2011) points out, there is also the unintended effect of playing into disempowering stereotypes of women in the Global South as victims.

I have argued elsewhere that a shortcoming of gender and climate change research is that an insufficient amount of thought seems to have been given to the cultural-discursive (or ideational) dimensions of climate change, or to the ways in which gendered environmental discourses shape dominant understandings of the issue (MacGregor, 2010, 2011). If the concept of gender is used as a 'lens' through which to interpret the cultural-discursive construction and framings of climate change, then new kinds of answers and arguments become possible. Theoretical lenses or perspectives act to foreground some things, while backgrounding others. Applying the feminist tool of a 'gender lens' allows a focus on the construction and performance of gender identities as well as how societies organise power, work, pleasure, resource distribution and knowledge along gender lines (Runyan and Peterson, 2014). This approach to gender analysis involves understanding power relations between men and women and the discursive constructions of the hegemonic masculinities and femininities that shape the way we interpret, debate, articulate and respond to social/natural/technological phenomena such as climate change.

Feminist environmental scholarship has a long tradition of deconstructing the binaries in Western thought. Val Plumwood's explanation of the concept of dualism, which she calls 'the construction of a devalued and sharply demarcated sphere of otherness' (1993: 41), has been influential in the development of a gender lens in feminist environmental thought. For most feminist environmental theorists, the binaries that structure dominant ideologies – good–bad, mind–body, public–private, reason–emotion, victor–victim – are part of the same logic that underpins the domination of women by men and the natural environment by the human species. Dismantling the binaries and devising strategies for escaping dualistic traps have been central to research and writing in this field.

Making use of these theoretical tools allows us to see that, in the affluent world, climate change has been shaped by a number of hegemonic discourses. At once a scientifically-defined problem and a challenge for capitalism, global warming has been conceptualised in a way that reflects dominant power structures which are themselves gendered. Ecological modernisation and 'green growth' (the successor to sustainable development) are two answers to the climate crisis that have been created by and for those with the most to gain. These are powerful frames because each works to set the global policy agenda and to inform local experiences and understandings. An example that has attracted the attention of feminist scholars is the discourse of environmental security that casts climate change as a serious existential threat requiring typically masculine militarised and top-down state responses (Detraz, 2011). Moving in the opposite direction, there are feminist criticisms of the discourse of overpopulation (and the need to control women's fertility) in response to climate threats (Hartmann, 2010). These discourses map neatly onto hegemonic gender roles and stereotypes, with security lining up on the masculinity side of the binary slash and reproduction lining up neatly with femininity. They are not neutral: casting masculine security as solution and feminised reproductivity as threat means that existing forms of privilege and disadvantage are sustained.

Evidence: vulnerable victims – resilient subjects

An example that is ripe for critical feminist consideration is the emergence of a 'vulnerability–resilience' dualism that now dominates climate policy at all levels. Writing for the Annex I Expert Group on the UNFCCC, Levina and Tirpak make the binary clear: '"vulnerability" seems largely to imply an inability to cope and "resilience" seems to broadly imply an ability to cope. They may be viewed as two ends of a spectrum' (2006: 15). With the focus of climate politics having shifted away from mitigation to adaptation, there has been a concomitant change in the framing of climate policy goals (Pelling, 2011). Where once the aim was to slow down or avert the crisis that will hurt the vulnerable, it is now incumbent on governments and citizens to be prepared for coping with – perhaps even to benefit from – what comes. Within the adaptation literature, the construction of the 'resilient subject' in opposition to the 'vulnerable victim' fits squarely within this value-laden framing. Arguably, the move from 'vulnerability reduction to resilience thinking' can be understood as a change in interests and objectives that valorises elite human ingenuity and 'leaves the poor and vulnerable behind' (Cannon and Muller-Mann, 2010: 633).

Feminist discourse analysis reveals that this move reflects the hegemonic masculinist framing of the crisis. Climate victims appear as passive figures in need of help to become self-reliant enough to cope in harsh conditions beyond their comprehension and control. People who are positioned as vulnerable to extreme weather and other forms of climatic destruction are both feminised and racialised. One can Google the phrase 'climate victim' and find a wealth of images of women in saris standing hip-deep in flood waters. The climate victim's opposite number is the resilient subject who possesses the mental, physical, and material resources to survive and thrive – to bounce back after – whatever nature throws his way. He is active and strong and knows how to make the best of a bad situation. A Google search of 'resilience' yields images of super-heroes and men in suits and news headlines about business opportunities and sporting prowess. In everyday life, and in the discourse of civil contingencies, climate resilience equates to 'being prepared' (like a boy scout), not caught out by (mother) nature. At the uppermost levels of global climate politics, white-Western-male geoengineers are celebrated as modern-day Baconian Supermen who can harness the powers of techno-science to control the very weather (Fleming, 2007; Fox-Keller, 1985).

Resilience increasingly supplants the softer and more modest goal of 'sustainability' as commitment to an ecological ethic of human-nature interdependence fades. Seen through a gender lens, the valorisation of climate resilience over human vulnerability is founded on scientistic and masculinist values; it naturalises neoliberal rationalities of governance and celebrates human exceptionalism. Not only does this simplistic binary ignore complexities on the ground, it also removes expectations of citizen resistance to the root causes of ecological crisis, thereby casting it as an inevitable and therefore non-political *fait accompli*. Moreover, it runs counter to a signature insight of ecofeminist philosophy that, as embodied and embedded beings, all humans of all genders are at all times vulnerable to changes in our surroundings and at some times resilient enough to adapt to change – just like the rest of 'nature' (Alaimo, 2009).

Plumwood's tools for escaping the dualistic traps set by the dominant discourse include challenging false choices and recognising the diversity and complexity of what has been 'othered' (1993: 60). Importantly, she advises against a 'strategy of reversal', where that which has been devalued becomes that which is celebrated. In the case of the victim–subject/vulnerability–resilience binaries that frame contemporary climate narratives, gender analysis involves dismantling the traps so that they may be better avoided. It should not be about showcasing women victims as 'the keys to climate action' (Robinson and Verveer, 2015) because doing so serves to entrench false and homogenising pictures of (certain groups of) women (Arora-Jonsson, 2011). Instead, theoretical tools can be used to contextualise gender norms and challenge power relations.[3] My point is that it is good feminist practice to be attentive to and critical of dominant discourses: critical analyses of how discourses are gendered help to steer clear of the traps they set.

Argument 2: Gender equality and environmental protection are linked

Gender balance, sometimes the result of the feminist tool called 'gender mainstreaming', has been linked with effective environmental decision-making: the better the representation of women, the higher the likelihood that environmental protection and precaution will be prioritised. The corollary, of course, is that when decisions are made by male-dominated groups, there is a much greater chance that environmental concerns will fall off – or never make it onto – the policy agenda.

McCright and Dunlap (2011) are two of a very few sociologists to study the effects of hegemonic masculinity on perceptions of climate change. They use the phrase 'the white male effect' (WME) to describe the observed tendency of white males to be less concerned with climate change and all manner of environmental risk than women and minorities.[4] Using US data from 2001–2010, McCright and Dunlap (2011) have found that conservative white males are significantly more likely than other Americans to deny climate change. Their findings are in line with a significant body of research that suggests that elite white men are more accepting of risks and therefore less precautionary than all other social groups (cf. Finucane *et al.*, 2000). A parallel might be drawn between this denial of climate risks and the hyper-masculine banking industry's failure to recognise the warning signs of the 2008 financial crisis (Enloe, 2013).

In the field of feminist economics it is considered relevant to investigate which people and what values lay behind the practice of economic decision-making.

Julie Nelson (2012), for example, has recently looked at the role of economists in US climate politics. She argues that the precautionary principle, formerly a pillar of sustainability and environmental ethics, is now dismissed or ignored by leading economists who advise on US climate policy – not only because of its interference with economic growth, but also because of the role played by hegemonic masculinity. After showing that all the top US climate advisors are men (e.g. Nordhaus, Toll, Weitzman) and giving evidence of their support for growth and geoengineering, Nelson writes: 'important voices in international negotiations – speak from a [stereotypically masculine] position that assumes that we are basically in control of our situation, and have no need for [stereotypically feminine] attitudes of care or caution' (2012: 3).

Studies of local environmental politics offer further evidence of the connection between gender balance and pro-environmental or pro-climate decisions. For example Buckingham *et al.*'s (2005) research on waste-management systems for the European Commission found that the local governments with the highest recycling rates had a higher percentage of women employees than average. They also tended to include fewer engineers and more decision makers from diverse professional backgrounds such as education and social work. In a study investigating Swedish municipal governments' responses to climate change, Dymén, Andersson and Langlais (2013) conclude that strong gender awareness is associated with a high level of climate change response. They note that gender balance in government working groups and committees is an important factor in this outcome. This evidence does not prove causation, but makes it possible to argue that there is a link between gender balance – or in other words, the lack of male domination of decision-making – and pro-environmental policies (Buckingham, 2010). Therefore, another good answer to the 'gender and climate change question' is that women-friendly politics and climate-friendly politics go hand in hand.

Evidence: women's political status and CO_2 emissions

Looking at the connection between the political status of women and climate change policy, the empirical evidence appears strong. Out of the 70 most developed countries in the world, only 18 reduced or stabilised their overall carbon emissions between 1990 and 2004 (UNDP, 2007 cited in Buckingham, 2010). It is striking that 14 of these 18 countries had a greater-than-average percentage of female elected representatives (ibid.). This is a potentially powerful argument that should be used more often as an alternative to the 'women are hurt by climate change' response. Efforts to tackle climate change are hindered by the lack of women's involvement in political processes.

Ergas and York (2012) use cross-national data from 160 countries and statistical regression models to assess the environmental consequences of gender inequality, or, in other words, effects of women's political status on carbon emissions per capita. Previous research has linked women's political status to environmental treaty ratification (Norgaard and York, 2005) and to political designation of protected land area within nations (Nugent and Shandra, 2009). But Ergas and York are the first researchers to demonstrate a quantifiable link between women's political status and CO_2 emissions. Even when controlling for such indicators of modernisation (such as democratic governance and position in the world system), countries where women have higher political status tend to have lower per capita emissions (2012: 974). On the basis of this finding they argue: 'improving women's status around the world may be an important part of efforts to curtail greenhouse gas emissions and prevent dramatic climate changes from undermining the long-term prospects of societies' (ibid.: 974).

It would weaken the argument if complications and contradictory evidence were overlooked. Taking the path-breaking research by Ergas and York as a starting point, Magnusdottir and Kronsell (2015) have examined the impact of institutional gender equality in Nordic states (Sweden, Denmark and Norway) on climate change policy. Although they can agree that there is a correlation between gender equality and climate-friendly policies (indeed, Scandinavian countries are global fore-runners on both counts), they found that 'a critical mass of women does not automatically result in gender-sensitive climate policy making, recognising established gender differences in material conditions and in attitudes towards climate issues' (2015: 308). What we can take away from their findings is, first, that equal gender representation does not necessarily uproot the deeply embedded norms, practices, and values of masculinised institutions. Second, that it would be interesting to study the effects of women's equal political participation on male policymakers' performance of masculinity. Even though a critical mass of women in Scandinavian politics has not had the kind of direct effects on climate policy that Ergas and York (2012) have observed, it may – in the long run – lead to 'an alternative expression of masculinity, one that is more climate-sensitive (Magnusdottir and Kronsell, 2015: 320). There is nothing wrong with coupling a good argument with a call for more research on an under-studied topic.

Argument 3: The transition to a post-carbon world requires gender justice

My third and final line of argument is that if they are to work, strategies for climate mitigation and adaptation and for the long-term transition to more environmentally sustainable societies need to resonate with a significant proportion of the population. To resonate, visions of a post-carbon world arguably need to be inclusive of diverse experiences and to respond to the everyday needs of men and women. They need to make life easier rather than heap more duties and greater sacrifices on people who are already struggling from time poverty and declining quality of life. The fact that the needs and experiences of the majority of the world's population have been ignored in climate change politics has led to the emergence of a global movement for climate justice.

The phrase 'no climate justice without gender justice' has become a popular slogan for civil society organisations participating in climate negotiations and protests, such as the WGC at the Conference of Parties (COPs) and participants at the People's Climate March in New York City in 2014 (Gorecki, 2014). It is well known that the concept of climate justice expresses a belief that those most responsible for climate change have a duty to ensure that those who are most hurt are treated with respect and are compensated for climate-related loss and damage – as a matter of human rights. But what do they mean by 'gender justice'? A review of the writing published in conjunction with the slogan suggests that the focus tends to be on climate change impacts on women as an impoverished and marginalised group, as growers of crops and feeders of families. However justifiable, the profound sense of injustice that lies behind the banner does not seem connected to the feminist theoretical understandings of gender justice that could lead to more robust answers to the 'gender–climate question'.

Nancy Fraser (2013) offers a succinct and useful explanation of gender justice that could be useful to feminist climate change work. It is a delineation that stems from her critique of, and related desire to move beyond, the polarised choice between equality (treating women like men) and difference (treating women differently insofar as they are different from men). She treats justice as a more complex concept than can be captured in simple words like 'equality'. It is, for her, best conceptualised as being comprised of seven distinct normative principles that must be achieved at once in order for gender justice to be realised. The first three principles (anti-poverty, anti-exploitation and income equality) refer to the need for basic material conditions; fair distribution of wealth lies at the core of a socialist-feminist worldview. Principles four and five seek to redress the unfair gender division of labour by asserting the need for leisure time equality and equal respect for a range of work types. The final two principles seek to ensure equal participation in all spheres: these are the anti-marginalisation principle and the anti-androcentrism principle. According to Fraser, '[f]ailure to satisfy any one of them means failure to realise the full meaning of gender justice' (2013: 116).

Fraser's account of gender justice was originally designed as a tool for evaluating different possible versions of the post-industrial welfare state, very much part of a Left socialist debate in the early 1990s. More than twenty years on, it is visions of a post-carbon sustainable society rather than any kind of welfare state that dominate left-green debate. Important for the present discussion, Fraser argues that a concept of gender justice is important not just for present struggles against the degradation of women but also for making arguments for a better and different future world. She writes:

> Normative theorizing remains an indispensable intellectual enterprise for feminism, indeed for all emancipatory social movements. We need a vision or picture of where were are trying to go and a set of standards for evaluating various proposals as to how we might get there.
>
> (ibid.)

Gender and climate change activist discourse that highlights harm to women may be counterproductive not only because it supports problematic pictures, but also because it lacks a positive future vision. Does feminist environmental theorising offer good/better answers for why gender justice is a necessary precondition for a sustainable, post-carbon society? If it does, then I would argue that this line of argument deserves more attention in feminist interventions in climate change politics. Admittedly, feminist post-carbon visions are still under development, but they contain valuable evidence to support the argument at hand.

Evidence: solidarity economy not green economy

It is worth explaining first why feminist critics are sceptical of contemporary green visions, such as UNEP's *Green New Deal* and *Green Economy*, the Rio +20 outcome report *The Future We Want*, and the future as seen by the Transition Towns Movement (TTM) (the 'fastest growing environmental movement in the global north' (Barry and Quilley, 2009: 1). What the first two visions share is an unexamined faith in technology to solve socio-economic and ecological problems; the fixation on renewable energy and paid job creation in the (male-dominated) engineering and manufacturing sectors suggest a lack of understanding of the social implications of a future driven by ecological modernisation. Christa Wichterich offers this explanation: 'the driving force behind a greened economy is growth, not redistribution. Gender receives very little attention, power relationships are not examined; instead, large corporations are praised time and again for their pioneering role...' (2012: 40). The TTM, on the other hand, demonstrates an apocalyptic fixation with civilisational collapse that seemingly excuses a dubious image of self-reliance and 're-localised' community life that bears no mark of feminist values. According to Bay, the TTM is, at base, about communities coming together

> to develop strategies for addressing climate change by seeking ways to reduce carbon emissions as well as making lifestyle changes around food production and *nanna technologies*[5] (traditional ways of preserving, conserving, reusing materials and resources using various methods and techniques that previous generations used to minimise waste)
>
> (n.d.: 3, emphasis added).

This post-carbon future looks a lot like the past, when livelihoods were secured and environments were mediated through rigid divisions of labour along gender lines. These green visions do very little to redistribute wealth or to foster a fair division of necessary labour and equal political participation. Questions about gender equality are never asked. 'Unpaid female social reproduction work is thus silently accepted and assumed to be infinitely available' (Bauhardt, 2014: 65). Like mainstream economics, they take for granted that people will be cared for, for free, in private, and with no need for social organisation or democratic debate. In this way, these green post-carbon visions must receive a failing grade from the Fraser school of gender justice.

The work of feminist environmental scholars is helpful in articulating alternative visions of what a *gender-just* and sustainable society should/could be like. This work draws on a substantial body of theorising dating back to the 1970s to apply key insights to the challenge of imagining a post-growth and post-carbon world. For example, the feminist economics concept of the 'solidarity' or 'caring economy' directly counters green economy and transition visions (see, e.g. Gibson-Graham, 2006; Bauhardt, 2014; and Alber, Cahoon and Röhr, this volume). This concept was developed in the 1990s by German feminist political economists as the antithesis to the orthodox (neoclassical and neoliberal) understanding of the economy. Advocates criticise mainstream economics for externalising reproductive activities in the home from the economy, a shortcoming that stems in part from a failure to take the gender division of labour into account (Biesecker and Hofmeister, 2010; Bauhardt, 2014). Their criticisms are echoed in alternative proposals to resolve the current economic/ecological crises. Importantly, these proposals add 'the care crisis' (or crisis of social reproduction) to the economic and environmental crises that have already been diagnosed in Western nations (and some emerging economies). The vision of a solidarity economy is a radical challenge to mainstream diagnoses and proposed cures. As Wichterich writes,

> Instead of a neoliberal concept of individual responsibility, our actions should be guided by principles such as sharing, redistributing, and revaluing labor, prosperity, as well as power; cooperation and solidarity should decrease social competitiveness. A caring economy means that the entire economy is to be turned right side up again, shifting from speculation to provision. The goal is to re-embed the economy in social and natural relationships, and to link global social justice with environmental and gender justice.
>
> (2013: 41)

How do such feminist principles and goals translate into concrete strategies for addressing climate change that might resonate with more people? Applying Fraser's principles, it is possible to imagine ways of organising social life so that wealth and work are shared, gender-based exploitation and marginalisation are reduced, and environmental impacts are low. In affluent country contexts, one concrete strategy is the creation and provision of collective, public sustainability services that replace free labour performed by citizens with public services that are paid for by corporate taxes (Halme *et al.*, 2004). Through collectivising the work involved in social reproduction (e.g. of food provision, laundry, childcare) and pro-environmental work (recycling, waste reduction, repair), material throughput and carbon emissions would be reduced at the same time as diminishing the demands on people's time. An example of a sustainability service that would tick all of these boxes is food delivery schemes, already in operation in many cities in the Global North. Subsided provision of locally sourced food (either raw ingredients or prepared meals) direct to the doorstep could accomplish the following: increase time for other activities, reduce resource use, minimise waste, increase health and well-being and create jobs. Instead of 'green jobs' being primarily geared towards the manufactured eco-friendly gadgets (electric cars and solar panels) for which need must be created, an ecofeminist definition of green jobs involves taking care of inevitable human needs (i.e. for care) with minimal material resources. Similar social, economic and ecological benefits could be created by opening public cafeterias serving affordable meals prepared in low-impact ways in schools and the workplace. Green political theorist John Barry (2012) gives the examples of libraries, laundries, and light rail as collectivised services that could maximise welfare and minimise environmental impact. Co-housing (as practiced in Germany and the Netherlands) that enables the sharing of work and property (e.g. white goods, cars, gardening equipment) through communal living could potentially contribute to increased leisure time, more convivial social relations, a blurring of stereotyped gender roles, and lower carbon emissions. More research is needed, however, in order to evaluate the extent to which co-housing can serve as a climate mitigation strategy at the same time as promoting gender justice in practice (Tummers, 2015).

What is important about these feminist insights into solidarity and gender justice is that they demand resistance to, and reversal of, the downloading of environmental labour/duty to individual citizens. Refusing to give the free subsidies to patriarchal capitalism provided by feminised work is a first step in this political strategy. Climate change impacts hurt women, but women's work plays a role in sustaining the unsustainable systems that cause climate change. As citizens, women need to call an end to this system – not only in the name of gender justice but also because it is ethically wrong for one species to threaten the future of life on Earth for all others. I have written about this idea elsewhere in reference to the concept of feminist green citizenship as a political response to the climate crisis (MacGregor, 2014).

Conclusion

It often seems that climate change is the 'new and exclusive frame' for thinking about the human–environment nexus. There is an assumption that understanding climate change, and how it is affecting contemporary societies, requires new concepts, models and methods. New technologies and different lifestyles are needed for mitigation and adaptation; a new name for the present epoch must be declared. It is worth remembering, however, that the climate is changing while many deeply entrenched social problems remain impervious to change. This is a good reason for why seeing global warming as a crisis facing humanity as a whole is inaccurate, and why the new concept of 'Anthropocene' has been challenged for ignoring that there are winners as well as losers in the process of planetary domination (see e.g. Di Chiro, forthcoming). Simplistic explanations, no matter who constructs them, are counterproductive. They are inevitable, however, when responses to existential threats are called for without taking time for critical questioning or open debate.

This chapter has offered what I hope are three well-supported arguments that demonstrate the value of feminist environmental theoretical insights about discourse, politics and normative visions. Many of these insights have been developed over decades of feminist work to understand human–environment relations through the lens of gender. They remain relevant in the twenty-first century context of climate change because there has been so little change in patriarchal ideologies, power relations, political structures and policy solutions. The project of understanding climate change through a gender lens has more evidence to draw upon than a Google search of 'gender and climate change' currently yields, and far more dimensions than UN policymakers seem to grasp (see, e.g. the UNFCCC's answer to the gender-and-climate-change question).[6] It would not be possible to address them all in a Gender Day. In addition to giving some examples, my aim has been to suggest answers to the 'what does gender have to do with climate change' question that move the debate onto more positive and productive ground. While the work of activists and spokespeople such as Mary Robinson has been invaluable in getting gender included in UNFCCC documents and in drawing attention to the material links between climate change and the suffering of women, it is important that theirs are not the only interventions.

Notes

1 For full interview, see www.democracynow.org/2015/12/10/cop21_treaty_draft_excludes_gender_equality (accessed 10 January 2017).
2 For examples of such volumes, see Lever-Tracy ed. (2010) (a *handbook* on climate change!) and Dryzek *et al.* (2013).
3 See Okali and Naess (2013) for a useful discussion of how gender and power relations and power can be incorporated when integrating gender concerns in climate change research and policy.
4 There is also evidence to suggest that male survey respondents say they know more about climate change than women, but when asked questions that would demonstrate their knowledge, they tend to score less well than their claimed expertise would justify (see European Commission, 2014; McCright, 2010).
5 In a rare gesture to gender, some people involved in transition towns refer to work traditionally performed by grandmothers as 'nanna technologies' (or 'going back to old practices' [Kroen, 2010 in Bay, n.d.]). It is nothing more than a clever gesture, however, as no critical reflection on whether it is right that women should specialise in these tasks (*whatever they might be* – perhaps canning and crochet?) is forthcoming (cf. Bay n.d.).
6 See the UNFCCC webpage: 'Gender and climate change: What is the connection and why is it important?' Available online at http://unfccc.int/gender_and_climate_change/items/7516.php (accessed 30 August 2016).

References

Alaimo, S. (2009) Insurgent vulnerability and the carbon footprint of gender, *Kvinder, Kon & Forskning NR (Women, Gender and Research)*. Special issue Gendering Climate Change 3–4: 2–35.

Arora-Jonsson, S. (2011) Virtue and vulnerability: Discourses on women, gender and climate change, *Global Environmental Change*, 21: 744–751.

Barry, J. (2012) *The Politics of Actually Existing Unsustainability.* Oxford: Oxford University Press.

Barry, J. and Quilley, S. (2009) The transition to sustainability: Transition towns and sustainable communities. In L. Leonard and J. Barry (eds), *The Transition to Sustainable Living and Practice.* Bingley: Emerald Publishing, pp. 1–28.

Bauhardt, C. (2014) Solutions to the crisis? The Green New Deal, degrowth and the solidarity economy: alternatives to the capitalist growth economy from an ecofeminist economics perspective, *Ecological Economics*, 102: 60–68.

Bay, U. (n.d.) 'Transition Towns: social causes of climate change and private lifestyle changes'. Monash University. Available online at www.tasa.org.au/wp-content/uploads/2008/12/Bay-Uschi.pdf (accessed 29 March 2014).

Biesecker, A. and Hofmeister, S. (2010) Focus: (Re)productivity: Sustainable relations both between society and nature and between genders, *Ecological Economics*, 69(8): 1703–1711.

Buckingham, S. (2010) Call in the women, *Nature*, 468(25 November): 502.

Buckingham, S., Reeves D. and Batchelor A. (2005) Wasting women: The environmental justice of including women in municipal waste management, *Local Environment*, 10(4): 427–444.

Cannon, T. and Muller-Mahn, D. (2010) Vulnerability, resilience and development discourses in context of climate change, *Natural Hazards*, 55: 621–635.

Detraz, N. (2011) Threats or vulnerabilities? Assessing the link between climate change and security, *Global Environmental Politics*, 11(3): 104–120.

Di Chiro, G. (2017) Welcome to the white (m)Anthropocene? A feminist-environmentalist critique. In S. MacGregor (ed.), *The Routledge Handbook of Gender and Environment.* London: Routledge.

Dryzek, J., Norgaard, R. and Schlosberg, D. (2013) *Climate Challenged Society.* Oxford: Oxford University Press.

Dymén, C., Andersson, M. and Langlais, R. (2013) Gendered dimensions of climate change response in Swedish municipalities, *Local Environment*, 18(9): 1066–1078.

Enloe, C. (2013) *Seriously! Investigating Crashes and Crises as If Women Mattered.* Berkeley: University of California Press.

Ergas, C. and York, R. (2012) Women's status and carbon dioxide emissions: A quantitative, cross-national analysis, *Social Science Research*, 41(4): 965–976.

European Commission. (2014) Attitudes of European Citizens towards the Environment. Special Eurobarometer 416. Available online at http://ec.europa.eu/public_opinion/index_en.htm (accessed 10 October 2016).

Finucane, M., Slovic, P., Mertz, C., Flynn, J. and Satterfield, T. (2000) Gender, race, and perceived risk: The 'white male' effect, *Health, Risk and Society*, 2(2): 159–172.

Fox-Keller, E. (1985) *Reflections on Gender and Science.* New Haven: Yale University Press.

Fraser, N. (2013) *The Fortunes of Feminism.* London: Verso.

Gaard, G. (2011) Ecofeminism revisited: Rejecting essentialism and re-placing species in material feminist environmentalism, *Feminist Foundations*, 23(2): 26–53.

Gibson-Graham, J.-K. (2006) *A Post-Capitalist Politics*. Minneapolis: University of Minnesota Press.

Gorecki, J. (2014) No Climate Justice Without Gender Justice: Women at the Forefront of the People's Climate March. Available online at www.thefeministwire.com/2014/09/climate-justice-without-gender-justice-women-forefront-peoples-climate-march/ (accessed 14 January 2016).

Halme, M., Jasch, C. and Scharp, M. (2004) Sustainable homeservices? Toward household services that enhance ecological, social and economic sustainability, *Ecological Economics*, 51: 125–138.

Hartmann, B. (2010) Rethinking the role of population in human security. In R. Matthew, J. Barnett, B. McDonald and K.L. O'Brien (eds), *Global Environmental Change and Human Security*. Cambridge, MA: MIT Press, pp. 193–214.

Lever-Tracy, C. (ed.) (2010) *Routledge Handbook of Climate Change and Society*. London: Routledge.

Levina, E. and Tirpak, D. (2006) Key Adaptation Concepts and Terms; Agenda document 1. OECD/IEA Project for the Annex I Expert Group on the UNFCCC Paris. Available online at www.oecd.org/environment/cc/36278739.pdf (accessed 19 March 2013).

MacGregor, S. (2010) A stranger silence still: The need for feminist social research on climate change, *Sociological Review*, 57(2): 124–140.

MacGregor, S. (2011) Researching gender and climate change: From impacts to discourses, *Journal of the Indian Ocean Region*, 6(2): 223–238.

MacGregor, S. (2014) Only resist: Feminist ecological citizenship and the post-politics of climate change, *Hypatia: Journal of Feminist Philosophy*. Special Issue on Feminist Philosophy and Climate Change, 29(3): 617–633.

Magnusdottir, G.L. and Kronsell, A. (2015). The (in)visibility of gender in Scandinavian climate policy-making, *International Feminist Journal of Politics*, 17(2): 308–326.

McCright, A. (2010) The effects of gender on climate change knowledge and concern in the American public, *Population and Environment*, 32(1): 66–87.

McCright, A. and Dunlap, R. (2011) Cool dudes: The denial of climate change among conservative white males in the United States, *Global Environmental Change*, 21(4): 1163–1172.

Nelson, J. (2012). Is dismissing the precautionary principle the manly thing to do? Gender and the economics of climate change. Research Note #013. Institute for New Economic Thinking. New York. Available online at http://ineteconomics.org/sites/inet.civicactions.net/files/Note-13-Nelson.pdf (accessed 6 June 2014).

Norgaard, K. and York, R. (2005) Gender equality and state environmentalism, *Gender and Society*, 19(4): 506–522.

Nugent, C. and Shandra, J.M. (2009) State environmental protection efforts, women's status, and world polity: A cross-national analysis, *Organization & Environment*, 22(2): 208–229.

Okali, C. and Naess, L.O. (2013) Making sense of gender, climate change and agriculture in Sub-Saharan Africa: Creating gender-responsive climate adaptation policy. Available online at www.futureagricultures.org/ (accessed 8 April 2016).

Pelling, M. (2011) *Adaptation to Climate Change: From Resilience to Transformation*. London: Routledge.

Plumwood, V. (1993) *Feminism and the Mastery of Nature*. London: Routledge.

Reed, M. and Christie, S. (2009) Environmental geography: We're not quite home – reviewing the gender gap, *Progress in Human Geography*, 33: 246–255.

Robinson, M. and Verveer, M. (2015) Women are the victims of climate change – and the keys to climate action. *The Guardian*, Tuesday 1 December 2015 12.15 GMT. Available online at www.theguardian.com/commentisfree/2015/dec/01/women-victims-climate-change-keys-climate-action (accessed 23 January 2016).

Runyan, A.S. and Peterson, V.S. (2014) *Global Gender Issues in the New Millennium* (4th ed). Boulder, CO: Westview Press.

Shabbir, N. (2015) Women and climate change injustice: Thoughts from the Paris talks. *The Guardian*, Thursday 10 December 2015 13.52 GMT. Available online at www.theguardian.com/global-development/2015/dec/10/women-injustice-climate-change-thoughts-from-the-paris-talks (accessed 13 January 2016).

Tummers, L. (2015) The re-emergence of self-managed co-housing in Europe: A critical review of co-housing research. *Urban Studies* [Published online before print 22 May 2015, doi: 10.1177/0042098015586696]

United Nations Framework Convention on Climate Change. (UNFCCC) Official website. Available online at http://unfccc.int/ (accessed 30 August 2016).

Walker, G. (2012) *Environmental Justice: Concepts, Evidence and Politics*. London: Routledge.

Wichterich, C. (2012) The future we want: A feminist perspective, *Heinrich Böll Stiftung Publication Series on Ecology*, 21. Transl. Sandra Lustig. Berlin: Heinrich Böll Stiftung.

3 Integrating gender issues into the global climate change regime

Karen Morrow

Introduction: (re)characterising climate change and evolving the UNFCC regime

Climate change was originally characterised in the United Nations Framework Convention on Climate Change 1992 (UNFCCC) (UN, 1992) (adopted at the United Nations Conference on Environment and Development (UNCED)) as a largely, scientific/technical issue, albeit with strong economic ramifications. In consequence it was to be tackled in the traditional manner, through an international treaty regime focused on nation states, in the same fashion as had been applied to other global problems, including those concerning the environment.[1] The recent notable success of the international community in developing a regime to address ozone depletion (Canan and Reichman, 2002)[2] that had been produced by applying the established state-centric approach to international lawmaking in a new environmental context was doubtless influential in this regard. In common with ozone depletion, climate change is both a scientific and a socially embedded issue, but, while there is undoubtedly commonality here, there are, however, important points of contrast between the two issues. In many ways the ozone regime arguably represents the optimum scenario to date for the adoption of an effective global environmental law instrument, benefiting from: rapidly maturing and compelling scientific consensus; involving a problem that could be addressed within comparatively narrow parameters by tackling a small number of implicated industrial sources of ozone depleting chemicals (ODCs); and the fact that alternatives to ODCs were both viable and attractive prospects for producers. These conditions do not however apply in the context of the contested science (though we have finally moved on from denial to disputing specifics) of the complex morass of climate change; with myriad sources (both industrial and domestic) of greenhouse gases to contend with; and costly, underdeveloped or even unobtainable alternatives. In addition, the multi-scalar impacts of climate change (ranging from global, through regional to local levels and cognisant of the linkages between them) and their highly particularised manifestations, add a further degree of complexity to developing a legal regime to address the issues.

In short, climate change constitutes a very different practical and regulatory proposition to any environmental law issue hitherto. That said, international law has been called upon to deal with just this sort of complex, cross-cutting, socially embedded phenomenon in the context of sustainability and the resultant need to broaden participation in the international polis that this engendered. Though there are significant commonalities here, the necessary conceptual connections were not made at the UNCED even though sustainability, too, emerged on the international legal stage in this context, in the form of the Rio Declaration and Agenda 21 (Morrow, 2017). Climate change, like sustainability, features elements of scale, complexity and ubiquity (in infiltrating the whole gamut of human activity) that by their very nature make addressing it effectively a collaborative endeavour for society as a whole. So, while states are (and in the context of the current global legal order will remain) the key players in the global climate change regime, garnering effective responses to complex cross-cutting environmental–social–economic problems of this nature cannot feasibly be achieved by them alone – or even in concert. The response to the pervasive threat of climate change has therefore come to be viewed as requiring the invocation of a much broader participation in a collaborative governance project. The use of the term 'governance' in this context is apposite as it reflects the fact that 'government' as normally understood does not in fact exist in international law and that the norm-making community in this context is increasingly understood as being comprised of a complex web of actors including, nation states; but also embracing intergovernmental organisations (IGOs); courts and tribunals; treaty bodies; and a whole range of civil society actors, to name only the most central players. However, these actors enjoy highly differentiated status and play widely varying roles in the governance arena: here discussion will focus on the still comparatively novel area of stakeholder participation as a means of incorporating bottom-up/grass-roots input into the global climate change regime as this shifted into the mainstream in the run-up to the twenty-first UNFCCC Conference of Parties (COP 21, Paris) in 2015 (Jordan and van Asselt, 2015), and in particular its application to women as a group. It will become clear that making the connection between gender and participation in the first place and operationalising it in the UNFCCC has proved far from straightforward (Morrow, forthcoming).

In principle, then, understanding climate change as a cross-cutting, multiscalar and socially embedded problem casts it in the guise of a sustainability issue, and allows us to draw on the possibilities that this offers for re-shaping the global polis. The approach of attempting to yoke bottom-up civil society participation to top-down state and IGO activity in international environmental law and policy has already featured, albeit to as yet limited practical effect, in this area. In fact lip-service was paid to the need to adopt an expanded approach to participation in the climate change regime from the outset with Art. 4.1.(i) of the UNFCCC (UN, 1992) placing broad (but vague) obligations on signatory states in this regard. Notably, no obligations were placed on the convention institutions to do likewise. Nonetheless, the need to undertake outreach to stakeholders to bolster the fledgling regime swiftly became something of an institutional priority – albeit somewhat haphazard in its execution. The UNFCCC machinery initially developed working relationships with a number of the major groups who had been identified as stakeholders in the context of Agenda 21, specifically: business and industry and environment NGOs (included from the outset); local government and municipal authorities (1995); indigenous peoples (2001); and research and independent organisations (2003) (UNFCCC, 2011). However not all of the Agenda 21 major groups were early inclusions and notably absent were: trade union NGOs (2008); young people (2009) (UNJFICYCC, 2010); and women (2011) all of which had to wait for a considerable time to gain official recognition within the UNFCCC regime (farmers and agriculture NGOs remain subject to provisional recognition) (UNFCCC, 2013d). This prioritisation of certain stakeholder groups above others is rather telling. Closer examination, then, reveals that the pursuit of participation being touted by the UNFCCC regime was a highly selective exercise. Rather than consistently augmenting the range of sanctioned inputs into the global climate change regime, the steer taken, privileging predominantly technical, commercial/industrial and to a lesser extent environmental perspectives, arguably served to entrench the initial predisposition of the UNFCCC approach (Morrow, forthcoming). With the exception of the inclusion of the indigenous peoples constituency (an instrumental move that was a political and practical necessity for developing UNFCCC regime programmes such as the Reducing Emissions from Deforestation and Forest Degradation (REDD)), there has been a notable tardiness in integrating more obviously socially-oriented/people-centred aspects of participation into climate change governance. This general failure to follow through widening participation in the central regime processes speaks to a failure to grasp the import of the social embeddedness of climate change. The failure to recognise the importance of gender in the UNFCCC regime is illustrative of this and also significant (and arguably particularly egregious) in its own right.

Gender, climate change and (an intransigent) UNFCCC regime

The UNFCCC's failure to engage with gender in the context of climate change is difficult to justify for a number of reasons. In the first place, gender is recognised as a salient factor in climate change impacts as a manifestation of the broader understanding that women are both agents of and affected by environmental degradation in highly particularised ways. Thus, in failing to engage effectively with gender issues, the UNFCCC machinery was actually hampering an important facet of its own regime functionality. As alluded to above, the complex nature and multi-scalar manifestation of the sundry issues posed by climate change make statecraft alone an inadequate response – societal engagement is required – and failing to adequately engage almost half of the global population, with relevant contributions to make to the debate, not least on mitigation and adaptation (Brody *et al.*, 2008), necessarily hampers this endeavour. Furthermore women, like men, are both drivers of, and vulnerable, to climate change but these states are experienced in gender differentiated ways – and a regime that needs must engage with these issues cannot expect or even hope to arrive at viable law and policy responses in ignorance of these factors and their impacts. In terms of vulnerability, climate change is the latest in the long list of environmental disasters that impact most severely upon the poorest in society (Svensen, 2009). Pervasive societal gender inequality functions to ensure that, in this already stricken group, women are both among those most subject to the adverse impacts of climate change and ranked amidst those worst positioned socially, legally and economically to respond to them (Brody *et al.*, 2008). In terms of agency and capacity to engage with the effects of climate change, women's experiences are also mediated by gender considerations that shape the knowledge, understanding and experience of climate change impacts and which will be considered further below.

Second, in its reluctance to take gender issues on board, the UNFCCC regime was out of step with the rest of the UN. The UNFCCC regime failed to grasp the salience of the UN's organisational commitment to gender main-streaming as a means of fostering a more systemic, holistic and above all effective engagement with gender issues than hitherto (Morrow, 2006). For present purposes, the aim of mainstreaming to enculturate gender considerations as a systemic priority is significant, in particular, as pursued through the goal of propagating women's participation in law and policy, as is the UNFCCC's lengthy failure to take this on board. While the mainstreaming incarnation of UN gender policy was emerging alongside the participatory re-orientation of the climate change regime in the mid- to late-1990s the UNFCCC's failure to engage with the issues must also be viewed in light of the fact that gender was already an established UN institutional priority well before this point (Morrow 2006).

Insofar as climate change is concerned a number of the UN's specialised gender bodies saw the relevance of gender to climate issues in advance of the UNFCCC – for example the Commission on the Status of Women (CSW, 2008) and the Committee on the Elimination of Discrimination Against Women (CEDAW, 2009). UN bodies tasked with non-gender specific responsibilities and acting across a range of areas also identified the absence of a gender dimension to the climate change regime as problematic in advance of the UNFCCC arriving at this position, notably the UN Environment Programme (Wamukonya and Skutsch, 2001); the UN Development Programme and the World Conservation Union (UNDP, IUCN *et al.*, 2009).

The UNFCCC's shortcomings in regard to gender appear even more marked when contrasted with its sibling regimes, also spawned at Rio: the Convention on Biodiversity (CBD); and the Convention to Combat Desertification (UNCCD) (agreed in 1994 but rooted in the UNCED) which were active on this front well in advance of the climate change regime. The discrepancy in approaches between the (now acknowledged as) interconnected Rio regimes was graphically demonstrated when, in preparation for the United Nations Conference on Sustainable Development in 2012 – (Rio+20) a joint liaison group (JLG) of their three executive secretaries produced a report on the importance of gender in implementing the long-term objectives of their respective regimes (CBD, UNCCD, UNFCCC, 2012). While the UNFCCC had by this time stated that it recognised in principle the need to align itself with the UN's institutional approach to gender (UNFCCC, 2012a) and its contribution to the JLG report showed some movement on gender, it was also evident that it was lagging behind the CBD and the UNCCD. For example, the JLG report refers to '16 mentions' of gender in the Durban COP 17 2011 outcomes; standing in marked contrast to extant policy commitments under the other Rio regimes (CBD, UNCCD, UNFCCC, 2012: 5). Welcome as this was, after such a tardy start, a lengthy game of catch-up was inevitable.

Women, ecofeminism and the environment

If the institutional machinery of the UNFCCC was slow to grasp the significance of gender to climate change, the same was not true for women themselves, who sought (albeit unsuccessfully) to constitute a women's NGO forum at the first FCCC Conference of Parties (COP 1, Berlin) in 1995 (Wamukonya and Skutsch, 2001). This was unsurprising, given the rise of a gendered understanding of environmental issues. The development of ecofeminism was one phenomenon that made an important contribution to articulating and interrogating environmental matters as gendered concerns more generally. For present purposes, it highlighted two issues that have come to be understood as centrally important: first, it posited a broad conceptualisation of the intersections between gender and environmental concerns; and second, it adopted an approach that fused theory and grass-roots activism (Morrow, 2013) – this is of particular importance as women tend to be active in grass-roots organisations to a much more marked degree than in formal political processes (Verchick, 1996; Morrow, 2013) and capturing this may offer one corrective to the dearth of women's representation in decision-making processes more generally (discussed further below). These elements were first applied to 'gender' in the law and development debate; then rolled out to other areas, such as sustainability; and are latterly emergent in the climate change debate (Morrow, 2017). These will now be briefly considered.

The social ecofeminist approaches that dominate cutting edge ecofeminist thought point to a particularised relationship between women and the environment, characterised by shared oppression at the hands of a patriarchal society and founded on women's reproductive functions, broadly understood. The latter concept refers to an expansive notion of those caring activities that are necessary to nurture life and support the family unit and for which women are usually primarily responsible. Carrying out this caretaking role by necessary implication extends the sphere of the 'personal' far beyond the indi-vidual, into families and communities at all levels; and beyond the human, to concern itself most immediately with host ecosystems but ultimately with the biosphere as a whole. Social ecofeminist approaches also seek to engage with the fact that other complex interrelating societal factors exist alongside/ in combination with gender and ecological contexts to position women dis-advantageously as environmental actors. These include (but are not limited to): race; indigeneity; class; sexual orientation; age; and disability; individu-ally or in various shifting combinations in different contexts, interacting to place women among those most disadvantaged by environmental degrada-tion (Brody *et al.*, 2008). So social ecofeminism can be viewed as seeking to engage with the full spectrum of the complexity of women's experiences as environmental actors and as thus regarding the web of relationships between women and the environment as socially embedded in many and intricately interconnected ways (Sandilands, 1999).

While much feminist scholarship focuses on disadvantage, social ecofem-inism also recognises that, in an environmental context, the skills, situated knowledge and experience garnered by women in their multiple caretaking roles may also, given credence, render them potentially powerful agents in tackling environmental malaise. The resulting nuanced understanding of the relationship between gender and the environment that ecofeminism offers can readily be extrapolated into valuable insights into the salience of gender to climate change, and to women not only as its victims but also as potent (if under-employed) drivers of change. Failing to effectively engage women in cli-mate governance both unnecessarily impoverishes our understanding of cli-mate change issues and prodigally disregards a hugely valuable pool of talent, expertise and creativity to tackle them in a way that we can ill afford to do.

If the approach adopted by ecofeminism to understanding women as envi-ronmental actors offers important insights as to the gendered impacts of cli-mate change upon and informing action to address them, the way in which this knowledge has been gathered and understanding gained is also poten-tially instructive in building a fresh approach to participation. Ecofeminist approaches contend that sustainable and effective governance of environmen-tal issues must integrate gender perspectives and the concerns of nature. In this context, gender perspectives themselves should accommodate all genders – traditionally female and male and (as our understanding of the complexity of gender grows and necessitates the rejection of dualism) intersex perspectives (which may align with either gender or neither). This offers a necessary correc-tive to the dominance of male perspectives in what remains a strongly patriar-chal area of societal endeavour. Ecofeminist approaches at base regard genders as equal but at the same time acknowledge difference (between genders and within them) as centrally important and as something that must be accommo-dated in both epistemology and governance processes. Given the continuing dearth of women's participation in established forms of government gener-ally (UN Women, 2015), securing women's participation in emerging areas of governance is both crucial in its own right and in improving gender representation in the system more generally.

Ecofeminism invokes a potent admixture of personal agency and civil society activism, enlisting individual and collective participation in environmental affairs and offering a specific incarnation of feminist engaged citizenship (Lister, 2003). The incorporation of grass-roots activism as an integral factor in ecofeminism adds a further dimension to its potential to contribute to participation in global environmental governance, in that it brings to the table the previously lacking 'bottom-up' perspective.

Making the case for the gender constituency in the UNFCCC regime

After the false start at COP 1 (Berlin) in 1995, gender issues fell into abeyance in the context of the global climate change regime for a number of years until resurrected as part of a wider engagement with gender in the run-up to the 2002 World Summit on Sustainable Development (WSSD) and in the Millennium Development Goals. Fuelled by this and informed in part by the type of insights discussed above and by the expectations generated by the UN's commitment to gender mainstreaming, women pressed for formal participant status in the climate change regime, seeking recognition for a gender constituency. The approach adopted for building the gender constituency and making the case for inclusion in the UNFCCC regime appears to owe much to ecofeminist thought and praxis.

The official quest for recognition for the gender constituency was initiated by the women's caucus running alongside COP 14 (Poznan) in 2008. The case had to be made that such a grouping could and would represent the diverse interests and expertise of women globally – a considerable challenge in itself. The approach adopted was to craft a coalition to orchestrate the campaign headed by an umbrella group comprising: Women in Europe for a Common Future (WECF); ENERGIA (International Network on Gender and Sustainable Energy); the Women's Environment and Development Organisation (WEDO) and GenderCC – Women for Climate Justice. This constituted a multi-regional network of women, gender activists and climate justice experts, committed to reaching out globally. Outside of elected bodies (and arguably often within them), 'representativeness' is a hugely contested issue and particularly so given the diversity of such an enormous and oversimplified category like 'women'. Ecofeminist approaches, however, eschew the essentialism that this could imply and have instead developed strategies through which to address the need to encapsulate diversity. These were applied to good effect in building the gender constituency in the context of climate change (Morrow, 2013). An inclusive, global approach to fostering active dialogue, using electronic communication, in particular, to strong effect, and interweaving technical expert opinion and grass-roots experience, were of central importance in consensus-building around core issues. This was clearly evident in the process of moulding the draft charter of the women's and gender constituency which functioned as both a rallying call and an opportunity to arrive at an agreed statement of values and intent. The Charter ascribed to a number of significant principles in terms of constituency membership and operation, including: democratic and participatory governance; respect for divergent positions; and broad, equitable and representative membership (GenderCC, 2011: Article 2). The objectives included in the Charter also sought to promote broad participation by women in the UNFCCC regime: prioritising making women's voices and experiences heard; and feeding women's views into ongoing climate change discourse (GenderCC, 2011: Article 3).

In light of this multi-faceted and ultimately persuasive campaign (and in support of its emerging institutional priorities, discussed above) the UNFCCC secretariat approved the women/gender grouping's application for provisional constituency status in 2009; full constituency status (comprising official observer status and associated participation rights) was awarded just prior to COP 17 (Durban) in 2011.

Women, leadership and the global climate change regime

Parallel developments in bringing gender to the fore in the global climate change regime included action aimed at boosting women's leadership capacities in order to enhance their participation in the regime – a move reflecting the need to address a specific manifestation of the more general lack of visibility and power in policy and decision-making processes (Verchick, 1996). This need for, and drive to secure, leadership roles for women had become a focal issue for women in the run-up to COP 16, Cancun in 2010, through a NGO-sponsored capacity-building and networking project run under the auspices of the Mary Robinson Foundation's Climate Justice Initiative (MRFCJ, 2013; Jackson, 2010). The project operated by establishing an informal women's leadership network, drawn from UN institutions, governments, civil society, and the philanthropic and private sectors. It was geared towards promoting: gender equality in UNFCCC sub-programmes; full participation for women in decision-making; and promulgating good practice. Women's leadership in the climate change regime also benefited from exploiting the fortuitous fact that women held the relevant ministerial portfolios in the host states for COPs 15 (Copenhagen), 16 (Cancun) and 17 (Durban) and they formed a troika in order to promote gender in the run-up to COP 17 (Sebastian and Ceplis, 2010).

The appointment of Christiana Figueres as UNFCCC's Executive Secretary in 2010 placed a woman as the UN's most senior official in the field and (while gender is only one matter of concern for her among many) she is on record as wishing to address continuing gender inequality in the climate change regime (UNCCNR, 2014). Another significant development at the highest level was the appointment of Mary Robinson, with her established track record as an enlightened advocate for gender and climate justice, as the UN special envoy for climate change. Important though such developments are, they appear to constitute the exception to the general rule and it is telling that in the run-up to COP 21, Paris 2015, of 28 people identified as key actors in the process, only 5 were women (King, 2015).

Progressing gender in the UNFCCC regime

There have been a number of encouraging developments with regard to gender within the UNFCCC regime in the wake of the grant of constituency status, in the attempt to rapidly operationalise its new-found institutional commitment to address gender issues. For example, since COP 18, Doha, 2012 a dedicated 'Gender Day' has become an established feature of the official UNFCCC COP meeting. COP 18 also saw the UNFCCC subsidiary body for implementation delivering a ground-breaking report on 'Gender and Climate Change' (UNFCCC, 2013c). Further positive change was apparent at COP 19, Warsaw, 2013 which included a workshop on gender balance in the UNFCCC (UNFCCC, 2013b) and showcased grass-roots good practice on addressing climate change (Momentum for Change, 2013). The new visibility for gender extends to its prominent presence on the main UNFCCC website.

More substantive developments include: the adoption of Decision 23/CP.18 on promoting gender balance in participation in both state delegations and the regime's constituent bodies (UNFCCC, 2012a). This was followed up at COP 20, Lima, 2014 in Decision 18/CP.20 by parties mandating the two-year Lima work programme on gender, with particular emphasis on: charting progress; capacity-building and training; initiating work on developing guidelines or other tools 'on integrating gender considerations into climate change-related activities under the Convention'; and mandating the appointment at a senior level of a UNFCCC gender focal point (UNFCCC, 2014b).

Commendably, the UNFCCC has also stated that it intends to bring gender to the fore in respect of the climate change regime's substantive activities, for example, the Clean Development Mechanism (CDM). However, at least in the early stages the institutional approach adopted has been at best somewhat cursory and at worst ill-informed, referencing gender mainstreaming on the one hand, while seemingly regarding it as something that can be tackled as an 'add-on' to the original gender-blind regime on the other (UNFCCC, 2012b).

Fuelling ongoing change

While Constituency status has ensured that gender is now at least firmly on the UNFCCC's institutional agenda, that, of course, does not guarantee substantive change. The opportunity to participate does not ensure influence. Other factors need to be brought into play, not least developing women's capacity to make full use of the opportunities that they have won (Brody *et al.*, 2008). More challengingly, there also needs to be a substantial and substantive shift in the regime's power dynamics – and this will encounter resistance on a number of fronts, both within the system where there will inevitably be conflicts with other more established regime priorities, and (in an international law context) among a substantial minority of signatory states.

Nonetheless, following on from and alongside official recognition for the gender constituency, the global climate change regime is beginning to act on gender, though the breadth and depth of this engagement remains to be seen. As evidenced by the visibility of gender in the outcomes of COP 21 in Paris in 2015, the signs are not encouraging. While gender equality and the empowerment of women are mentioned in the preamble to the Paris Agreement,[3] coverage in the substantive provisions is limited to single mentions in Article 7 – adaptation, and Article 11 – capacity-building (UNFCCC, 2015). It seems that, given the amount of work that needs to be done and the rate of progress thus far, for the foreseeable future, women will need to continue to press for progress. The UNFCCC regime's own monitoring activities have already played an important role in generating and disseminating baseline research on gender representation (or the lack of it) within the machinery of the UNFCCC (UNFCCC, 2013a) and in dramatising the need for change. The ongoing work of the MRFCJ is also making a particularly important contribution in this area, also using a statistical evidence base to make the case for 'gender balance', recommending that targets be set (no less than 40% and no more than 60% for either gender) for both regime bodies and state delegations and that sanctions be imposed for non-compliance (MRFCJ, 2013). Monitoring, reporting and valuation, rendering gender imbalance within the UNFCCC visible and verifiable and providing a yardstick to evaluate progress or the lack thereof is arguably as significant in this area as it is in the substance of the emerging climate regime (The Paris Aide Memoire 2015).

Conclusion: gender as an established institutional priority in the UNFCCC?

While the UNFCCC is now, albeit belatedly, making all the right noises in terms of embracing gender, it remains to be seen how thoroughgoing its institutional commitment actually is. In some areas we can see improvement: in terms of raw numbers, whereas women comprised 29.4 per cent in total of state delegations at COP 18, Doha 2012, with active encouragement by the UNFCCC regime, an overall improvement to 36 per cent was evident at COP 19, 2013 (UNFCCC, 2014a). However the situation with regard to

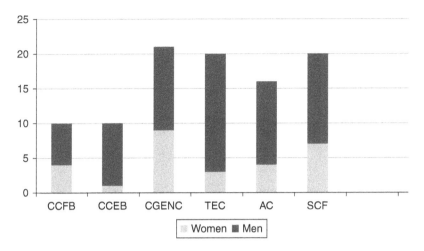

Figure 3.1 Women's representation in selected[a] UNFCCC constituted bodies 2014
CCFB – Compliance Committee Facilitative Branch
CCEB – Compliance Committee Enforcement Branch
CGENC – Consultative Group of Experts on National Communications
TEC – Technology Executive Committee
AC – Adaptation Committee
SCF – Standing Committee on Finance
[a]Only those bodies where the gender composition changed are shown.

gender in the UNFCCC's own constituent bodies reflects a diluted version of even this minimal progress. Changing the composition of the latter is in many ways more demanding than the former: it requires that women, still underrepresented in state delegations, are then selected for influential/leadership positions within the regime structure (see Figure 3.1; drawn from UNFCCC, 2014a).

As Figure 3.1 illustrates, despite recent initiatives, women remain (often significantly) in the minority in all of the UNFCCC constituted bodies. Furthermore, since 2013 only one body features significant improvement in women's representation – the CCFB (up 29%) with some improvement in the SCF (up 10%); improvements of 5 per cent or less in the TEC and the AC; and more worryingly a reduction in women's representation of 10 per cent in the CCEB and 9 per cent in the CGENC. Five further bodies showed no change (drawn from UNFCCC, 2014a). This confused picture, compounded by limited reference to gender in the Paris Agreement (UNFCCC, 2015), hardly represents substantial progress and leaves us to hope that the dawn on gender balance in the global climate change regime is merely progressing slowly and not, in fact, false.

Notes

1 See, for example, the Convention on Trade in endangered Species of Wild Flora and Fauna 1970 (CITES). Available online at www.cites.org/eng/disc/text.php (accessed 5 August 2015).
2 The Vienna Convention for the Protection of the Ozone Layer, 1985. Available online at http://ozone.unep.org/en/Treaties/hb_treaties_decisions-fbb.php?sec_id=155 (accessed 5 August 2015) and the Montreal Protocol on Substances that Deplete the Ozone Layer 1987 (as amended). Available online at http://ozone.unep.org/en/Treaties/hb_treaties_decisions-fbb.php?sec_id=5 (accessed 5 August 2015).
3 UNFCCC Draft Paris Agreement 12/12/2015. Available online at www.cop21.gouv.fr/wp-content/uploads/2015/12/l09r01.pdf (accessed 16 December 2016). The agreement is open for signature on 22/04/2016–21/04/2017 (art 20).

References

Brody, A., Demetriades, J. and Esplen, E. (2008) *Gender and Climate Change: Mapping the Linkages*. Brighton: BRIDGE. Available online at www.bridge.ids.ac.uk/reports/Climate_Change_DFID_draft.pdf (accessed 14 August 2015).

Canan, P. and Reichman, N. (2002) *Ozone Connections: Expert Networks in Global Environmental Governance*. Saltaire: Greenleaf Publishing.

CBD, UNCCD, UNFCCC. (2012) The Rio Conventions: Action on Gender. Available online at http://unfccc.int/resource/docs/publications/roi_20_gender_brochure.pdf (accessed 23 June 2014).

CEDAW. (2009) Statement of the CEDAW Committee on Gender and Climate Change, (adopted at the 44th session of CEDAW 20 July to 7 August, New York 2009) Available online at www2.ohchr.org/english/bodies/cedaw/docs/Gender_and_climate_change.pdf (accessed 14 August 2015).

CSW. (2008) Gender Perspectives on Climate Change, 52nd session of the Commission on the Status of Women, 2008. Available online at www.un.org/womenwatch/daw/csw/csw52/issuespapers/Gender%20and%20climate%20change%20paper%20final.pdf (accessed 14 August 2015).

GenderCC. (2011) Charter of the Women's and Gender Constituency under the UNFCCC. Available online at www.gendercc.net/fileadmin/inhalte/Dokumente/UNFCCC_conferences/Constituency/Women_Gender_Constituency_Charter_final.pdf (accessed 13 June 2014).

Jackson, S. (2010) (Mary Robinson Foundation (Climate Justice)/Realizing Rights (The Ethical Globalization Initiative)) 'Women's Leadership on Climate Justice: Planning for Cancun and Beyond' 17 September 2010. Available online at www.mrfcj.org/pdf/Meeting_Report_Womens_Leadership_on_Climate_Justice_17Sep2010.pdf (accessed 10 June 2014).

Jordan, A. and van Asselt, H. (2015) 'Can grassroots climate action save the planet?' Responding to Climate Change, 10 August 2015. Available online at www.rtcc.org/2015/08/10/can-grassroots-climate-action-save-the-planet/ (accessed 10 August 2015).

King, E. (2015) Who's Who in the world of climate diplomacy? Responding to Climate Change, 28 July 2015. Available online at www.rtcc.org/2015/07/21/whos-who-in-the-world-of-climate-change-diplomacy/#sthash.i7jVF41l.gmse (accessed 11 August 2015).

Lister, R. (2003) *Citizenship: A Feminist Perspective* (2nd edn). Washington Square, NY: New York University Press.

Mary Robinson Foundation – Climate Justice (MRFCJ). (2013) 'The Full View: Advancing the goal of gender balance in multilateral and intergovernmental processes'. Available online at www.mrfcj.org/pdf/2013-06-13-The-Full-View.pdf (accessed 7 August 2014).

Momentum for Change. (2013) Momentum for Change: Women for Results – sponsored by the Rockefeller Foundation. Available online at http://unfccc.int/secretariat/momentum_for_change/items/7318.php (accessed 13 June 2014).

Morrow, K. (2006) 'Not so much a meeting of minds as a coincidence of means: Ecofeminism, gender mainstreaming and the UN', *Thomas Jefferson law Journal*, 28(2): 185–204.

Morrow, K. (2013) 'Ecofeminism and the environment: International law and climate change', in M. Davies and V. Munro (eds), *The Ashgate Research Companion to Feminist Legal Theory*. Farnham: Ashgate, 377–394.

Morrow, K. (2017): In S. MacGregor (ed.), *Handbook on Gender and Climate Change*. London: Routledge.

Sandilands, C. (1999) *The Good-Natured Feminist: Ecofeminism and the Quest for Democracy*. Minneapolis: University of Minnesota Press.

Sebastian, J. and Ceplis, D. (CARES) (2010) Perspective on Gender and Climate Change at Cancun. Available online at www.aic.ca/gender/pdf/Gender_and_Climate_Cancun.pdf (accessed 17 August 2015).

Svensen, H. (2009) *The End is Nigh: A History of Natural Disasters*. London: Reaktion Books Ltd.

The Paris Aide-Memoire 2015 France/Peru. (2015) First informal ministerial consultations to prepare COP21 Paris, 20–21 July 2015 Aide mémoire produced by France and Peru. Available online at www.minam.gob.pe/somoscop20/wp-content/uploads/sites/81/2014/10/Aide-m%C3%A9moire-Paris-July-Informals.pdf (accessed 11 August 2015).

UN. (1992) Framework Convention on Climate Change (FCCC). Available online at http://unfccc.int/resource/docs/convkp/conveng.pdf (accessed 5 August 2015).

UNCCNR (United Nations Climate Change News Room). (2014) Christiana Figueres on Gender and Climate, 4 August 2014. Available online at http://newsroom.uat.unfccc.int/unfccc-newsroom/christiana-figueres-on-gender-and-climate/ (accessed 17 August 2015).

UNDP, IUCN CCGA. (2009) *Training Manual on Gender and Climate Change*. Available online at https://cmsdata.iucn.org/downloads/eng_version_web_final_1.pdf (accessed 13 August 2015).

UNFCCC. (2011) Non-governmental organization constituencies. Available online at https://unfccc.int/files/parties_and_observers/ngo/application/pdf/constituency_2011_english.pdf (accessed 7 August 2015).

UNFCCC. (2012a) Decision 23/CP.18 on 'Promoting gender balance and improving the participation of women in UNFCCC negotiations and in the representation of Parties in bodies established pursuant to the Convention or the Kyoto Protocol' FCCC/CP/2012/8/Add.3. Available online at http://unfccc.int/files/bodies/election_and_membership/application/pdf/cop18_gender_balance.pdf (accessed 7 August 2014).

UNFCCC. (2012b) CDM and Women. Available online at http://unfccc.int/resource/docs/publications/cdm_and_women.pdf (accessed 26 June 2014).

UNFCCC. (2013a) Report on Gender Composition. UNFCCC/CP/2013/4. Available online at http://unfccc.int/resource/docs/2013/cop19/eng/04.pdf (accessed 7 August 2014).

UNFCCC. (2013b) Workshop on Gender, Climate Change and the UNFCCC Workshop on Gender, Climate Change and the UNFCCC. Available online at

http://unfccc.int/files/adaptation/application/pdf/in_session_workshop_agenda_
web.pdf (accessed 11 August 2013).

UNFCCC. (2013c) Subsidiary Body for Implementation 'Gender and Climate
Change' FCCC/SBI/2013/L.16. Available online at http://unfccc.int/resource/docs/
2013/sbi/eng/l16.pdf (accessed 7 August 2014).

UNFCCC. (2013d) Contact Details and Constituency Information. Available online at
https://unfccc.int/files/meetings/warsaw_nov_2013/application/pdf/see_brochure_
contact_cop19cmp.pdf (accessed 17 August 2015).

UNFCCC. (2014a) Report on Gender Composition FCCC/CP/2014/7. Available
online at http://unfccc.int/resource/docs/2014/cop20/eng/07.pdf (accessed 17
August 2015).

UNFCCC. (2014b) Lima work programme on gender Decision 18/CP.20 FCCC/CP/
2014/10/Add.3. Available online at http://unfccc.int/resource/docs/2014/cop20/eng/
10a03.pdf#page=35 (accessed 17 August 2015).

UNFCCC. (2015) Paris Agreement FCCC/CP/2015/L.9/Rev.1. Available online at
www.cop21.gouv.fr/wp-content/uploads/2015/12/l09r01.pdf (accessed 16 December
2016).

UNJFICYCC. (2010) UN Joint Framework Initiative Children. Youth and Climate
Change: 'Youth Participation in the UNFCCC Negotiation Process: The United
Nations, Young People, and Climate Change'. Available online at http://unfccc.int/
cc_inet/files/cc_inet/information_pool/application/pdf/unfccc_youthparticipation.
pdf (accessed 5 August 2015).

UNWOMEN. (2015) Facts and Figures: Leadership and Political Participation.
Available online at www.unwomen.org/en/what-we-do/leadership-and-political-
participation/facts-and-figures (accessed 13 August 2015).

Verchick, R.R.M. (1996) 'In a greener voice: Feminist theory and environmental
justice', *Harvard Women's Law Journal*, 19: 23.

Wamukonya, N. and Skutsch, M. (2001) 'Is there a gender angle to the climate
change negotiations?' Available online at www.unep.org/roa/amcen/Projects_
Programme/climate_change/PreCop15/Proceedings/Gender-and-climate-change/
IsthereaGenderAngletotheClimateChangeNegiotiations.pdf (accessed 10 June
2014).

4 Gender justice and climate justice

Building women's economic and political agency through global partnerships

Patricia E. Perkins

Introduction

Climate justice – attention first to the needs of those most marginalized by climate change – is a feminist issue, and women are leading climate justice movements worldwide. Poverty, healthcare, education, work and political agency are all gendered, because women have less access and control over money, basic services and livelihoods, making women more likely to be negatively impacted by the world's changing climate in all countries. At the same time, their caring responsibilities, experiences and skills often make women experts on the best way of using limited resources to address climate-related challenges. Removing barriers to women's political agency and leadership is crucial for reducing women's vulnerability to climate risks. This implies fundamental changes in governance, prioritizing care for others and the world, and transformation of the economic, social and political systems that have produced climate change.

This chapter examines bottom-up strategies which are emerging in the face of climate change, especially with regard to their gendered impacts and women's agency. The details of each particular community's situation – ecological, social, political – are vitally important. How do communities organize socially and politically to meet biophysical and weather-related changes that affect their livelihoods? How are the needs of the most vulnerable addressed? How does women's engagement at the grass roots, and women's political leadership, facilitate equity in these initiatives? Since climate change affects women everywhere, global communication, networking and knowledge-sharing are vital to improve women's ability to face the impacts and lead in finding solutions.

I have been involved with collaborative climate justice partnerships through international projects – the Sister Watersheds project with Canadian and Brazilian partners (2002–2008) and a Climate Change Adaptation in Africa project with partners in Canada, Kenya, Mozambique and South Africa (2010–2012) – as well as green community development initiatives in marginalized Toronto neighbourhoods, and other networking projects. This work has demonstrated the wide applicability of local-level efforts in vulnerable communities to address equity challenges by developing strategies and materials for increasing the knowledge, interest and engagement of local residents on water-related and climate change issues, focusing in particular on women and youth. I have seen how collaboration between university researchers and community activists/organizers can generate fruitful synergies, strengthen educational outreach, build skills, and foster global and local networking. Women's approaches, leadership, local knowledge and indigenous understandings of power, land, and human–environment relationships are beginning to transform governance in many places worldwide.

A major theoretical frame for this analysis is feminist political ecology: the understanding that "the same dynamics that produce unequal access to resources or disproportionate vulnerabilities to environmental changes are often key components of social and political difference" (Buechler and Hanson, 2015: 7–8). Also theoretically relevant to gender and climate justice is Nobel Economics medallist Elinor Ostrom's work on polycentric (multiple-scale) governance and its benefits in addressing global environmental change by reducing opportunistic behavior, distributing benefits at many scales, building trust and reciprocity, and facilitating progressive policy leap-frogging or inspiration and sharing from one area to another (Ostrom, 2014: 120–121).

In the growing literature on gender and climate change, various perspectives on women's roles are represented. Women are often seen as victims, particularly needy or requiring special protection, or perhaps as virtuous because they consume less (due mostly to their poverty) or do the bulk of vital environmental and physical/social reproduction work (due to discrimination and patriarchy). Women's environmental leadership (since they and their children bear the brunt of many impacts and their work is essential to remediating damage) is sometimes noted approvingly. At times, there is even recognition of women's theoretical and practical contributions to building new kinds of socio-economic systems in which commons are managed sustainably, food and work are shared, and forward-looking collective priorities can flourish. In this chapter, I try to distinguish these various perspectives on women's roles and seek out glimpses of the fundamental transformations which may address climate justice over the long term.

The chapter concludes by noting some commonalities in these stories from the Global North and South, and some ways in which communication, solidarity and mutual reinforcement can strengthen and inspire feminist climate justice activism.

Gender justice and climate justice

"Climate justice" is defined by activists in various ways, which progressively involve an intersectional understanding that the impacts of climate change are grounded in gender, class and race (Climate Justice Now!, 2013; Nagel, 2015; Kaijser and Kronsell, 2014; Godfrey and Torres, 2016; Black, 2016; IRIS,

2009) and necessitate a radical transformation of existing political, social and economic systems (Klein, 2014; Black et al., 2014; LaDuke, 2016; Salleh, 2009; Kaufman, 2012; Gibson-Graham, 2006; Bond and Dorsey, 2010).

Women are increasingly familiar with climate change at the grass-roots level, because of its impacts on their paid and unpaid working lives, yet they usually have subordinate or limited roles in government, and limited political agency to address the causes of climate change (Perkins and Figueiredo, 2013; Habtezion, 2013). This is the case worldwide: a 2012 report found that "women's involvement in climate change decision-making at national, European and international levels is still low" and that women are a low proportion of graduates in scientific and technological fields deemed important for climate change response (EIGE, 2012: 3).

However, women's economic and social contributions, both paid and unpaid, are centrally important in times of climate change – traditionally female roles such as childrearing, care, skills transmission, education and community work (Perkins, 2013a; Perkins, 2013b; ILO, 2009). As Elinor Ostrom noted, these sorts of skills and roles are centrally important for equitable commons governance so that disputes can be resolved and irresponsible pollution and resource depletion prevented. Ostrom's empirical research showed that successful commons governance institutions tend to evolve in stable, caring, land-based communities with long-standing conflict-resolution mechanisms and interdependent living arrangements that allow each community member a place and a voice (Ostrom, 2010). An example of the link between commons, women's roles and climate justice comes from Honduras:

> Time and again, experience has shown that communities fare better during natural disaster when women play a leadership role in early warning systems and reconstruction. Women tend to share information related to community well being, choose less polluting energy sources, and adapt more easily to environmental changes when their family's survival is at stake. Women trained in early warning disaster reduction made a big difference in La Masica, a village in Honduras that, unlike nearby communities, reported no deaths during Hurricane Mitch in 1998. Integrating gender perspectives in the design and implementation of policies and laws also helps meet the gender-differentiated impacts of environmental degradation – shortage of water, deforestation, desertification – exacerbated by climate change.
>
> (ILO, 2009: 3)

In South Africa, the "Million Climate Jobs Campaign" sponsored by COSATU, the Confederation of South Africa Trade Unions, was initiated in 2011 and calculated that more than 3 million new jobs could be created in a Just Transition (union-based framework for the shift to a more sustainable economy). Environmentally friendly production processes are often more labor-intensive and thus create jobs; attention and policies are needed to make sure these jobs are available to both men and women, with equal pay for equal work and adequate training for all (COSATU, 2012: 42–44; Masterman-Smith, 2011).

A 2012 ILO presentation on "How can women benefit from green jobs?" states that 80 percent of the jobs related to green technologies will be created in the secondary sector (industry/manufacturing) and that women will likely lose out in terms of training opportunities and new skills for these jobs. The obstacles for women's access to green jobs are mainly the same as in the traditional "brown economy": lack of access to education, finance, decision-making, skills, and discrimination. But rather than ignoring or accepting them, these barriers can be addressed through targeted support for women's business entrepreneurship, gender mainstreaming in green jobs, job training and education for women (especially in science and technology fields), affirmative action, and increasing women's access to productive resources (Wintermayr, 2012: 19–20).

Equipping women to do the work of building more sustainable economies has many spin-off benefits (Karlsson, 2007). The Grameen Shakti (GS) microloans initiative in Bangladesh has trained more than 5,000 women as solar PV technicians and maintenance workers and has installed more than 100,000 solar home energy systems in rural communities in Bangladesh (ILO, 2009: 4). Brazil's "one million cisterns" program in the drought-plagued Northeast includes special courses for women who learn concrete construction skills in order to qualify for jobs building water-retention tanks (Mulheres Pedreiras, 2013: 1). Hundreds of women have been trained in this program since 2003 – along with many more men – as a result of concerted activism and organizing by women's groups in Brazil and their international partners and supporters (Morães, 2011: 141–166). In both of these cases, women are learning new, transferrable skills and gaining economic opportunities through their involvement in the "green transition" of their local communities.

Indigenous women in Canada, whose activist leadership against fossil fuel extraction and pipelines is well-recognized (Manuel, 2015: 211; Thomas-Mueller, 2014; Perkins, 2017), draw clear connections between environmental racism in First Nations communities, climate change, gender-based violence, colonialism, and the need for fundamental economic transformation (Awâsis, 2014; Coats, 2014). Says Cree anti-tarsands activist Melina Laboucan-Massimo:

> Violence against the earth begets violence against women. I think when we don't deal with both of them we're not ever really going to resolve the issue of the colonial mind and the colonial mentality and the values of patriarchy and ... capitalism that essentially exploit the land and exploit our women.
>
> (Gorecki, 2014)

These examples demonstrate the importance of moving from a "women as needy and vulnerable" to a "women as contributors and leaders" frame in the context of gender and climate justice. The gendered injustices of climate change, which intersect with race and class-based injustices, include not only the differential impacts of the changing climate on women's livelihoods, working lives, unpaid work and bodies, but also their political agency and participation in the transition to a more sustainable socio-economic future. Women are not only "victims" of climate change, they are also key theoreticians, stakeholders and actors producing social, political and economic transformation (Alston and Whittenbury, 2013; Cohen, 2017). This has tremendous potential to help revolutionize the problematic structures of capitalist global economies and decision-making that have generated climate change, to lead towards a more sustainable future (Salleh, 2009; Godfrey and Torres, 2016; Klein, 2014).

The following section outlines some detailed strategies and experiences with grass-roots activism, networking, and global partnerships to advance women's climate justice leadership. As Ostrom noted, layering of multiple nested governance functions, which she termed "polycentricity," is a characteristic principle of successful commons governance, in the context of local appropriateness (Ostrom 2009: 89–90). If we consider the global climate justice movement as a commons, or common cause (see Buckland, 2015), how can activists and academics build polycentric movements for climate and gender justice in a world where growing inequality is a corollary to global climate change?

Community-based responses to climate change, North and South

Feminist political ecology offers a number of insights. First, while global distributional inequities certainly heighten the impacts of climate change, this is not just a problem of the Global South. Intersectional, global analysis and action are needed. Second, since each local situation is complex and different, and women's degrees of power vary widely, local-level actions must serve as the basis for building political movements to advance women's views and economic interests, and acknowledge their social contributions in times of climate change. Finally, to link scales and begin to address the fundamental system changes that are crucial for sustainability, women's solidarity and networking as part of new polycentric governance systems offer the possibility of disrupting unjust political hierarchies and pernicious vested interests, in order to build a more equitable and sustainable future.

The following cases provide a few examples of how this is happening in communities across the globe.

Sister Watersheds: equity on São Paulo's watershed committees

Brazil has a progressive watershed management system which requires participation by civil society representatives on watershed committees, but low-income people and women in particular are underrepresented. Watershed committees are formed "so that water users can collectively help to decide issues of allocation, infrastructure and regulation at the watershed level" (Hinchcliffe et al., 1999; Perkins, 2004). However, low-income local residents and especially women often are not motivated to become involved in these processes, and the participatory engagement rhetoric is increasingly criticized as out of touch with practice (Roledo, 2016; Lemos et al., 2010).

The Sister Watersheds project (2002–2008) linked universities and NGOs in Canada and Brazil in developing strategies and materials for increasing the knowledge, interest and engagement of local residents on water-related issues, focusing on low-income neighbourhoods in São Paulo, and in particular on low-income women. It included student exchanges, research, community engagement and knowledge-sharing in local communities and nearby universities. Conceptualized by progressive Brazilian environmental educators Dr Marcos Sorrentino and Larissa da Costa of the Ecoar Institute for Citizenship, a leading environmental education NGO based in São Paulo, the project's design evolved throughout its implementation by organizers at Ecoar and York University in Toronto.

The project developed and tested training programs by conducting workshops led by its local NGO partners with more than 1,450 participants, approximately two-thirds of them women, and by partnering with other community organizations to present content on topics related to environmental education and watershed management. For example, staff from Ecoar contacted groups of elementary school teachers, public health extension agents, and other community-based workers and provided in-service training for them about water and health, basic ecology, and public policy questions related to water in their local communities. The various training programs were shaped and modified to be specifically appropriate for groups of women, children, youth, health agents, school groups, teachers, film/culture/music/arts organizations and Agenda 21/environmental education groups. The workshops focused on water management, environmental education, community development and democratic participation, with particular emphasis on gender and socio-economic equity.

The focus and methods for each set of workshops were developed collaboratively by Ecoar's community organizers along with the members and leaders of each group of participants, to meet their interests and needs. For example, workshops for local health agents emphasized sanitation, infant and child diseases, nutrition, and ways of spreading health education throughout each neighbourhood community. In this way, the project's organizing and training built on women's socio-economic roles, both unpaid (as mothers, carers, community activists) and paid (as health agents, nurses, cleaners) to increase women's confidence about what they knew and how they could share and use their knowledge for improved watershed governance, for example, by participating in technical committees within the watershed governance structure, attending public meetings, running for office themselves, voting for or working to support public officials with progressive positions on urban water issues. The methodologies, techniques and materials developed for these workshops and training programs – made freely available to other organizations through publications and websites – contributed to the capacity of project partner organizations and individual staff members and students to continue related work on watershed policy issues into the future.

The curriculum materials and techniques developed by the project were tested and fine-tuned in more than 220 workshops designed and led by project staff, student interns, and university exchange students in three watersheds – two in Brazil and one in Canada – where university campuses are located near low-income residential areas. The more than 1,500 Brazilian workshop participants were potential participants in Brazil's watershed committees, as civil society representatives/organizers and as community members whose knowledge and sensitivity on water issues was heightened by the project's educational initiatives. In Canada, most participants were youth attendees at after-school and summer-camp programs run by a local social service organization; they focused on the recreational and community-building aspects of the local watershed, centred on a creek that divides the low-income neighbourhood from the York University campus – one of the most polluted streams in Canada. The day the children donned hip-waders to sample the stream and examine the water for invertebrates, in a workshop led by citizen-science animators at a local conservation organization, they cheered each time they could find something moving under the microscope.

The outreach materials developed by the project include an illustrated Manual on Participatory Methodologies for Community Development containing a set of workshop activities and background materials for participatory community environmental education programs and training sessions focusing on water and gender equity issues; an illustrated guide with practical exercises focusing on urban agroecology; a full-colour socio-environmental atlas which brings together ecological, hydrological and social information about one local watershed in a series of interactive maps; a video about the history and environment of one of the watersheds; a publication outlining Agenda 21 activities in schools; and several blogs and websites with materials and discussion-starters on watershed topics, as well as a book and many journal articles, masters' papers, and other academic publications contributing to the literature on participatory watershed education in Brazil and in Canada. When Brazilian organizers from Ecoar visited the low-income Jane-Finch neighbourhood near York University in Toronto to see what kind of water-related challenges exist in Canada, their arts and music-based participatory energy was a real hit with local youth, most of whom are first-generation newcomers to Canada. Canadian social service organization staff and York students received some great practical training in how to run a phenomenal workshop that day.

This project helped both its university and NGO participants to bridge the gap between academic and community-based methods of environmental education. Graduate exchange students studied and contributed to local training programs; faculty members wrote about the theoretical and practical benefits of public participation in watershed management; NGOs supervised students who received academic credit for their community-organizing work; professors led local watershed governance structures; innovative methods for environmental education were shared internationally. This collaboration allowed new perspectives on water management to evolve, with benefits for all participants' training/education programs. The University of São Paulo, York University and Ecoar developed dozens of new partnerships with other community organizations as a result of this project. Students, both in Brazil and in Canada, played a crucial role in developing the linkages between academic institutions and community-based NGOs. Both locally and internationally, students sought out community organizations for their research and field experiences, and shared the results of their work with both academic and non-academic audiences. The student exchanges of this project thus fueled its interdisciplinary and educational bridging contributions.

This project thus exemplifies all three of the feminist political ecology principles noted above, as well as the beginnings of a polycentric approach: local-level knowledge and priorities served as the foundation for larger-scale political action both in São Paulo and Toronto; intersectional analysis of the links among gender, ethnicity, economic status, environmental quality, and political agency were central to the project's focus; and sharing the specificities of each watershed's situation among the international participants allowed solidarity and increased political awareness to grow. Watersheds in fact are a natural model for polycentric networking, since their branching structure and interrelated upstream-downstream interdependencies facilitate multi-scale understanding and communication.

Climate change and water governance in Durban, Maputo and Nairobi

Building on the Sister Watersheds project, the opportunity for a second water-based environmental organizing and education partnership arose, this one more explicitly engaged with climate justice, linking universities and civil society organizations from three African countries.

According to the Intergovernmental Panel on Climate Change (IPCC),

> Africa is one of the most vulnerable continents to climate change and climate variability. This vulnerability is exacerbated by existing developmental challenges such as endemic poverty, limited access to capital, ecosystem degradation, and complex disasters and conflicts.
>
> (IPCC, 2007)

Income inequality in South Africa, Mozambique and Kenya is among the largest in the world; in all three countries, equity struggles related to water are growing in social, political and ecological significance, which is both a symptom and a cause of urban vulnerabilities related to climate change.

Democratic mediation of equity conflicts related to water, and sustainable long-term management of water resources in the face of climate change, requires public participation, in particular by low-income marginalized women – the experts on water availability and use.

"Strengthening the role of civil society in water sector governance towards climate change adaptation in African cities – Durban, Maputo, Nairobi" was a three-year project (2010–2012) with ten partner organizations (one university and two civil society organizations in each of the three cities, plus York University in Toronto). Its activities included collaboration between students, NGOs and academics as well as community-based research and environmental education. Project partners included the following community-based organizations in Africa: the Kilimanjaro Initiative (KI) and Kenya Debt Relief Network (KENDREN) in Nairobi; Women, Gender and Development (MuGeDe) and Justiça Ambiental (JA) in Maputo; and Umphilo waManzi (Water for Life) and the South Durban Community Environmental Alliance (SDCEA) in Durban. The University of Nairobi (Nairobi), Eduardo Mondlane University (Maputo), and the Centre for Civil Society at the University of KwaZulu-Natal (Durban) provided academic research coordination and student supervision for this project.

The project focused on low-income areas of each city, as these are most severely affected by periodic flooding and other climate change impacts and included training and research sponsorship for students and faculty in the partner universities; support for community-based research, workshops in low-income communities and secondary schools, curriculum and materials development, and skills development within the partner NGOs; training of environmental educators and organizers; contributions to the pool of experienced and qualified community workers in each country; strengthening of all the partner institutions' capabilities to carry out international projects; and contributions to the international literature and professional knowledge concerning water issues, environmental education techniques, and community organizing for improved civil society involvement in governance. The networks built extended from local and community-based linkages through regional and national-level policy groupings to international academic and policy networks on civil society, watershed management and governance.

The political process of policy development and implementation depends on the interchange between civil society groups, researchers generating information on current realities, and government.

One objective of this project was to build partnerships between academics and community activists at the local level in each city. Spanning educational differences and often class and ethnicity, this type of partnership encourages and allows the partner NGOs to reflect on and analyse their activities and to document "learning" more systematically than they are often able to do, by bringing student researchers into the NGOs as collaborators/interns. The partnership also encourages universities to be more pragmatic about teaching and research, and to "field-test" approaches towards community organization, equity and capacity-building. Students committed to the project's goals of fomenting participatory engagement by local people in municipal water decision-making are given practical opportunities to develop their skills, as a way of hastening each city's climate change preparedness. Partly because most of the students and civil society activists involved were women, this project contributed to the integration and meaningful participation of women in formal decision-making processes within their own organizations and communities, as well as helping to build their political agency and understanding of the systemic aspects of climate change. More fundamentally, the project's modeling of gender analysis (keeping track of the representation by gender in workshops and meetings; questioning the gender impacts of project initiatives, water and education conditions, government climate change policies, political access, etc.; seeking out women and girls as participants and interviewees) helped train all partners about gender-sensitive methodologies and priorities and the need for them.

Specific examples of how climate change responses combine well with gender-aware community organizing include the following:

- The Kilimanjaro Initiative (KI), a youth-focused NGO, upgraded a sports field in Nairobi's Kibera slum, on the banks of the Nairobi River, which helps prevent housing from being flooded during extreme weather events. In addition, KI organized community forums on sustainable water management and environmental education, as well as community and river clean-ups. Young women's leadership is central to their organizing, and they prioritize women's access to sport, safety for women and girls in public spaces, and gendered water-related issues (Bilal, 2014).
- In Durban, women activists from Umphilo waManzi and the South Durban Community Environmental Alliance coordinated "learning journeys" where government officials visit low-income neighbourhoods to hear about local women's experiences with flooding, sanitation and other types of climate change stresses, allowing them to bring these views into policy discourse.
- Maputo University environmental education students worked with intermediate school youth on after-school activities related to climate change. Most participants – teachers, community organizers, students – were women and girls. The project thus created opportunities for their education, skills development and political engagement that provided a special emphasis on women's environmental leadership.

While there was less balance between Canadian and African civil society/ university partners (and thus in the sharing of information and knowledge) than in the Sister Watersheds case, this project facilitated exchanges among the African partners that led to insights for the students and activists involved, as well as their communities. After working on local civil society organizations' initiatives in their own cities, several students from Nairobi and

Maputo spent a few months in Durban hosted by activist organizations, to witness how their water-related issues and strategies were similar, and also very different. Canadian students with experience in Toronto water issues and organizing also were able to spend time in Nairobi and Durban to learn about their challenges and how civil society groups were addressing them.

This project, too, thus was grounded in the feminist political ecology principles of intersectional analysis and action, local level priorities as the basis for broader political intervention, and women's solidarity and networking to build multi-scale global systems. Several of the students involved in this project were able to travel to continue their studies in Canada or at other partner universities, and supportive information exchanges on water issues and climate justice continue to be shared among the partners.

Green Change: climate change, urban renewal and jobs in Toronto

The process of working on these international projects has affected my own approach and understanding of the need for climate justice research and action within Canada. Multi-scale, intersectional injustices exist in Canada as they do globally; polycentric structures are required to combat these injustices; building effective political movements must start with grass-roots action; women, especially indigenous young women, are leading in theory and practice.

In Toronto, Canada, the effects of climate change are being noted particularly through high amounts of summer rainfall and sudden storms with intense winds and heavy rain, which seem to be becoming more frequent (Todd, 2011). Higher amounts of rainfall stress the aging urban water/sewer infrastructure, resulting in sewage overflows into Lake Ontario (Binstock, 2011; City of Toronto, 2011). The City has launched basement flooding programs to prevent water backups during rainstorms. The increasing numbers of extremely hot days in the summer have led the City to develop a "cooling centres" program where those without air conditioning can come to public libraries, community centres and other communal spaces which offer extended hours on very hot days.

From an equity perspective, both the weather extremes and the resulting policies have disproportionate negative effects on low-income Toronto residents, particularly women, since they are more likely than higher-income people to occupy basement apartments and to be renters, not house-owners (who can, in contrast, benefit from government infrastructure subsidies). The wage gap between men and women in Ontario has worsened in recent years; women working full-time earn 24 percent less than their male counterparts, and women are twice as likely to be low-income earners. The Ontario poverty rate for women increased from 25 percent of single women in 2007 to 32 percent by 2011 (McInturff, 2014: 16–17). Canadian women over the age of 75 have twice the poverty rate of men in cities – 30.6 per cent to 15.3 per cent (Canadian Council on Social Development, 2007: 11). Lower-income people and women, especially seniors, are also more likely to depend on public transit (which is often disrupted during storms) and on public beaches and parks for recreation; they are less likely to have air conditioning in their homes; and they are more likely to have health conditions which are severely exacerbated by age and heat, such as diabetes, high blood pressure and heart conditions. These equity implications have been noted in some reports (Khosla, 2013; Duncan, 2008; McInturff, 2014; Callaghan et al., 2002; Block, 2010, Canadian Women's Foundation, 2015), but have yet to be analysed or emphasized in official publications or policy frameworks.

One particular low-income neighbourhood in Toronto has borne the brunt of several recent extreme weather events. It is the Jane-Finch neighbourhood, located in northwest Toronto near York University. Due to dense urban growth and university development, the area has become increasingly built-up over the past decade, which has increased surface runoff. During an intense storm in August 2005, more than 150 mm of rain fell in the area, and normally-placid Black Creek became a rushing torrent which washed out its culvert under Finch Avenue, a local arterial roadway, leaving a gaping 50-meter-wide hole. Construction of a new bridge for Finch Avenue took six months and cost more than $3 million. During this time, public buses and commuter traffic had to be rerouted through the York University campus, causing delays and problems for the university and local residents alike. This was a graphic example of how extreme weather events – which are increasing in frequency due to climate change (Toronto Environment Office, 2008) – in conjunction with aging infrastructure, urban sprawl (including campus development), and increasing rapid rainfall runoff, can have costly and traumatic effects on everyone in the watershed. It also shows how organizing at the local level nests with and responds to global-level climate impacts and effects, and how intersectional analysis is fruitful: just as in global comparisons, the low-income neighbourhoods like Jane-Finch in Toronto, and low-income people – those whose consumption is least responsible for carbon emissions – are those least able to protect themselves from climate impacts.

Since 2009, women leaders in the Jane-Finch community have instigated, fundraised for, and brought about many environmental changes, including the planting of a community garden, an eco-friendly Earth Hour event at a local school, tree planting for Earth Day 2010, and a celebration of Earth Day 2011 in a space that was secured as a Centre for Green Change. At the Centre for Green Change – a community training facility modeling green building techniques – local residents and youth who are concerned about the protection of the environment can become engaged in the process of green change and mobilize others, while they increase their own knowledge and skills, initiating individual and collective actions towards building a healthy, safe, prosperous, and environmentally friendly neighbourhood. The vision for the Centre includes a Pathways to Green Jobs Program, to educate and promote environmental stewardship, green jobs, and eco-entrepreneurship, building on successful work of the Carpenters Union's CHOICE Apprentice

Program and incorporating an environmental training component. The Centre for Green Change works with residents to create and to innovate and to use their hands and creative minds while protecting the community's natural resources. The Centre is expanding the number of long-term, high-quality green jobs for local residents, especially for youth. Led first by local environmental activist Rosemarie Powell, and later by Clara Stewart-Robertson, the project has won awards and accolades, including the 2010 Toronto Green Award in the Community category and the 2011 Urban Leader Award for Imagination, and has forged many partnerships with other conservation agencies, environmental groups, unions and community organizations across the city. Community activist Majora Carter was honoured at a Green Change fundraiser, and most of those involved have been young women.

Reasons for the project's success include its focus on connecting with the community directly and its emphasis on local skills and needs. The project aims to identify the skills and knowledge already present within community members – such as gardening, food processing, construction, forestry – and to empower them to use these assets for green change. The level of skills and knowledge in the Jane-Finch community is significant. Many neighbourhood residents are recent immigrants to Canada, and many of them come highly trained and experienced from their home countries, but are unable to break the barrier of employment in Canada. Becoming a "Green Change Agent" provides respect and recognition.

Green Change has also organized local residents to address planning and transportation issues, pipeline controversies, and other environmental challenges in the neighbourhood. Through public meetings and workshops, this form of community organizing educates while strengthening the voice and confidence of marginalized community members so that their views are taken more seriously in public decision-making. Since the large majority of those working with Green Change as organizers, student volunteers and community members are women, this project indicates how women's solidarity, networking and training help to improve governance and build polycentric institutions.

Conclusion: women's networking for climate justice

Beyond the huge differences in the situations outlined above, where I've worked with academic and civil society partner groups on initiatives linking community development, urban water governance, climate justice, environmental education and promoting gender equity, there are also a number of recurring common themes. Women everywhere are the dynamos behind most environmental organizations and activism, especially local issue-based activism (Perkins, 2013a). Where marginalized communities are particularly impacted by environmental hazards such as fossil fuel extraction, processing, and pipelines (as, for example, in South Durban, São Paulo, and Jane-Finch/Toronto), local movements for climate justice are compelling and difficult for higher-level governments to quash; their connections with other such initiatives regionally and internationally are building strong polycentric movements to further climate justice in opposition to the fossil fuel economy. In this movement-building, women's solidarity and networking are crucial.

Water and sanitation are another focus of women's organizing and activism worldwide, and indeed in all of the situations I've been involved with. Because of gendered social roles in both paid and unpaid work, and water's use as both a factor and means of production in cleaning, cooking, agriculture, child and elder care, and healthcare, women's expertise and voice on water issues are strong and hard to deny, even where infrastructure and technical questions tend to dominate policy discussions (as on Brazil's water governance committees). Women's safety, while obtaining water, accessing urban sanitation, and making use of water resources for the benefit of the entire community, is a particular priority in women's water governance organizing at all levels.

Environmental education, another priority for women in Toronto, Maputo, Nairobi, Durban, and São Paulo, can mean formal and informal, child and adult education; education grounds climate justice organizing in local examples of global phenomena. The projects I have described mobilized academic research funds to support grass-roots education and community organizing as well as networking among academic and civil society partners across boundaries of class, ethnicity, and nationality. Each in its own small way, helped make possible women's solidarity, intersectional analysis and action as the basis for polycentric governance and movement-building.

Community-based education and organizing are fundamental to creating the conditions for local knowledge to be shared and utilized, through equitable democratic participation. Building inclusive governance structures and strengthening the role of civil society, especially of women, in water governance and environmental education are essential components for addressing climate change vulnerability and fostering resilience and sustainability in urban centres as well as rural areas (Fordham et al., 2011; Vincent et al., 2010). Non-governmental organizations in the Global South have vast expertise in such initiatives, which is potentially transferable to other places across the Global South and also to some in the Global North.

Community-based environmental education initiatives, which are relevant and interesting for local residents and increase their job opportunities, knowledge of watershed and other local environmental issues, understanding of basic political and ecological principles, and confidence to express and act on their views, can serve as the basis of a climate change intervention approach which is progressive, constructive and democratic. This, in turn, increases the resilience and sustainability of watershed and climate change decision-making processes. It also lays the groundwork for community organizing and extension of the environmental education activities to larger constituencies in local areas affected by climate change. Such grass-roots initiatives – and the global sharing of ideas on how to design and implement them, freely available for adaptation in other places – stand in contrast to top-down climate change adaptation mechanisms controlled from the Global North and mainly by men.

Climate justice – addressing the impacts of climate change on the poorest first – is a powerful imperative at every level, from the local to the global. Civil society groups worldwide, mostly led by women, are using online and in-person networking tools as appropriate in each particular situation to share ideas on how to educate their communities, promote climate justice, obtain funding, press politically for policies that address the needs of marginalized people, and if ignored by governments, take action themselves. This bottom-up movement builds resilience and contributes to strong polycentric governance in the face of the social and political repercussions of extreme weather events and other impacts of climate change.

References

Alston, M. and Whittenbury, K. (eds), (2013) *Research, Action and Policy: Addressing the Gendered Impacts of Climate Change*. Dordrecht/Heidelberg/New York/London: Springer.

Awâsis, S. (2014) Pipelines and resistance across Turtle Island. In T. Black, S. D'Arcy, T. Weis and J.K. Russell (eds), *A Line in the Tar Sands: Struggles for Environmental Justice*. Toronto: Between the Lines, pp. 253–266.

Bilal, S. (2014) Interview in *A soccer field in Kenya* (video, directed by Sara Marino). Available online at http://ccaa.irisyorku.ca/publications/video-a-soccer-field-in-kibera/ (accessed September 4, 2016).

Binstock, M. (2011) *Greening Stormwater Management in Ontario*. Toronto: Canadian Institute for Environmental Law and Policy (CIELAP). Available online at http://cielap.org/pub/pub_greeningstormwaterman.php (accessed June 23, 2011).

Black, T. (2016) Race, gender, and climate injustice: dimensions of social and environmental inequality. In P. Godfrey and D. Torres (eds), *Systemic Crises of Global Climate Change: Intersections of Race, Class and Gender.* New York: Routledge.

Black, T., D'Arcy, S., Weis, T. and Russell, J.K. (2014) *A Line in the Tar Sands: Struggles for Environmental Justice*. Toronto: Between the Lines.

Block, S. (2010) *Ontario's Growing Gap: The Role of Race and Gender*. Ottawa: Canadian Centre for Policy Alternatives. Available online at http://ywcacanada.ca/data/research_docs/00000140.pdf (accessed September 4, 2016).

Bond, P. and Dorsey, M.K. (2010) Anatomies of environmental knowledge and resistance: Diverse climate justice movements and waning eco-neoliberalism, *Journal of Australian Political Economy*, 66(December): 286–316. Available online at http://search.informit.com.au/documentSummary;dn=833845077924067;res=IELBUS> ISSN: 0156-5826 (accessed May 4, 2011).

Buckland, K. (2015) The common(s) denominator: Oil and water on a common river, *EJOLT Report*, 23(September): 63–73. Available online at www.ejolt.org/wordpress/wp-content/uploads/2015/09/climate-justice-report.pdf (accessed September 5, 2016).

Buechler, S. and Hanson, A. (2015) *A Political Ecology of Women, Water, and Global Environmental Change*. New York; London: Routledge.

Callaghan, M., Farha, L. and Porter, B. (2002) *Women and Housing in Canada: Barriers to Equality*. Centre for Equality Rights in Accommodation. Available online at www.equalityrights.org/cera/docs/CERAWomenHous.htm (accessed September 4, 2016).

Canadian Council on Social Development. (2007) Age, gender and family: Urban poverty in Canada, 2000. Available online at www.ccsd.ca/images/research/UPP/PDF/UPP-AgeGenderFamily.pdf (accessed September 4, 2016).

Canadian Women's Foundation. (2015) Fact sheet: Women and poverty in Canada. Available online at www.canadianwomen.org/sites/canadianwomen.org/files//Fact%20Sheet%20-%20WOMEN%20%26%20POVERTY%20-%20Dec%2014%202015_1.pdf (accessed September 4, 2016).

City of Toronto. (2011) WWFMP – Going for the flow. Available online at www.toronto.ca/water/protecting_quality/wwfmmp/about.htm (accessed May 8, 2011).

Climate Justice Now! (2013) Definition of Climate Justice. Available online at www.climate-justice-now.org/em-cjn/mission/ (accessed November 28, 2013).

Coats, E. (2014) What does it mean to be a movement? A proposal for a coherent, powerful, indigenous-led movement. In T. Black, S. D'Arcy, T. Weis and J.K. Russell (eds), *A Line in the Tar Sands: Struggles for Environmental Justice*. Toronto: Between the Lines, pp. 267–278.

Cohen, M.G. (ed.), (2017) *Gender, Climate Change and Work in Rich Countries*. London; New York: Routledge.

COSATU – Confederation of South African Trade Unions. (2012) A just transition to a low-carbon and climate resilient economy: COSATU policy on climate change, a call to action. [pdf] Confederation of South African Trade Unions. Available online at www.cosatu.org.za/docs/policy/2012/climatechange.pdf (accessed November 28, 2013).

Duncan, K. (2008) *Feeling the Heat: Women's Health in a Changing Climate*. Canadian Women's Health Network. Available online at www.cwhn.ca/en/node/39416 (accessed September 4, 2016).

EIGE – European Institute for Gender Equality. (2012) Review of the implementation in the EU of area K of the Beijing Platform for Action: Women and the Environment. Gender equality and climate change. [pdf] European Institute for Gender Equality. Available online at http://eige.europa.eu/sites/default/files/Gender-Equality-and-Climate-Change-Report.pdf (accessed November 28, 2013).

Fordham, M., Gupta, S., Akerkar, S. and Scharf, M. (2011) Leading resilient development: grassroots women's priorities, practices and innovations. GROOTS International, Northumbria University School of the Built and Natural Environment and United Nations Development Programme. Available online at www.undp.org/content/dam/aplaws/publication/en/publications/womens-empowerment/leading-resilient-development–grassroots-women-priorities-practices-and-innovations/f2_GROOTS_Web.pdf (accessed September 4, 2016).

Gibson-Graham, J.K. (2006) *A Postcapitalist Politics*. Minneapolis: University of Minnesota Press.

Godfrey, P. and Torres, D. (eds), (2016) *Systemic Crises of Global Climate Change: Intersections of Race, Class and Gender*. New York; London: Routledge.

Gorecki, J. (2014) 'No climate justice without gender justice': women at the forefront of the People's Climate March. *Feminist Wire*, September 29. Available online at www.thefeministwire.com/2014/09/climate-justice-without-gender-justice-women-forefront-peoples-climate-march/ (accessed September 4, 2016).

Habtezion, S. (2013) Overview of linkages between gender and climate change. *Gender and Climate Change Training Module 1.* New York: United Nations Development Programme. Available online at www.undp.org/content/dam/undp/library/gender/ Gender%20and%20Environment/TM1_Africa_GenderClimateChange_Overview. pdf (accessed June 21, 2015).

Hinchcliffe, F., Thompson, J., Pretty, J., Gujit, I. and Shah, S. (eds) (1999) *Fertile Ground: The Impacts of Participatory Watershed Management.* London: Earthscan/ IT Publications.

ILO. (2009) Green jobs: improving the climate for gender equality too! Available online at www.ilo.org/wcmsp5/groups/public/@dgreports/@gender/documents/publication/ wcms_101505.pdf (accessed November 28, 2013).

Intergovernmental Panel on Climate Change (IPCC). (2007) Climate Change 2007: Synthesis Report. Available online at www.ipcc.ch (accessed May 4, 2011).

IRIS. (2009) Report of the First International Ecojustice Conference: How will disenfranchised peoples adapt to climate change? Strengthening the Ecojustice movement. Institute for Research and Innovation in Sustainability, York University, Toronto. Available online at www.irisyorku.ca/wp-content/uploads/2010/11/Ecojustice-Conference_final.pdf (accessed December 2, 2011).

Kaijser, A. and Kronsell, A. (2014) Climate change through the lens of intersectionality, *Environmental Politics*, 23(3): 417–433.

Karlsson, G. (ed.), (2007) Where Energy is Women's Business. National and Regional Reports from Africa, Asia, Latin America and the Pacific. [pdf] Available online at www.energia.org/cms/wp-content/uploads/2015/04/06.-karlsson_csdbook_lores. pdf (accessed September 4, 2016).

Kaufman, C. (2012) *Getting Past Capitalism: History, Vision, Hope.* Lanham, MD: Lexington Books.

Khosla, P. (2013) Women's poverty in cities. National Network on Environments and Women's Health and Toronto Women's Call to Action. Available online at www.twca.ca/ wp-content/uploads/2013/02/Women_Poverty_in_Cities.pdf (accessed September 3, 2016).

Klein, N. (2014) *This Changes Everything.* NY/Toronto: Random House, Knopf.

LaDuke, W. (2016) *The Winona LaDuke Chronicles: Stories from the Front Lines in the Battle for Environmental Justice.* Ponsford, MN: Spotted Horse Press.

Lemos, M.C., Bell, A.R., Engle, N.L., Formiga-Johnsson, R.M. and Nelson, D.R. (2010) Technical knowledge and water resources management: A comparative study of river basin councils, Brazil, *Water Resources Research*, 46(6), June. Available online at http:// onlinelibrary.wiley.com/doi/10.1029/2009WR007949/full (accessed September 3, 2016).

Manuel, Arthur, (2015) *Unsettling Canada: A National Wake-Up Call.* Toronto: Between the Lines.

Masterman-Smith, H. (2011) Green jobs and a just transition for women workers? Available online at www.ntwwc.com.au/uploads/File/OWOL%20conference%20 papers/Masterman-Smith.pdf (accessed November 28, 2013).

McInturff, K. (2014) Ontario's gender gap: Women and jobs post-recession. Canadian Centre for Policy Alternatives. Available online at www.policyalternatives.ca/ sites/default/files/uploads/publications/Ontario%20Office/2014/03/Ontario's%20 Gender%20Gap%20FINAL.pdf (accessed September 4, 2016).

Morães, A. (2011) Gendered waters: the participation of women in the program 'one million cisterns' in the Brazilian semi-arid region. PhD dissertation. University of Missouri.

Mulheres Pedreiras: Curso vai capacitor agricultoras para a construção de cisternas de placas, (2013) *Agricultura Familiar e Agroecologia.* Available online at http:// aspta.org.br/2013/11/mulheres-pedreiras-curso-vai-capacitar-agricultoras-para-a-construcao-de-cisternas-de-placas/ (accessed November 28, 2013).

Nagel, J. (2015) *Gender and Climate Change: Impacts, Science, Policy.* London: Routledge.

Ostrom, E. (2009) Beyond markets and states: polycentric governance of complex economic systems. Nobel Economics Prize lecture, December 8. Available online at www.nobelprize.org/nobel_prizes/economic-sciences/laureates/2009/ostrom_lecture. pdf (accessed October 29, 2013).

Ostrom, E. (2010) Polycentric systems for coping with collective action and global environmental change, *Global Environmental Change,* 20: 550–557.

Ostrom, E. (2014) A polycentric approach for coping with climate change, *Annals of Economics and Finance,* 15(1): 97–134.

Perkins, P.E. (2004) Participation and Watershed Management: Experiences From Brazil,' paper presented at the conference of the International Society for Ecological Economics (ISEE), Montreal, Canada, July 10–14.

Perkins, P.E. (2013a) Environmental activism and gender. In D. Figart and T. Warnecke (eds), *Gender and Economic Life.* Northampton, MA: Edward Elgar, pp. 504–521.

Perkins, P.E. (2013b) Green community development and commons governance. In USSEE (United States Society for Ecological Economics), University of Vermont, Burlington, June 9–12.

Perkins, P.E. (2017) Canadian indigenous female leadership and political agency on climate change. In M.G. Cohen (ed.), *Gender, Climate Change and Work in Rich Countries.* London and New York: Routledge.

Perkins, P.E. and Figueiredo, P. (2013) Women and water management in times of climate change, *Journal of Cleaner Production.* Special issue on Women, Water, Waste, Wisdom, and Wealth. Available online at www.sciencedirect.com/science/article/pii/S0959652612001011 (accessed November 28, 2013).

Roledo, C. (2016) Governança da água: um estudo sobre a gestão e a qualidade da água da sub-bacia hidrográfica do Rio Una (São Paulo). Master's dissertation, University of São Paulo, Faculty of Public Health.

Salleh, A. (2009) *Eco-sufficiency and Global Justice: Women Write Political Ecology.* London/New York/Melbourne: Pluto Press/Spinifex.

Thomas-Mueller, C. (2014) The rise of the native rights-based strategic framework: our last best hope to save our water, air, and earth. In T. Black, S. D'Arcy, T. Weis and J.K. Russell (eds), *A Line in the Tar Sands: Struggles for Environmental Justice.* Toronto: Between the Lines, pp. 240–252.

Todd, A. (2011) Climate change and water governance in the Greater Toronto Area. MES paper, York University.

Toronto Environment Office (2008) Ahead of the Storm… Preparing Toronto for Climate Change. Available online at www1.toronto.ca/City%20Of%20Toronto/ Environment%20and%20Energy/Our%20Goals/Files/pdf/A/ahead_of_the_storm. pdf (accessed 4 September 2016).

Vincent, K., Wanjiru, L., Aubry, A., Merson, A., Nyandiga, C., Cull, T. and Banda, K. (2010) *Gender, Climate Change and Community-based Adaptation.* New York:

United Nations Development Programme. Available online at file:///Users/esperk/ Downloads/Gender%20Climate%20Change%20and%20Community%20Based%20 Adaptation%20(2).pdf (accessed September 4, 2016).

Wintermayr, I. (2012) How can women benefit from green jobs? An ILO approach. *European Parliament, Women's and Gender Equality Committee.* [pdf] Available online at www.europarl.europa.eu/document/activities/cont/201203/20120301ATT39684/ 20120301ATT39684EN.pdf (accessed November 28, 2013).

5 Gender and urban climate change policy

Tackling cross-cutting issues towards equitable, sustainable cities

Gotelind Alber, Kate Cahoon and Ulrike Röhr

Introduction

When world leaders came together on the first day of COP 21, which two weeks later culminated in the Paris Agreement, the scene was strikingly familiar. Heads of state – predominately male – shook hands and posed for the cameras, emphasising the role of their country in the global plan to avoid dangerous climate change by limiting global warming to well below 2°C. Notably, of the 23,000 government representatives sent to negotiate a new climate agreement, less than 20 per cent were women – a considerable drop from previous COPs, which had seen a slow but steady increase in women's participation.[1] The outcome of COP 21 was widely celebrated as a breakthrough, yet strong commitments from states were lacking in the final text, as were meaningful references to human rights and gender equality. Reading between the lines, critics observed that it seems increasingly unlikely that the ambitious goal of keeping global warming below 1.5–2 degrees could be met within the necessary timeframe and, worryingly, that states were not willing to ensure that future climate action would be undertaken in a fair and equitable manner.

In light of the slow pace of international climate negotiations, it is unsurprising that an ever-growing number of sub-national actors and stakeholders are engaging in climate action. Cities, in particular, have been identified as central to climate responses, given the urgency of limiting global greenhouse gas emissions and dealing with the unavoidable impacts of climate change in urban areas. More and more cities around the world are undertaking climate action planning and implementing local strategies and policies, yet so far little consideration has been given to the integration of gender dimensions. Indeed, until recently, research on the linkages between gender and climate change has largely focused on climate impacts and adaptation, often in rural settings. The specific challenges which emerge in an urban context remain largely unaddressed, highlighted by the very limited number of case studies or resources providing examples of gender-responsive urban climate policies.

This gap at the urban level stands in contrast to the recent progress made on gender and climate change in other areas, including in international and national climate policy. While gender advocates and women's organisations have been highly critical of the lack of gender awareness in the UNFCCC process in the past, a broad range of gender-related issues were addressed in the lead up to the Paris Agreement, coming down to lengthy discussions between Party delegates about the need to ensure gender equality as a cross-cutting principle for climate action and the placement of gender and rights language throughout the text. Following the COP 18 'Gender Decision' and the 'Lima Work Programme on Gender',[2] Parties have shown a growing willingness to consider the gender dimensions of climate change – 160 countries included a reference to either 'women' or 'gender' in their Nationally Determined Contributions, which will set the path for national-level climate action post-2020.

Even if implementation on the ground still remains challenging (see Bee, this volume), these developments are testament to the evolving nature of climate policy and can arguably be seen as a reflection of the deepening understanding that climate change, as a complex issue, will need to be met with comprehensive and multifaceted responses. For cities, too, COP21 was seen as a crucial moment for their role to be acknowledged and consolidated, evidenced by a strong show of presence from the local government and municipal authorities constituency (LGMA) and city networks. In Paris, hundreds of mayors from around the world signed the Paris City Hall Declaration during the Climate Summit for Local Leaders at Paris City Hall on 4 December. These actions were largely symbolic, yet the growing presence of cities in the climate negotiations, and the increasing commitment of cities in the Global South, suggest a broadening of the international climate agenda – moving away from a narrow focus on the national level. This can potentially open up space for new approaches, such as the channelling of climate finance to urban climate policy efforts and the involvement of civil society actors at local level.

In this chapter, we will argue that urban climate policy is emerging as a key site of action from a gender-perspective. Examining the challenges faced by cities in the context of climate change through a gender lens, we will highlight several issues that remain underexplored, including the profound inequality that exists in urban areas and the need to recognise the role of 'care' and gendered labour in a city's economy. In an effort to move from analysis towards finding solutions, we will explore the extensive options available to cities to reduce their considerable contributions to global greenhouse gas emissions, enhance the resilience of their populations and reduce disaster risks in a gender-responsive manner. Finally, we provide a number of recommendations for how urban climate policy can be used as a transformative tool to work towards more equitable, resilient and sustainable cities.

Urbanisation, inequality and climate challenges

Urbanisation presents a growing challenge closely linked to climate change and the actions taken to address it. At present, 54 per cent of the world's population lives in urban areas (UN DESA, 2014). With population growth and rapid urbanisation, this figure is expected to increase to over 64 per cent by 2050, representing an additional 2.5 billion people living in urban areas, predominantly in rapidly growing cities in Africa and Asia (UN DESA, 2014).

This trend goes hand in hand with increasing inequalities, largely because in developing countries, urban growth processes are primarily driven by low-income populations (George and McGranahan, 2013). Moreover, in both high- and low-income countries, larger cities are characterised by higher levels of inequality than their smaller counterparts (Behrens and Robert-Nicoud, 2014; Mattoo and Subramanian, 2010), to the extent that the divide between the privileged and underprivileged in most cities is essentially as large as the global divide between industrialised and low-income countries (UN-Habitat, 2008). The higher system complexity of large cities is often connected with increasing inequality, because greater specialisation of skills leads to segregation according to wealth, income, status and power (Hodgson, 2003). This discrepancy becomes particularly apparent in the spatial inequality easily visible in cities, where a small percentage of the population has access to more space for generous housing, mobility and recreational activities, while the majority is crowded together in densely-populated and often most polluted areas. The chronic lack of low-income housing in most growing cities has resulted in the multiplication of inadequate makeshift dwellings, or slums. In addition to income inequality and marginalisation, the urban poor face unequal access to formal employment, but also unequal access to water, sanitation and the rule of law (George and McGranahan, 2013). This is often compounded by the concentration of infrastructure investments – governed by the interests of a public sector commandeered by private interests and the real estate market – in middle- and upper-class residential areas (George and McGranahan, 2013).

Spatial inequality, and urban inequality more broadly, is a relevant factor for climate change, both in terms of the causes of climate change and vulnerability to its impacts. The poorest groups, such as slum dwellers, have much smaller carbon footprints than the wealthy, who participate in overconsumption and lead carbon-intensive lifestyles. Despite having contributed the least to the problem, the poor are often the hardest hit and face considerable obstacles in adapting to the impacts of climate change. Dense human settlements or informal settlements are particularly exposed to climate hazards and are often located in landslide or flood prone areas (UNDP, 2008). Furthermore, rapid and unplanned urbanisation can place pressure on already strained infrastructure and services, upon which urban populations are often highly dependent. Adopting a climate justice lens when looking at human settlements therefore reveals that 'climate change is not simply happening to cities, but rather is being produced through the city and in turn serving to reproduce or challenge existing forms of uneven development and urban inequality' (Bulkeley, Edwards and Fuller, 2014: 34).

Cities will face considerable challenges in addressing these issues of poverty and marginalisation alongside worsening climate impacts. Approximately 360 million urban residents live in coastal areas less than 10 metres above sea level and are vulnerable to flooding and storm surges (Moser and Satterthwaite, 2008). Fifteen of the world's 20 megacities are at risk from rising sea levels and coastal surges, and while these are largely concentrated in developing countries, in Europe, 70 per cent of the largest cities also have areas that are particularly vulnerable to rising sea levels (World Bank, 2010: 16). Severe heatwaves are also expected to become more frequent and intense with climate change, affecting urban areas most strongly. During particularly severe heat waves in 2002 and 2003, between 35,000–70,000 people are estimated to have died in Europe (Dhainaut *et al.*, 2003; Pirard *et al.*, 2005; World Bank, 2010). The victims were disproportionally elderly citizens – in particular elderly women – in urban areas, highlighting the social vulnerability of certain groups, both in the Global North and South. A growing body of research outlines the potentially disastrous impacts of climate on settlements and infrastructure, with loss of life and assets, negative health impacts, salinisation of water sources, water shortages, high food prices, food insecurity and disruption to livelihoods and city economies (World Bank, 2010). Learning to cope with and address such situations is thus of upmost importance to local governments and it is unsurprising that a growing number of cities are already developing comprehensive climate responses.

At the same time, cities have a clear role to play in tackling global emissions. According to estimations provided by UN Habitat, between 60 and 70 per cent of energy-related greenhouse gas emissions are released in cities, and further growth is predicted in light of urbanisation trends (UN-Habitat, 2011). In some sectors, such as urban development and spatial planning, urban policymakers are able to shape settlements and draw on synergy effects from the combination of mitigation and adaptation by prioritising low-carbon housing and mobility. While the options for equitable urban climate policy will be more fully explored later in this chapter, it is worth noting that local governments are in a good position to integrate social issues with climate action in order to challenge existing patterns of inequality and marginalisation.

Addressing gender dimensions of climate change in urban contexts

Research on the social dimensions of climate change suggests that climate change impacts will serve to further exacerbate existing inequalities. It is perhaps nowhere more obvious than in cities that a range of factors interact to produce social inequality – including gender, but also income, class, race, ethnicity, age, education and so on. Caretaking roles are also highly relevant for daily urban life – childcare responsibilities, for example, are generally more relevant to transport needs and mobility patterns than are many other factors. The framework of intersectionality can provide a useful tool for understanding how these social categories overlap and are mutually reinforcing. Scholars have argued that an intersectional analysis shows that the way individuals relate to climate change depends on their positions on context-specific power structures based on social categorisations, at the same time helping to avoid oversimplification based on stereotypes or static categories (Kaijser and Kronsell, 2014). Adopting a gender lens to urban poverty, for example, reveals that female-headed households are more likely to experience poverty and marginalisation and to a greater degree of severity (UN DESA, 2010). Similarly, while households are often taken as the standard unit of analysis, there is reason to believe that considerable intra-household disparity can exist between men and women within the family and the household, in terms of consumption, cash income, control over earnings and expenditure (UN DESA, 2010).

These examples serve to highlight that poverty is only part of the picture. Research on the linkages between gender and climate change over the past decade has highlighted that socially-constructed gender roles and identities and underlying power relations affect the way women and men experience and respond to climate change (Skinner, 2011). In cities women are said to enjoy greater social, economic and political opportunities than their rural counterparts, yet there is considerable evidence to suggest that barriers to women's empowerment remain widespread in urban areas (UN-Habitat, 2013). Contributing factors include persistent legal and sociocultural discrimination, the 'gender pay gap' and the gendered division of labour, as well as gender-based violence and sexual harassment, both in the home and in public spaces. While the full extent of these disparities cannot be explored in great length here, certain aspects are particularly relevant in the context of rapidly worsening climate change and increasing urbanisation. There are many specific conditions in cities that interact with climate change, or could be exacerbated both by climate impacts, with gendered implications.

As mentioned above, a key consideration is the gendered division of labour which persists in all countries in the world and which remains a core underlying element of every city's economy. This gendered division characterises both the paid labour force and the unpaid 'care economy'. While men's labour is largely concentrated in 'productive/income-generating work', women undertake the major role in 'reproductive, unpaid labour, which includes routine domestic chores as well as more specialised care work' (Chant, 2013: 13). As a result, women are more likely to be given the primary responsibility for family care, including the provision of food, caring for children, elderly and sick family members (Miranda, 2011). In the event of climate change, and in particular in post-disaster situations, this workload can increase, resulting in an even greater care burden for women. Challenges such as a lack of food security, and insufficient access to energy, mobility and water services, can make the task of caring for family members more difficult and time-consuming (Chant, 2013: 19).

From a gender perspective, a consideration of time use is therefore crucial, given that across all regions of the world, women spend on average between three and six hours on unpaid care activities, while men spend between half an hour and two hours. Although there are regional variations, overall women spend two to ten times more time on unpaid care work than men (ActionAid, 2013; see also OECD, 2014), who subsequently have more leisure time. This kind of unpaid labour is generally under-recognised and valued less than 'productive' labour and is still almost completely neglected in conventional economic statistics and assessments of urban economic activities.

Subsequently, women in rural and urban areas experience what is sometimes referred to as time poverty, alongside other gendered labour market inequalities. While it has been argued that rural women are often more 'time poor' than those living in cities, studies of urban prosperity and poverty highlight that stark contrasts often exist between women's inputs to and benefits from the accumulation of wealth in the cities (Chant, 2013). Indeed, although urbanisation is often associated with greater access to employment opportunities, lower fertility levels and increased independence for women, the gendered division of labour has significant implications for women's ability to actively take part in the labour market and the type/quality of employment opportunities available to them (OECD, 2014). In cities in the Global South, for example, research on urban poverty has highlighted that poorer women are compelled to combine unpaid care activities with paid activities to survive in cash-based economies, often from a very young age (Chant, 2013; Tacoli, 2012). In many cases, this involves work in the lowest-paid formal and informal sector activities with long working hours and travel times, in addition to time spent on existing unpaid domestic labour and care work (Chant, 2013; see also Beyazit, 2015 and Jirón and Imilan, 2015). In general, women are more likely to be involved in informal employment and also for a longer period of time than men; in many parts of the world – although predominantly in the Global South – women's informal economic activities are commonly based at home, as a result of constraints on women's spatial mobility arising from moral and social norms, as previously mentioned, and due to the demands placed on women by reproductive ties (Chant, 2013: 15). Statistics on home-based work are difficult to come by, yet estimates suggest that women make up the vast majority of the world's estimated 100 million home-based workers. More than half of these workers are found in South Asia, where the share of women is around 80 per cent (WIEGO, 2007). Domestic-based income-generating options are especially limited in nature and earning potential for female slum dwellers, yet those living and working in informal settlements face the significant risk of losing their source of income alongside homes in the event of a flood or other climate-related disasters (Alber, 2010).

In urban areas, service deficiencies and associated time burdens limit income generation among women, given that access to services, including childcare and health services, water and transport, are often decisive for women's labour force participation (Chant, 2013). Disruptions to services due to climate change serve to further exacerbate existing challenges, limiting the capacity of poorer women and other vulnerable groups to adapt and make necessary changes.

Rural–urban migration, too, is an increasingly relevant consideration in the context of climate change, with the use of migration as a coping factor already shown to have clear gender impacts. While many studies and development programmes have focused on male out-migration in rural communities, more recent research highlights that when faced with natural disasters and a diminishing resource base, women may seek to migrate as well, usually to urban centres (IOM, 2008: 34). The International Organisation for Migration (IOM) argues that while lone women migrants will face similar challenges to their male counterparts in finding employment, affordable housing, and accessing social services, they may have a more difficult time in urban areas due to gender-based discrimination (IOM, 2008). These difficulties are largely specific to local contexts, yet existing research highlights certain gendered migration patterns or trends, such as that of rural women from poor or peasant families migrating to cities often ending up in poorly paid-jobs or the informal sector due to their lack of education and skills, as has been the case in India and Peru (Chindarkar, 2012; IOM, 2008). Similar observations have been made in the Philippines where women from the fishing communities, who were grappling with the harsh impacts of climate change, migrated locally into cities to work as domestic help for affluent families (UNFPA, 2009: 3).

Migration can take place both as a result of slow-onset climatic changes, such as drought and salinisation, and in the wake of climate-related disasters and severe weather events. However, in practice, climate change can rarely be isolated as a sole cause for migration, making it 'difficult, if not impossible, to stipulate numbers' for future migration flows (Kolmannskog, 2008). Nevertheless, it is clear that climate change has an impact on the movement of persons, and that the number of displaced persons and migrants are expected to rise in the future, potentially including the displacement of entire populations and with clear impacts for urbanisation (Kälin and Schrepfer, 2012). Forced migrants, especially forced to flee quickly from climate events, are at greater risk of health concerns, sexual exploitation, human trafficking and sexual and gender-based violence (IOM, 2008). Rapid, unplanned migration can also result in a dramatic growth in informal settlements. Rural–urban migration has also been linked to a rise in female-headed households within cities, as well as, more generally, to a diversification of households among low-income groups and demographic mobility with urban settings (Chant, 1998; Tacoli and Mabala, 2010). Policymakers will therefore face the challenge of changing demographics due to climate change alongside direct climate impacts.

As previously highlighted, informal settlements and low-income populations are exposed to greater climate risk due to geographical vulnerability, high density and weak structures, yet limitations to adequate housing also have considerable gender dimensions. Gender continues to be a major axis of discrimination in housing access and in particular housing ownership, for example, although recent research suggests that this is apparently less entrenched in urban than rural areas, with a tiny handful of cities even having achieved gender parity in house ownership (Chant and McIlwaine, 2015). Nevertheless, estimates show that women represent less than 15 per cent of land and property owners worldwide and data gathered from 16 low-income urban communities in developing countries showed that only one-third of owner-occupiers were female (Chant, 2013). The proportion of women living in slums is also considerable. For example, women make up 80 per cent of the slum dwellers in the Tondo district of Manila in the Philippines – the largest slum in South East Asia (Brot für die Welt 2009 in Alber, 2010). In the Global North, too, women often face limitations in terms of access to affordable, quality housing. Low-quality, energy inefficient housing has been identified as a factor which contributes to fuel poverty. According to the World Health Organization (WHO), in some European countries, up to 30 per cent of the population cannot afford sufficient fuel for heating to maintain an adequate indoor temperature (Pye and Dobbins, 2015; Rehfuess, 2006). Women, especially single mothers and elderly women, are more likely to experience fuel poverty due to gendered income differentials and subsequently lower pensions.

Similarly, although women are in the majority among urban farmers in many cities around the world, they tend to carry out subsistence-based farming activities, whereas men play a greater role in urban food production for commercial purposes (Hovorka *et al.*, 2009). Studies on gender and urban agriculture subsequently highlight the need to identify and address potential gender dimensions linked to food security, in order to avoid entrenching existing gender differentials (Hovorka *et al.*, 2009).

These examples serve to highlight that gender dimensions are relevant in light of a range of different climate challenges in urban areas, warranting further consideration and with consequences for most key sectors within cities – from housing to transport and energy.

Addressing gender dimensions of urban climate policies

While this is by no means an exhaustive exploration of the gendered implications of climate change in urban areas, it is also important to consider how policy responses can address these challenges and, importantly, to minimise further negative impacts. Given the relevance of time-use from a gender perspective, for example, policies which have the (unintended) effect of increasing the work burden of family care-givers can be detrimental, as is the case with certain energy-saving measures, the utilisation of biomass-based heating systems or waste separation (Carlsson-Kanyama and Lindén, 2007; Organo *et al.*, 2013). As feminist scholars and gender experts have criticised, climate responses have tended to be gender-blind and androcentric,[3] with the result that male perspectives are favoured or adopted without question, neglecting women's identities, attitudes and behaviour or regarding these as deviations from the 'norm' (Jarvis, 2009; Roehr *et al.*, 2008). Cities are no exception, and although some local governments have made efforts to conduct gender mainstreaming or address gender dimensions in urban planning and budgeting, this has generally not been extended to climate policies and measures (Alber, 2010).

However, taking social considerations – including gender – into account in policy responses can allow for the different needs and priorities of city inhabitants to be addressed. An extensively researched example is transport, which has both clear gender dimensions and significant relevance for mitigation efforts in cities.

In both the Global North and South, statistics on transport use reveal differences in the way men and women use transport, including trip patterns – men are more likely to commute and/or to travel for leisure, while women tend to work closer to, or at home and make shorter trips, more often, for work, shopping and childcare purposes (Breengaard *et al.*, 2007; European Institute for Gender Equality, 2012; Tran and Schlyter, 2010). In general, women travel shorter distances and have more limited access to motorised means of transport, making them more dependent on public transport services (GIZ, 2015; Tran and Schlyter, 2010). These differences provide some explanation for men's higher energy consumption, which, combined with their higher income, spending and food preferences, results in a larger average carbon footprint compared to that of women. Both adaptation and mitigation strategies can benefit from such insights, which can help to identify specific vulnerabilities or needs, key target groups and priorities for action.

Furthermore, the considerable 'gender gap' which continues to exist in leadership and decision-making can be particularly apparent at the local level, where women are often underrepresented in local decision-making structures or bodies and relevant sectors such as transport and energy. An analysis by the Climate Alliance of European Cities (2005) indicated that in positions relevant to climate change policy, such as urban energy and transport planning, the share of women is considerably lower than men, particularly in executive positions. Arguably, addressing the underrepresentation of women in these fields is more a matter of promoting the right of women to have an equal voice than necessarily resulting in improved understanding of gender issues (given that gender balance does not automatically reflect in greater gender awareness), yet it is crucial to ensure that a diversity of opinions and preferences is heard and recognised in the context of climate responses.

Gender can play a role when it comes to preferences for mitigation options, for instance, research in Europe shows that women are more likely to reject high-risk technologies and more willing to make changes to their own behaviour and consumption (European Institute for Gender Equality, 2012). Surveys in urban areas also reveal that residents of cities have various priorities and needs which are often gender-differentiated, in terms of access to resources, services and infrastructure, and different capabilities to make changes. Due to their lower socio-economic status, women are on the whole less likely to own their own house, or as rental tenants they are less likely to be in the position to make decisions relating to major investments such as heating systems, renovations or household appliances (European Institute for Gender Equality, 2012). However, though women often lack the opportunity to carry out energy efficiency-related improvements or investments, it has also been observed that they are often more prepared and willing to make changes (European Institute for Gender Equality, 2012).

It is worth noting, however, that this does not necessarily mean that women are more 'climate-friendly' – in many cases they merely lack the means to generate a large carbon footprint, as a result of more limited access to energy and transport services, or financial means for consumption. Heightened awareness of the 'risks' of climate change and technologies is presumed to be a reflection of socially constructed roles and expectations, such as that women 'care' for the fate of future generations, rather than any inherent green tendencies that come naturally to women (Hemmati and Roehr, 2007; MacGregor, 2010).

Thus, without meaning to generalise or over-simplify the categories of 'men' and 'women', a gender analysis does reveal that policies need to respond to the needs of both women and men and speak to the social context in which they are implemented in order to avoid further entrenching existing inequalities. It remains to be seen how truly transformative change, such as an energy transformation which brings together supply and demand, could serve to challenge deeply entrenched societal patterns, including the separation of reproduction and production (Kanning *et al.*, 2016). Gender differentiation can therefore help to develop a more nuanced understanding of how climate change impacts upon urban populations, as well as how greenhouse gas emissions occur at the level of the end user, ideally allowing policymakers to tailor their climate policy interventions and develop effective and equitable responses –with positive impacts on gender relations, and ideally, transformative potential.

From gender-responsive to gender-transformative: the culture of care as a prerequisite for a low-carbon city

While mainstreaming gender considerations into climate policy is an important step in its own right, the necessary transition towards an equitable low- or zero-carbon society would arguably require more than just adding gender to existing climate responses. It certainly requires more than introducing resource-efficient technologies and 'greening' production, which is currently the main focus of national and local climate policy strategies in Europe and other parts of the world. In order to be truly transformative, climate policies need to address the root causes and drivers of climate change, at the same time as contributing to making society more equitable. There is good reason to look to cities as potential drivers of these kinds of change. Cities are not only a main actor in climate change mitigation, they are the level of authority closest to citizens – their policies directly impact the lives and livelihoods of city inhabitants. Cities thus have the ability to either stimulate or limit the innovation of citizens and transformative initiatives. From a gender perspective, fostering a culture of care in cities could bring about the crucial shift from gender-responsive to gender-transformative climate policies.

'It's the economy, stupid' – but what about the care economy?

The main driver of climate change is the economy, both in terms of production and consumption. As feminist scholars (Biesecker and Gottschlich, 2013; Donath, 2000; Gottschlich *et al.*, 2014; Kuhl, 2012) have argued, existing green or sustainable economy approaches consistently fail to break the neoliberal paradigm of growth as a prerequisite for development. Contrary to popular belief, in middle- and high-income countries economic growth does not typically lead to more prosperity, but rather to a constant increase of consumption (genanet, 2011). Decoupling economic growth from the consumption of resources has proven to be a limited or even unsuccessful strategy to date (genanet, 2011). The so-called 'rebound effect' has meant that emission reductions achieved by producing goods in a more energy and resource efficient way, or by using more energy efficient appliances, are often counteracted by increases in consumption. While the extent of the rebound effect is still subject to scientific debate, it is clear that increases in energy efficiency typically result in more modest energy savings than expected (Sorrell *et al.*, 2009; Sorrell, 2007).

An unwavering belief in economic growth is largely incompatible with the principles of care for other human beings, an awareness of social inequality, and the need to care for future generations and nature (genanet 2012). Material wealth does not automatically result in well-being and satisfaction for all. Improving equality is a key precondition for an improvement of social relations, which are also necessary for a sustainable society, as Wilkinson and Pickett (2009) have highlighted. Societies in high income countries lack the urgently needed fundamental shift to a 'less is more' lifestyle that requires us to ask what exactly should be produced, which products are necessary and for whom, and whether the social and ecological consequences of the production are socially acceptable (genanet, 2011: 5). It is not only about more efficient production; equally important is the idea of living more sufficiently. Thus, restructuring the economic system primarily along ecological lines, while largely omitting justice issues, is not enough to create sustainable ways of living and planning for climate change mitigation and resilience.

Feminist critiques of the economy, which also examine the notion of a green economy, assert that the invisible and unvalued or undervalued part of the economy – the care-economy – is in actual fact the foundation of the whole economy, and as such should be a major point of reference in all sustainable economy approaches. Moreover, the term 'care' is not necessarily simply limited to human–human relationships. Feminist ecological economics is based on an understanding of care as a productive force, rather than a 'reproductive' one, analogous to the productivity of nature. Care is more than just the action of caring for others or ourselves, it is about 'caring for the values of freedom, equality, and justice' (Tronto, 2013: xii), as well as the natural world and our place in it.

Tronto's (2013) idea of a 'caring democracy' makes the hidden dimensions of care a central concern for politics. The reorganisation of the economy with an orientation towards care as a supportive, socio-ecological transformative principle thus becomes a question of democracy. Within this approach we can identify strong linkages between environment and climate change debates with feminist economy and care approaches. Care as an activity and a culture is a prerequisite for sustainable development in general, and in particular for a sustainable economy and for climate change mitigation and resilience.

What does a culture of care mean?

Care is not just an activity (caring for), it is also a practice that encompasses ethical, emotional and relational dimensions (caring with). It is therefore both a set of values and a series of concrete practices. Care can be defined as a principle for the regulation of the societal relationship to nature and a mindful ('careful') approach to nature. Moreover, care can be regarded as an ethical and political principle, which provides a means of drawing attention, first to the connection between a practicable ethics of care and the socio-political and structural background and second, to address the unequal power relations which mean that care is devalued (Gottschlich *et al.*, 2014).

A caring society should be one in which care pervades all major societal institutions from the global level to the local level. Additionally, a caring society urges people to be aware of asymmetrical relations and dependencies that shape both individuals' lives and society. Based on the experience of everyday life and the care economy, feminist approaches have described the special quality of caring. This quality is expressed in shouldering responsibility for others and making a conscious commitment to other people, to society at large, and to nature.

There is an urgent need for a process of re-thinking and re-shaping the responsibility for caring. Expressed generally, caring can be perceived as a group of activities that includes

> everything that we do to maintain, continue, and repair our 'world' so
> that we can live in it as well as possible. That world includes our bodies,

ourselves and our environment, all of which we seek to interweave in a complex, life-sustaining web.

(Tronto, 1993)

This would require promoting the ethics and attitude(s) of care in our whole societies, so that care-giving/care-receiving is seen not only as a remedy for those who tend to fall out of the system, but as a means of ensuring that our society, economies, politics, governance systems become intrinsically caring. For that we need a *culture* of care, in which education and civil society play a crucial role.

Based on this understanding, the following general principles can make an indispensable contribution to sustain livelihoods and lead to a reorganisation of economy in a sustainable and caring way. A caring economy:

- focuses on the needs of people,
- aims at facilitating life of nature and human beings, and ensuring a good life for everyone,
- is embedded in its social-ecological context,
- needs to be tolerant towards errors and reversible in order to allow a turnaround if needed (costly, high-risk technologies like nuclear power or the use of genetically modified organisms are an example of the opposite)
- anticipates long-term consequences,
- acts thoughtfully, slowly and is transparent in terms of time and space.

(Biesecker *et al.*, 2000)

Cities provide important opportunities to re-shape the economy, for example by regulation, incentives or subsidies. To become low-carbon and climate resilient, cities therefore would need to take up these opportunities and make the transition towards a 'caring city' (see box below). The results of an online survey on environmental protection and care in women-led companies reveal that they are willing to do more for the environment and to prevent climate change, yet at the same time, the respondents also emphasised the difficulties of combining self-employment with care responsibilities – but also the greater flexibility offered in terms of hours (Röhr, 2013). Policymakers could learn lessons from such initiatives run by women who face the daily reality of the care crisis, facing both the environment and people. Of course, it is easier to change cultural systems and react to conditions in small enterprises, compared to large ones or to large systems like cities, yet it is crucial to find ways of scaling up social innovations and to make them the rule, rather than the exception.

The ethical principle and culture of care provides us with a starting point for a critique of the existing system and a chance for change towards a sustainable future, as well with answers to many of the questions relating to what a caring city could look like: what this would mean for administration, for infrastructures, for services, and so forth.

A caring city – need for change

Transformation processes must be supported and accompanied by political measures in cities.

- With respect to time: A gender-just, socially sanctioned reduction in working hours that is self-determined by each person creates individual and societal freedom from the need for economic growth. Being 'time wealthy' is both the aim of, and the prerequisite for, a resource-light life.
- With respect to (infra)structure: On the one hand, the allocation of public goods such as water, energy, education, culture and sport facilities, mobility and health services, must take into account social aspects, justice and care. On the other hand, infrastructure must be developed with social use in mind and facilitates the solidarity economy at the local and regional level (such as community supported agriculture, donation markets or 'freecycling', and cooperatives). A prerequisite for this would be the transparency of political processes, as well as the political will to provide structural support for democratically supported transformations.
- With respect to democracy: new models of participation must be found that take into account various interests and represent real opportunities for public influence.

Towards solutions in practice

In order to address the comprehensive challenges posed by climate change, climate policy needs to cover all sectors of production and consumption. For mitigation, this includes energy supply and demand, mobility and transport, and waste management; and for adaptation, it encompasses but is not limited to, water demand and supply, sanitation, conservation and enhancement of ecosystems, and disaster risk reduction. Urban climate policies therefore require a comprehensive and integrated approach, centred around urban planning, to lay the path for a transition towards low-carbon and resilient communities. Just as steps to 'green' the economy are not far-reaching enough to address the climate crisis, it is also evident that investments in large-scale technologies, like more efficient power stations or dams for flood protection, do not represent comprehensive enough solutions at local level. A range of policies and measures is needed, including those which target people and change their behaviour, ranging from demand-oriented management and planning in the energy, transport and water sectors and a redistribution of urban space to accommodate non-motorised transport, through to measures to improve energy efficiency alongside promoting decentralised renewables, avoid and recycle waste, improve food security and build resilient settlements.

However challenging such policies may be, they can yield considerable benefits for the population, including improved air quality and health, job creation and cost savings, more effective allocation of resources, enhanced resilience to other hazards than climate variability, and an overall enhanced liveability of cities. Moreover, if implemented in a gender-responsive and socially fair manner, the distributional effects of these benefits will be more equitable and inequalities will not be aggravated by climate change. Policies are also likely to be more acceptable, viable and effective.

In order to realise these benefits, mainstreaming is needed both for climate change and gender equality, as both are cross-cutting issues. Even if cities have not yet adopted an ambitious comprehensive climate strategy, they can make efforts to pursue a gender-just, climate-proof and low-carbon development strategy in their day-to-day decision-making and management. Mainstreaming of climate and gender policies must work in two directions: policies on equal opportunities and gender need to take climate change into consideration, for instance, by anticipating future climate variability and taking the impacts into account, whereas climate change policies need to integrate gender dimensions in terms of needs, preferences, capacities, and potential impacts on people. 'Cross-mainstreaming' is therefore a challenging endeavour, requiring changes and learning processes in urban governance and institutional settings.

Moreover, achieving the transformation of existing cities is a long-term task, requiring longer planning horizons than are usually foreseen in politics. Built environments have an inherent inertia. They reflect the social and economic systems at the time of their construction, but remain for a long time, even after these systems have changed. This is relevant for both gender and climate issues: characteristics of male dominated and fossil fuel intensive urban structures and infrastructures include, for example, the segregation of housing from the workplace, and public transport geared towards the needs of commuters, rather than the complex and manifold trips of those who combine care work and employed work – in most cases, women. Neglecting to address gender issues may thus either constrain the mobility of women in urban areas, or lead to higher greenhouse gas emissions, if they are in the position to opt for private motorised transport.

While policymakers often have good intentions, cities often struggle to meet the challenge of cross-mainstreaming gender and climate policy, sometimes due to opposition or a lack of resources. The role of active civil society and community-based organisations in pushing for stronger action is therefore crucial. However, these groups are not necessarily gender-aware or knowledgeable about gender issues. The 'transition towns' movement, for example, aims to move the urban agenda forward by promoting low-carbon cities, yet gender equality has not been part of their agenda and gender issues have only recently been taken into account, primarily with regard to gender balance (Ward, 2014). Thus, in order to serve as drivers for more progressive, transformative policies, environmental organisations and other civil society actors need to devote time to, and conduct training on, gender issues. Conversely, women's groups with their considerable knowledge of gender issues and human rights often do not have the capacity or knowledge of climate change to effectively influence urban climate strategies. Yet, these actors are in particular potentially able to strongly promote gender-responsive climate policy, once it becomes apparent that this policy area can provide a crucial opportunity to achieve tangible benefits and positive equality outcomes for their constituencies. This is the case when adaptation actions are geared towards enhancing resilience of marginalised groups, for example, or when mitigation measures involve promoting accessible, safe and affordable public transport systems.

Integration of gender into the recommended policy cycle

While striving for a fundamental transformation towards sustainable cities in the widest definition of the term, the limitations posed by the somewhat tedious process of day-to-day policymaking in cities should be acknowledged. In order to start changing visions into reality, a step-by-step approach is therefore advisable, following the policy cycle model that is part of most recommended methodologies for urban climate policy. Based on our experience of working with cities, we will highlight where, and how, gender issues need to be integrated into the various steps of a generic policy cycle (Alber, 2015).

Throughout the entire cycle, the participation of communities and civil society should be ensured, from the setting of targets, collection of information, analysis and decision-making through to implementation, monitoring and evaluation. As previously mentioned, women tend to be underrepresented in participatory processes, even at community and neighbourhood level (see, e.g. Das, 2014; Figueiredo and Perkins, 2013). Appropriate provisions and arrangements are therefore needed to ensure that women can participate equally in planning and articulate their needs, preferences and opinions. A range of resources and tools can be used to guarantee that women's voices are heard, including guidebooks developed by cities (see, e.g. Senatsverwaltung für Stadtentwicklung Berlin, 2011; Urban Development Vienna, 2013).

Commitment: First of all, it is crucial to get the objectives right. Although it is not necessarily part of the established policy cycles, cities should enter into a commitment to address climate change, particularly given that in most countries, climate policy is not necessarily part of the cities' obligations.[4] At this stage, there is no need to set quantitative short- and mid-term targets, but long-term visions are important. Moreover, rather than pursuing a one-dimensional strategy, narrowly focused on greenhouse gas emissions, a multidimensional approach is advisable, which involves striving for a low-carbon, climate-resilient, equitable, gender-just, inclusive and caring city.

Problem analysis: This step generally includes conducting greenhouse gas inventories and climate change vulnerability assessments. In order to take gender into account, data should be disaggregated wherever possible. In most cases, this is not possible for energy-related greenhouse gas emissions, which are only available for the household level, but not for individuals (see

Alber, 2010). It is therefore useful to conduct surveys and include questions on attitudes and preferences. As for vulnerability, most assessments are only broken down to the community level. For the most vulnerable communities, it is advisable to carry out community-based vulnerability assessments which look at intra-household dynamics and give special attention to the potentially most vulnerable groups, such as female-headed households (Ahmed, 2010).

Strategy development and priority setting: To address gender inequalities, priority for adaptation should be given to building resilience, measures to improve food security such as urban gardening, climate proofing of settlements, enhancement of water infrastructure and services, and gender-responsive disaster risk reduction, based on the Sendai Framework, for example (United Nations Office for Disaster Risk Reduction, 2015). Gender-responsive mitigation must respond to needs while limiting or cutting GHG emissions, for instance by providing access to low-carbon energy and mobility services for women, or offering gender-sensitive advice on energy efficiency. To learn about the needs of marginalised populations, cities need to consult with community-based organisations and women's organisations, drawing on the wealth of methodologies available to ensure equal participation. Moreover, specific actions should seek to identify the constraints and needs of women and other vulnerable groups, such as exploratory walks (UN-Habitat, 2012).

Policy formulation and adoption of action plan: Once a draft action programme has been designed, the portfolio of policies and measures should be subject to a gender scan in order to identify critical elements that affect gender equality (Alber, 2015). Questions to ask include: Does the policy or measure concern one or more target groups and will it have an effect on the daily life of the population or on specific groups? Does it affect gender differences in regard to rights, resources, participation, values and norms?

If the policy or measure is likely to affect certain groups, a 'Gender Impact Assessment' should be conducted for the planned actions in order to reveal relevant inherent gender issues and make adjustments if necessary. This kind of assessment can be used to analyse the impact of planned projects and pro-grammes on women and men, as well as on gender relations more broadly. Again, questions to ask include: Does the action affect women and men differently? Might it lead to positive or negative impacts on gender equality? Special attention should be given to the gender factors explained above, for example, who is responsible for family care, how does the measure impact care work, does it increase household chores? How can care work be taken into consideration without reinforcing gender stereotypes? If cost increases are involved, how do they affect different consumer groups and what is their impact on captive customers whose alternatives are constrained? If incentives are provided, who will serve to benefit (for instance, only house-owners, or also tenants)? Further questions should identify the need for, and availability of data and knowledge to properly assess the impacts of the measure on gender equality, and the stakeholders to be involved in the development of the initiative, that is, whether additional experts and groups need to be consulted.

Implementation, monitoring and evaluation: For monitoring, feasible ways must be found to gather sex-disaggregated data on changes in emissions and vulnerability, and on positive and negative socio-economic impacts of policies. A number of cities, including Berlin, Vienna, Brussels, Mexico City, Cochabamba and Bamako already use gender-responsive budgeting in order to determine if public spending reaches women to the same extent as men and whether gender equality is promoted. Gender budgeting can create enabling policy frameworks, build capacity and strengthen monitoring mechanisms to support accountability to women. It should be applied to climate policy-related expenditures, in order to track the allocation of resources and analyse gender-differentiated impacts.

Conclusions and recommendations

The gender dimensions of climate policy are widely acknowledged. Research suggests that addressing these dimensions can make climate policy more effective at all levels, including the urban level, and improve its acceptance and viability, as well as contribute to gender equality. Cities should therefore pursue an integrated approach and enter into commitments to work towards becoming low-carbon, climate-resilient, equitable, gender-just and inclusive, as well as caring in the broad sense explained above.

Carbon intensive economies and societies, mal-adaptation as well as gender inequality, have shaped cities, and thus are all deeply rooted in urban layouts and institutional structures. To overcome the inbuilt inertia, transformative processes are needed involving institutional change and a redistribution of power and resources. Rather than adapting women and the underprivileged to the city, cities must adapt to their needs and listen to the voice of women's and community-based groups. This encompasses a call for social, rather than merely technical innovations.

National governments can assist cities by providing guidance on mitigation and adaptation strategies and policies, as well as offering incentives to stimulate action or make climate finance available. In this context, they should provide methodologies that integrate gender considerations. City networks, thematic networks and international agencies which offer guidance for local governments to address climate change should address gender issues as well, and recommend gender-responsive approaches and tools. Multilateral and bilateral climate finance should apply criteria that involve gender-balanced participation and gender-sensitive priorities and approaches. In order to put more vigour to these activities, efforts to integrate gender into international agreements, in particular in the UNFCCC process, should also be strengthened and expanded to cover urban climate policy.

Furthermore, more research is needed to better understand the linkages between gender and climate change in urban areas. This includes addressing the lack of sex-disaggregated data on vulnerability to the impacts of climate change as well as on the drivers of climate change, such as mobility and energy consumption, and on the socio-economic impacts of policies. Yet, while data help to highlight gender dimensions and differences, this is not sufficient alone. An analysis of the underlying causes is required, in order to effectively address inequality and discrimination, at the same time as responding to climate change.

Notes

1 GenderCC (2016) 'Women's participation at the climate negotiations'. Available online at www.gendercc.net/genderunfccc (accessed 12 December 2016).
2 The Doha gender decision (Decision 23/CP.18) is available at https://unfccc.int/ files/bodies/election_and_membership/application/pdf/cop18_gender_balance.pdf and the Lima Work Programme on Gender (Decision 18/*CP*.20) at https://unfccc. int/files/meetings/lima_dec_2014/decisions/application/pdf/auv_cop20_gender.pdf (both accessed 1 June 2016).
3 'The term "androcentrism" is understood to embrace certain patterns of thought, observation and action in regard to political, economic, scientific and societal issues. These patterns place men and maleness at the center or deem them to be the yardstick and standard while seeing women and femaleness as a "peculiarity", as a deviation from the standard' (Roehr *et al.*, 2008: 17).
4 Since the early 1990s, however, emerging city networks have encouraged cities to adopt voluntary commitments to address climate change. Examples include the Climate Alliance (www.climatealliance.eu) in Europe, ICLEI's CCP campaign (archive.iclei.org/index.php?id=10829) and more recently, the C40 cities (www.c40. org). These networks offer guidance and tools and allow for sharing of experiences.

References

ActionAid. (2013) Making care visible. Available online at www.actionaid.org/sites/ files/actionaid/making_care_visible.pdf. (accessed 31 August 2016).

Ahmed, A.U. (2010) *Reducing Vulnerability to Climate Change: The Pioneering Example of Community-based Adaptation.* Centre for Global Change, in association with CARE Bangladesh, Dhaka.

Alber, G. (2010) Gender, Cities and Climate Change (thematic report prepared for Cities and Climate Change Global Report on Human Settlements 2011*)*. UN-Habitat, Nairobi, Kenya.

Alber, G. (2015) Gender and Urban Climate Policy. Gender-Sensitive Policies Make a Difference. GIZ, UN-Habitat, GenderCC, Bonn and Eschborn. Available online at http://gendercc.net/resources/gendercc-publications.html (accessed 31 August 2016).

Behrens, K. and Robert-Nicoud, F. (2014) Survival of the fittest in cities: Urbanisation and inequality, *The Economic Journal*, 124(581): 1371–1400.

Beyazit, E. (2015) Are wider economic impacts of transport infrastructures always beneficial? Impacts of the Istanbul Metro on the generation of spatio-economic inequalities, *Journal of Transport Geography*, 45: 12–23.

Biesecker, A. and Gottschlich, D. (2013) Wirtschaften und Arbeiten in feministischer Perspektive – geschlechtergerecht und nachhaltig? in S. Hofmeister, C. Katz and T. Mölders (eds), *Geschlechterverhältnisse und Nachhaltigkeit. Die Kategorie Geschlecht in den Nachhaltigkeitswissenschaften.* Opladen: Barbara Budrich, pp. 178–189.

Biesecker, A., Mathes, M., Schön, S. and Scurell, B. (eds) (2000) *Vorsorgendes Wirtschaften: auf dem Weg zu einer Ökonomie des guten Lebens.* Bielefeld: Kleine.

Breengaard, M.H., Christensen, H.R., Oldrup, H.H., Poulsen, H. and Malthesen, T. (2007) TRANSGEN- Gender Mainstreaming European Transport Research and Policies: Building the knowledge base and mapping good practices. Co-ordination for Gender Studies, University of Copenhagen. Available online at http://koensforskning. soc.ku.dk/projekter/transgen/eu-rapport-transgen.pdf/. (accessed 31 August 2016).

Bulkeley, H.A., Edwards, G.A.S. and Fuller, S. (2014) Contesting climate justice in the city: Examining politics and practice in urban climate change experiments, *Global Environmental Change*, 25: 31–40.

Carlsson-Kanyama, A. and Lindén, A.L. (2007) Energy efficiency in residences: Challenges for women and men in the North, *Energy Policy*, 35: 2163–2172.

Chant, S. (1998) Households, gender and rural-urban migration: Reflections on linkages and considerations for policy, *Environment and Urbanization*, 10: 5–22.

Chant, S. (2013) Cities through a 'gender lens': A golden 'urban age' for women in the global South? *Environment and Urbanization*, 25: 9–29.

Chant, S. and McIlwaine, C. (2015) *Cities, Slums and Gender in the Global South Towards a Feminised Urban Future.* New York: Routledge.

Chindarkar, N. (2012) Gender and climate change-induced migration: Proposing a framework for analysis, *Environment Research Letters*, 7(2) 025601.

Climate Alliance of European Cities. (2005) *Climate for Change. Tools for Promoting Women in Executive and Management Positions.* Frankfurt am Main: Climate Alliance of European Cities.

Das, P. (2014) Women's participation in community-level water governance in urban India: The gap between motivation and ability, *World Development*, 64: 206–218.

Dhainaut, J.-F., Claessens, Y.-E., Ginsburg, C. and Riou, B. (2003) Unprecedented heat-related deaths during the 2003 heat wave in Paris: Consequences on emergency departments, *Critical Care*, 8: 1–2.

Donath, S. (2000) The other economy: A suggestion for a distinctively feminist economics, *Feminist Economics*, 6(1): 115–123.

European Institute for Gender Equality. (2012) Review of the Implementation in the EU of area K of the Beijing Platform for Action: Women and the Environment Gender Equality and Climate Change. Available online at http://eige.europa.eu/ sites/default/files/documents/Gender-Equality-and-Climate-Change-Report.pdf (accessed 31 August 2016).

Figueiredo, P. and Perkins, P.E. (2013) Women and water management in times of climate change: Participatory and inclusive processes, *Journal of Cleaner Production*, 60: 188–194.

genanet (ed.) (2011) Green Economy: Gender_Just! Towards a resource-light and gender-just future. Discussion Paper. Available online at www.genanet.de/fileadmin/ user_upload/dokumente/Care_Gender_Green_Economy/G3_discussion_paper_ en.pdf (accessed 31 August 2016).

genanet (ed.) (2012) Sustainable economic activity: Some thoughts on the relationship between the care economy and the green economy. Available online at www. genanet.de/fileadmin/user_upload/dokumente/Care_Gender_Green_Economy/ G3_Hintergrundpapier_Care_Gottschlich_EN.pdf. (accessed 31 August 2016).

GenderCC (2016) Women's participation at the climate negotiations. Available online at www.gendercc.net/genderunfccc (accessed 12 December 2016).

George, M. and McGranahan, G. (2013) The legacy of inequality and negligence in Brazil's unfinished urban transition: Lessons for other developing regions, *International Journal of Urban Sustainable Development*, 5(1): 7–24.

GIZ. (2015) Policy Advice for Environment and Climate Change. Available online at www.paklim.org (accessed 31 August 2016).

Gottschlich, D., Roth, S. and Roehr, U. (2014) Doing Sustainable Economy at the Crossroads of Gender, Care and the Green Economy. Debates – Common Ground – Blind Spots, CaGE-Texts 4/2014. Berlin.

Hemmati, M. and Roehr, U. (2007) A huge challenge and a narrow discourse, *Women & Environments International Magazine*, 74/75: 5–9.

Hodgson, G.M. (2003) Capitalism, complexity, and inequality, *Journal of Economic Issues*, 37(2): 471–478.

Hovorka, A., Zeeuw, H. and de Njenga, M. (eds) (2009) *Women Feeding Cities – Mainstreaming Gender in Urban Agriculture and Food Security*. Rugby, UK: Practical Action Publishing.

IOM. (2008) Migration and climate change, IOM Migration Research Series (MRS) 31. International Organization of Migration. Available online at www.iom.cz/files/Migration_and_Climate_Change_-_IOM_Migration_Research_Series_No_31.pdf (accessed 31 August 2016).

Jarvis, H. (2009) *Cities and Gender*. London; New York: Routledge.

Jirón, P. and Imilan, W.A. (2015) Embodying flexibility: Experiencing labour flexibility through urban daily mobility in Santiago de Chile, *Mobilities*, 10: 119–135.

Kaijser, A. and Kronsell, A. (2014) Climate change through the lens of intersectionality, *Environmental Politics*, 23: 417–433.

Kälin, W. and Schrepfer, N. (2012) Protecting People Crossing Borders in the Context of Climate Change. Normative Gaps and Possible Approaches. University of Bern. Available online at www.unhcr.org/4f33f1729.pdf (accessed 30 August 2016).

Kanning, H., Mölders, T. and Hofmeister, S. (2016) Gendered energy – Analytische perspektiven und potenziale der geschlechterforschung für eine sozial-ökologische gestaltung der energiewende im raum, *Raumforschung und Raumordnung*, 74: 213–227.

Kolmannskog, V. (2008) *Future Floods of refugees. A Comment on Climate Change, Conflict and Migration*. Oslo, Norway: Norwegian Refugee Council.

Kuhl, M. (2012) The Gender Dimensions of the Green New Deal – an analysis of policy papers of the Greens/EFA New Deal Working Group. A study commissioned by the Greens/EFA. Available online at https://europeangreens.eu/sites/europeangreens.eu/files/GND%20social%20dimension%20adopted.pdf (accessed 29 August 2016).

MacGregor, S. (2010) Gender and climate change: From impacts to discourses, *Journal of the Indian Ocean Region*, 6(2): 223–238.

Mattoo, A. and Subramanian, A. (2010) Equity in Climate Change: An Analytical Review, World Bank Policy Research Working Paper Series.

Miranda, V. (2011) Cooking, caring and volunteering: Unpaid work around the world. OECD Social, Employment and Migration Working Papers, No. 116, OECD Publishing.

Moser, C. and Satterthwaite, D. (2008) Towards pro-poor adaptation to climate change in the urban centres of low- and middle-income countries, Climate Change and Cities Discussion Paper 3. IIED.

OECD. (2014) Cities and Climate Change. Policy Perspectives. National governments enabling local action. Available online at www.oecd.org/env/cc/Cities-and-climate-change-2014-Policy-Perspectives-Final-web.pdf (accessed 31 August 2016).

Organo, V., Head, L. and Waitt, G. (2013) Who does the work in sustainable households? A time and gender analysis in New South Wales, Australia, *Gender, Place & Culture*, 20(5): 559–577.

Pirard, P., Vandentorren, S., Pascal, M., Laaidi, K., Le Tertre, A., Cassadou, S. and Ledrans, M. (2005) The impact of the 2003 heat wave on daily mortality in England and Wales and the use of rapid weekly mortality estimates, *Euro Surveillance*, 10(7).

Pye, S. and Dobbins, A. (2015) Energy poverty and vulnerable consumers in the energy sector across the EU: Analysis of policies and measures. Available online at https://ec.europa.eu/energy/sites/ener/files/documents/INSIGHT_E_Energy%20 Poverty%20-%20Main%20Report_FINAL.pdf (accessed 31 August 2016).

Rehfuess, E. (2006) *Fuel for Life: Household Energy and Health*. Geneva: World Health Organization.

Roehr, U., Spitzner, M., Stiefel, E. and von Winterfeld, U. (2008) Gender justice as the basis for sustainable climate policies. A feminist background paper. genanet, German NGO Forum on Environment and Development, Berlin, Bonn.

Röhr, U. (2013) FrauenUnternehmen Green Economy. Ergebnisse einer Online-Befragung. Available online at www.genanet.de/projekte/fuge.html (accessed 31 August 2016).

Senatsverwaltung für Stadtentwicklung Berlin. (ed.) (2011) *Gender Mainstreaming in Urban Development. Berlin Handbook*. Berlin: Kulturbuch-Verlag.

Skinner, E. (2011) *Gender and Climate Change. Overview Report*. Brighton, UK: BRIDGE, Institute of Development Studies.

Sorrell, S. (2007) The rebound effect: An assessment of the evidence for economy-wide energy savings from improved energy efficiency. Available online at www.ukerc. ac.uk/programmes/technology-and-policy-assessment/the-rebound-effect-report. html (accessed 31 August 2016).

Sorrell, S., Dimitropoulos, J. and Sommerville, M. (2009) Empirical estimates of the direct rebound effect: A review, *Energy Policy*, 37(4): 1356–1371.

Tacoli, C. (2012) *Urbanization, Gender and Urban Poverty: Paid Work and Unpaid Carework in the City*. London; New York: Human Settlements Group, International Institute for Environment and Development; Population and Development Branch, United Nations Population Fund.

Tacoli, C. and Mabala, R. (2010) Exploring mobility and migration in the context of rural-urban linkages: why gender and generation matter, *Environment and Urbanization*, 22(2): 389–395.

Tran, H.A. and Schlyter, A. (2010) Gender and class in urban transport: The cases of Xian and Hanoi, *Environment and Urbanization*, 22(2): 139–155.

Tronto, J.C. (2013) *Caring Democracy Markets, Equality, and Justice*. New York: New York University Press.

Tronto, J.C. (1993) *Moral Boundaries: A Political Argument for an Ethic of Care*. New York: Routledge.

UN DESA. (2010) The World's Women 2010. Trends and Statistics. United Nations.

UN DESA. (2014) World Urbanization Prospects. United Nations. Available online at https://esa.un.org/unpd/wup/Publications/Files/WUP2014-Highlights.pdf (accessed 1 June 2016).

UNDP. (2008) Human Development Report 2007/2008. Fighting Climate Change: Human Solidarity in a Divided World. Basingstoke: Palgrave Macmillan.

UNFPA. (2009) State of World Population 2009 – Facing a Changing World: Women, Population and Climate. New York, NY.

UN-Habitat. (2008) State of the World's Cities 2008/2009. Harmonious Cities.

UN-Habitat. (2011) Cities and Climate Change: Global Report on Human Settlements, 2011. [Nairobi]; London; Washington, DC: UN-Habitat; Earthscan.

UN-Habitat. (2012) Promising Practices on Climate Change in Urban Sub-Saharan Africa. Nairobi.

UN-Habitat. (2013) State of Women in Cities 2012–2013. Gender and the Prosperity of Cities.

United Nations Office for Disaster Risk Reduction. (2015) Sendai Framework for Disaster Risk Reduction 2015–2030. UNISDR.

Urban Development Vienna. (ed.) (2013) Manual for Gender Mainstreaming in Urban Planning and Urban Development, Vienna, Austria: Werkstattbericht.

Ward, F. (2014) Is gender an issue in Transition? Transition Culture. Available online at https://transitionnetwork.org/blogs/rob-hopkins/2014-10/gender-issue-transition (accessed 31 August 2016).

WIEGO. (2007) Informal Workers in Focus: Home-based Workers. Fact Sheet. Women in Informal Employment: Globalizing and Organizing, Cambridge, MA.

Wilkinson, R.G. and Pickett, K. (2009) *The Spirit Level. Why More Equal Societies Almost Always Do Better*. London: Penguin.

World Bank. (2010) Cities and climate change: An urgent agenda. Washington, DC: World Bank.

6 Natures of masculinities

Conceptualising industrial, ecomodern and ecological masculinities

Martin Hultman

Introduction

As gender scholars dealing with environmental issues are all very aware, men are the big problem. Especially white, middle-class, middle-aged, fairly rich men who travel too much, eat too much meat and live in energy intensive buildings. The truth is that if we quantitatively analyse per capita emissions and per capita ecological footprints, it is these particular men who are the problem (Räty and Carlsson-Kanyama, 2010). At the same time it is this category of men who dominate climate change, both the research and the politics (Anshelm and Hansson, 2014) as well as the denial of it, which has been referred to as 'the conservative white male effect' (McCright and Dunlap, 2015). This analysis is of course nothing new, put forward by ecofeminist research for a long time (e.g. Mies and Shiva, 1993; Buckingham, 2004).

This chapter takes this knowledge to further conceptualise historically situated and contemporary enacted forms of masculinities in rich, western countries with high per capita emissions. It introduces the configurations of 'industrial masculinities', 'ecological masculinities' and 'ecomodern masculinities'. I will discuss how industrial masculinities portray Nature as bits and pieces and work with it accordingly as well as how ecomodern masculinities are able to depict nature as alive and in need of care, but only if it fits with neoliberal market mechanisms and end-of-pipe technologies. Then, I discuss the possibility of ecological masculinities enacted with care, interconnectedness and the need for small-scale and localisation of resources. These configurations of masculinities display the entanglements of discourses in actors, and of actors in discourses of contemporary environmental politics. Exploring different configurations of masculinities might shed further light on how gender identities are constructed. In so doing, this text both elaborates on new concepts of masculinities and broadens our understanding of material cultural formations in the present form of global politics.

Gender and environment

Gender as an analytic source of inspiration opens up the possibility of interdisciplinary analysis that includes research on identity issues as well as bodies. There has been important research regarding issues such as women's activism and how women are affected by industrialisation and colonialisation in forms of, for example, GMO-politics, forestry, hydropower dam-building and mining, all of which display the gender–technology–environment nexus (e.g. Shiva, 2006; Arora-Jonsson, 2013; Haraway, 1988; Alaimo, 2010). The contribution of these fields is especially prominent and enlightening in relation to women affected by environmental injustices (e.g. Neumayer and Plümper, 2007) and women's environmental activism (e.g. MacGregor, 2013). The majority of the research centred on gender and environment has thus been carried out in poor nations with low-carbon emissions per capita (Arora-Jonsson, 2014).

Gender-related research in the field of environmental politics has gradually become more important, even though feminist research in this field still has a significant untapped potential (Buckingham, 2010; MacGregor, 2009). One of the potentialities is to elaborate on the male aspect of environmental politics, especially the question of how different masculinities enhance or influence environmental issues from the Global North (Gaard, 2014). While there has been long and inspiring ecofeminist research into gender roles and gender inequity in relation to environment and development goals, there has been little concern with constructions of masculinities to examine how masculinities are embedded in and through environmental policy (Connell and Pearse, 2014). This gap is curious, but not unexpected as hegemonies tend to present themselves as the 'normal'. It is thus troublesome, not least because men have dominated environmental politics heavily for many years, but much needed to conceptualise in a period in history in which the current hegemony has created catastrophic climate change.

Masculinities and environment made visible

The environmental historian Carolyn Merchant's classic book *The Death of Nature: Women, Ecology, and the Scientific Revolution* (1980; see also White, 1967) identified nature-destructive masculinities as a prime cause of environmental problems. That perspective has since been somewhat lost, taken for granted or even suppressed (Gaard, 2014). This lacuna is not only unfortunate, because of the importance men have in environmental politics, but also unforeseen because one of the first studies in which the influential concept of 'hegemonic masculinities' was used dealt with men and masculinities in environmental social movements (Connell, 1990). Raewyn Connell's study discussed how certain powerful ideas and practices within, for example, the environmental political field, were sustained through actions connected to hegemonic masculinities (which has similarities to what I call industrial masculinities in this chapter, more about that later). Since then, this concept has played a central, though sometimes contested, role in masculinities studies (Barrett, 1996; Bird, 1996; Connell and Messerschmidt, 2005; Connell and Wood, 2005). I will not be able to describe the development of masculinities studies in its entirety in this chapter, but I do agree with Hearn *et al.* (2012), Messerschmidt (2012), Gottzén and Mellström (2014), and Christensen and Jensen (2014) that it is important to find new subject positions beyond the normative and binary gender order. This chapter, then, is a humble attempt to research bodies as part of wider political structures (e.g. Haraway, 1988; Hearn *et al.*, 2014). My suggestion is to look at different, always-in-the-making, masculinities that are within and part of wider political structures which embody the nature of knowledge, materiality, power and meaning (Laclau and Mouffe, 1985; Barad, 2007). Theoretically, this chapter is thereby based on the plural idea of situatedness, understanding men as an unstable changeable entity, but at the same time a researchable category of different configurations.

A large field called masculinities studies has, from the 1990s onwards, been evolving around the issue of different configurations of materiality, values and practices among men. Few scholars, however, have so far been interested in continuing the analysis of masculinities and environment that Merchant and Connell started. There are some historical studies made on, for example, colonialisation in India (e.g. Sramek, 2006) and the environmental movement of the 1960s (e.g. Rome, 2003; Melosi, 1987; Hazlett, 2004), but few studies are trying to understand contemporary masculinities in regards to energy and environment questions (Connell and Pearse, 2014). This might have to do with research funding strategies as well as the fact that most researchers in the environmental field are men and may be unwilling to shed light on their own practice (Magnusdottir and Kronsell, 2014). Rural studies reviewing life in the countryside are somewhat of an exception (e.g. Brandth, 1995; Bell, 2000; Campbell and Bell, 2000; Emel, 2001), but they are concerned with geographically specific practices. Within Ecophilosophy we find a notable dissertation by Paul Pulé. His historical overview of the discussions within Deep Ecology, Ecological Feminism and Social Ecology, and the strong standpoint he introduces of recognising the caring element in men and his proposal for changing masculinities, are important reading (Pulé, 2013). We also find some contributions to this discussion within the broad field of what can be called Environmental Sociology. Quite a few quantitative studies show different consumption patterns of women and men in the same income bracket, demonstrating that men have a far heavier ecological footprint than women (e.g. Räty and Carlsson-Kanyama, 2010) combined with less care for environmental issues and lower membership in environmental organisations (e.g. Grasswick, 2014). Qualitative studies reveal the masculine connection to technological fixes (e.g. Alaimo, 2009; Hird, 2015; Hultman, 2013). Of late there are also analyses which connect global environmental issues with masculinities and discuss, for example, climate change, values, class and power (Fleming, 2007; McCright and Dunlap, 2015; Anshelm and Hultman, 2014b) as well as from another perspective, discuss the favourable link between state environmentalism and gender equality (Norgaard and York, 2005). Highly relevant to the discussion about global climate change is the research with a focus on extractive industries. These studies focus on micro-practices such as values on oil-rigs and in fracking communities (e.g. Filteau, 2015). Within Ecocritical literature studies the contributors to Mark Allister's anthology *Eco-man. New Perspectives on Masculinities and Nature* (2004) are of interest. This collection of essays has the ambition of showing the social construction of the environment in relation to masculinities. More specifically it shows different forms of masculinities in nature–culture experiences and explores the potential of living positively with nature. Allister's book promised another opening for discussions about different forms of masculinities within environmental political fields. But once again these ideas failed to find fertile soil and masculinities research continues to lack discussion on environmental issues, just as the environmental political field largely fails to engage with masculinities studies (Dymén *et al.*, 2013).

Conceptualising masculinities with case studies of climate change, environmental history and energy politics

This conceptual chapter is based on more than a decade of research into three different empirical studies of climate change, environmental history and energy politics respectively, in which I have had the opportunity to make in-depth case studies. First, this conceptual chapter about masculinities is based on an extensive study of global climate change debates published in the book *Discourses of Global Climate Change* (Anshelm and Hultman, 2014b). A set of 3,500 articles found in the database Artikelsök, using keywords such as climate change was studied in this book. The database contains articles published in all Swedish national newspapers, all major regional newspapers and the vast majority of magazines. As this material was compiled and arranged chronologically, read through, and sorted with discourse analytic tools (Anshelm and Hultman, 2014b), it revealed the dominance of industrial masculinities. Second, my PhD thesis completed in 2010 explored energy politics between 1978 and 2005, from perspectives of environmental history and cultural studies. I grounded my analysis on extensive reading of over 2,000 journal articles, archive studies and interviews (Hultman, 2013). I draw on this PhD work especially when discussing 'ecomodern masculinities'. The final empirical material that underpins this chapter and the configurations of masculinities that I discuss is a study of ecological sustainable entrepreneurship. This study is based on over forty interviews and field work in New Zealand and Sweden (Hultman, 2014b, 2014c, 2016). The conceptualisation of 'ecological masculinities' is explained with examples from that study.

Industrial masculinities

Industrial masculinities is the first example of masculinities in the field of environmental politics that I will discuss, and it has a strong foothold in industrial modernisation as shown by, for example, Merchant (1980, 1996; see also White, 1967). What I term industrial masculinities has in previous research been coined 'heroic masculinity' (Holt and Thompson, 2004), 'breadwinner masculinities' (Stacey, 1990), 'hypermasculinities' (Parrott and Zeichner, 2003), or 'cowboy masculinities' (Donald, 1992) when the studies involved have dealt with, for example, everyday consumption, work, sport, movies or farming. While each of these terms has advantages, I have chosen industrial masculinities because of its clear connection to the industrial modern discourse. This term also makes a broader connection to societal hegemonies beyond a particular role for men within that society. Industrial masculinities contains values from engineering and neo-classical economics, favouring large-scale and centralised energy technologies and the practice of patriarchy. Examples can be drawn from large-scale hydropower, nuclear power plants and fossil fuel technologies (Öhman, 2007). It is primarily distinguished by a separation of humans from Nature. Further industrial masculinities value Nature only as a resource for human extraction (White, 1967). Carolyn Merchant identified a kind of western masculinity that accuses others of religious fervour while using faith as a basis for its embrace of an industrial society. Since the so-called period of Enlightenment, a separation between man/woman and culture/nature has been created with the purpose of establishing the dominance of men/culture as rulers over women/nature. Merchant detects an important change from organic metaphors of nature, which were dominant up until the sixteenth century in Europe, to mechanical metaphors, until eventually nature was regarded as not having any intrinsic values, more or less dead as stated in the title of her book, although possible to make use of when trying to simulate a human-made Eden on Earth. This shift coincided with the rise of industrial-scale operators who viewed nature as a resource in, for example, the mining, energy and timber sectors (Merchant, 1980).

Connell characterised the men who dominated and ran industrial modernisation until the 1990s as representing a hegemonic form of masculinity (Connell, 1990), which was an important factor in influencing energy and environmental politics in industrialised society. It was this form of masculinities that men and women within the environmental movement in the 1970s and 1980s needed to challenge as they tried to strengthen an alternative worldview (Connell, 1990). There was a strong dominant belief that industrial modernisation, with its large-scale engineering focus, centralisation, and fossil/nuclear-based economy, would take care of its own environmental problems (Nye, 1994; Cohn, 1997). There was a strong faith among the elites that economic growth and a rationalisation of production could fulfil both energy and environmental policy requirements. When pollution created problems such as smog, the solution was to build higher chimneys (McNeil, 2001). Industrial modernisation was presented simultaneously as a cause of environmental problems and as a tool for overcoming them (Anshelm, 2010).

Climate scepticism as example of industrial masculinities

In the twenty-first century, this form of industrial masculinity forms the basis for climate scepticism. When industrial modernisation once again was truly challenged in the climate change debate, industrial masculinity, this time in the form of climate sceptics, reappeared on the environmental political scene (Anshelm and Hultman, 2014a). Climate scepticism is shaped by a small group of men who embody industrial masculinity. McCright and Dunlap (2003) identify the neo-conservative political movement in the US as a central actor which is in turn influenced by a small group of "dissident" or "contrarian" scientists who lend their credentials and authority to conservative think tanks. It is well recognised that in order to maintain an illusion of intense controversy, industries, special interest groups, and public relations firms have manipulated climate science and exploited the US media. Climate scepticism however, is not a social movement, but a project of a few influential men (Oreskes and Conway, 2010). In research based on Gallup surveys in the US, McCright and Dunlap (2011), who take gender into account, have found a correlation between self-reported understanding of global warming and climate change denial among white men with conservative values. This suggests that climate change denial is a form of identity-protective cognition; or rather, part of an identity process. In the US, as in Sweden, this scepticism is articulated by a small, homogenous group of, almost exclusively, men and conservative think tanks. These men have successful careers in academia or private industry, strong beliefs in a market structured society, and a great mistrust of government regulation. A few voices return again and again with virtually the same dystopian predictions. The sceptics' arguments are strengthened by references to the authority of their own titles and positions found in a variety of natural science academic disciplines and thereby demonstrate a general belief in the positivistic industrial modern science underpinning these disciplines. In relation to climate science, but only here, these sceptics adopt a constructivist position, presenting it as a mix of science and politics so entwined that they can no longer be distinguished (Anshelm and Hultman, 2014b).

In Sweden, sceptics have connections to associations where representatives of business, and scientific and technological research meet. One clear example is of Per-Olof Eriksson, a former board member of Volvo and former president of Sandvik. He wrote an article in the leading Swedish business paper *Dagens Industri*, declaring his doubts that carbon emissions affect the climate and stating that the Earth's average temperature has risen due to natural variations (Eriksson, 2008). Another example is Ingemar Nordin, professor of philosophy of science, who posited that the IPCC's selection and review of scientific evidence was consistent with what politicians wanted. Nordin claimed that politics shaped basic scientific research, and that scientists who produced politically acceptable truths were awarded funding (Nordin, 2008). Professors in economy Marian Radetzki and Nils Lundgren claimed that the IPCC deliberately constructed their models "in an alarmist direction" using feedback mechanisms that gave the impression that significant climate change was taking place (Radetzki and Lundgren, 2009). Subsequently 15 Swedish professors in various natural science disciplines, all men, proclaimed themselves as climate sceptics (Einarsson *et al.*, 2008).

Instead of understanding climate sceptics as anti-science or anti-political, I argue that it is important to understand how their masculine identity has been shaped by the hegemony of industrial modernisation and how this figuration is co-constructed with their challenges to climate science. This rationality of domination over nature, instrumentality, economic growth, and linearity has been hegemonic throughout the industrial modern era (Merchant, 1996). Climate sceptics believe that only through continuing modernisation and economic growth through industrialisation – what these men have done their whole lives – can the world's problems be solved. This is where climate sceptics use patriarchal arguments and claim to be defending the well-being of future generations, poor people and endangered species. They, the ones who made industrialisation possible, assert that they are the only ones capable of solving today's global problems (Anshelm and Hultman, 2014). Industrial masculinities are figurations that evaluate nature as made up of pieces to use, man as the chosen dominator, and engineering as the method of creating wealth for all humans. The mere talk about a vulnerable earth transformed by anthropogenic emissions is handled with denial or strong scepticism by those enacting industrial masculinities.

Ecological masculinities

In contrast to industrial masculinity is ecological masculinities. The terminology comes from, on the one hand, my own work on ecological discourse, as well as previous research on masculinities which has used similar, but not the same, terminology (Pulé, 2013; Gaard, 2014). Discussing this figuration is important, as Greta Gaards put it "neither ecocriticism, nor men's studies, nor queer ecologies, nor (to date) ecofeminism has offered a theoretical sophisticated foray into the potentials for eco-masculinities" (Gaard, 2014). I will give examples of ecological masculinities from my research into ecological sustainable entrepreneurship, so called ecopreneurship.

Industrial masculinity was challenged in the mid-1960s by what I term 'ecological masculinities'. The environmental movement has argued that the risks associated with modern industrial energy technologies, emissions, and pollutants, demand societal change. In contrast to dominating practices and ideas, a number of environmental activists and public intellectuals formulated a vision including arguments for small-scale technologies, the decentralisation of power, and criticism of economic growth as a measure of prosperity, as well as the need to develop renewable energy sources (Anshelm, 2000; McNeil, 2003). Visions of another society were formulated and practiced throughout the 1980s which challenged the dominant modern industrial energy and environmental politics. The transformative power of the ecological discourse was evident in public opinion, the election of Green Party members to parliamentary assemblies in different parts of the world, debates in the mass media, new regulations, and small-scale renewable energy projects (Hultman, 2014a). On the surface it may appear similar to romantic ideals and conservative currents that have existed since the beginning of industrial modernity. However, there was one major difference: the green wave not only proposed a recycling of traditional technologies and traditional values, it also created a practice of intentional communities creating an alternative modernity. A practice with a long history (see Tummers, this volume). These change agents not only shut themselves off from society, but also created alternative projects. Their models and experiments were part of a mighty international peak in environmental consciousness in the mid-1980s (Hultman, 2014a). During this period, changed forms of masculinities of a more caring, humble and sharing sort were presented as being more appropriate in an ecologically sound society. These masculinities, created among other places within the environmental movement, challenged the industrial modern hegemony (e.g. Connell, 1990; Jeffords, 1993). Part of this ecological discourse is what I call 'ecological masculinities' which I have found while researching transitional ecopreneurship.

Ecopreneurs as examples of ecological masculinities

One ecopreneur practising ecological masculinities is Pete Russell who lives in an eco-village on Waiheke Island, New Zealand, and runs a distribution network of organically home-grown food called Oooby. He explains how he supports the local economy by shortening the supply chain, thereby challenging the bulk food market controlled by supermarkets: "the people, who we are supporting are those making another food system possible." His mother inspired him to be part of a local-economic system which exchanged goods and services, rather than use cash: "Mom always has been a grass root person, living in a mud brick house." By starting a business making pastry he became aware of the "myopia of the globalized food system". A pivotal moment for Pete was becoming a father and moving to Waiheke Island where he had conversations with people, which opened up his eyes to how he was connected in the world. He'd realised that he was part of a destructive capitalism that he would never have signed up to if knowing it, instead "as human animals, we need to fulfil a sense of place, familiar with the space, relationships you evolve". Having a rootless modern lifestyle, moving from one place to another, never allows relationships to evolve. Pete says that his "real sense of fulfilment is about having a meaningful role to play within a group of people, a community of people that really matter". Pete declares that we need to create "Localized cells with basic requirements. We need to move towards a more harmonious, peaceful world" and in his daily life he wants to contribute to that.

Common to the calls for a different politics and ecological masculinities is the implication that extensive social structural change is needed and that this is not something that could be achieved through voluntary, individual con-sumer choices and market solutions. It requires a powerful public engagement and politicians who assume long-term responsibility for the biosphere, even if it means interference in citizens' consumption habits and behaviour.

Another of the ecopreneurs interviewed in New Zealand is the architect and Māori entrepreneur Rau Hoskins who is the CEO of designTRIBE. Hoskins is well known in New Zealand as a catalyst for an emerging Māori group of architects and as a prominent Māori spokesperson. Hoskins' inspiration and how he practices architecture is connected to a holistic worldview on how to integrate new technologies with old traditions: "connected to alternative practice, to energy, to waste, and to water." Hoskins talked about a way of sharing and caring in a distributed system in which the resources in his land are understood as common:

> "The land up there has had an ancestral spring. That has remained con-stant, and good quality water goes to a holding tank and is then dis-tributed around to the other houses. Each house has its own tank in a distributed network."

For Hoskins, Eurocentric architecture alienates humans from the environment. One of his primary goals with architecture is "Making people spend more time outdoors. Our ancestors only retreated inside when it was dark". For him, permaculture is a Western way of understanding the interconnectedness with the earth in the same way as the Māori do: "permaculture and Māori have synergy [...] Māori and greens come together so they can be a significant block".

One of the important aspects of ecological masculinities is the transi-tional agency directed towards creating a sustainable society. As Greta Gaard explains:

> Patriarchy has shaped most contemporary industrial capitalist cultures, so eco–masculinities would need to recognize and resist the identity shap-ing economic structures of industrial capitalism, its inherent rewards based on hierarchies of race/class/gender/age/species/sex/sexuality, and its implicit demands for ceaseless work, production, competition, and achievement.
>
> (2014: 232)

Dave Jordan is a farmer's boy who has embarked on a journey of transition. His concern about the amount of litter and pollution on the land and in the waterways led to him identifying hemp as a solution to those problems. Hemp has been used throughout history as fuel, textile, medicine, oil and so forth; it has similar properties to oil but without the polluting emissions from fossil fuels (Deitch, 2003; Hopkins, 1951). Growing up on a farm, he began his working life as a nature guide in the south island of New Zealand. One defining moment for him was when he stood up at a meeting with the New Zealand tourist industry to proclaim that the image of 100 percent pure NZ was a fraud, something he got to know not least through family experience of deaths from illnesses such as cancer, likely from using pesticides. NZ is not as clean as he had previously thought and from then on he began looking for solutions. His brother inspired him to learn about hemp, and he got to know how "hemp and mankind [sic] fit together as a glove. It is also a poor man's plant." Dave says he is working the farm "together with my partner as equals, all day around". Hemp could have been part of the industrial revolution but was stopped by the oil industry and punitively regulated (Deitch, 2003; Hopkins, 1951). Dave believes that he creates something special when growing hemp, and the value of getting feedback is fantastic for him: "people come and gives us hugs and thank us". Dave expresses ecological masculinity by caring for humans and non-humans alike. He has chosen a livelihood that is part of nature's own way of working and his hemp production contributes to a more sustainable world, making him a transitional agent.

These examples give glimpses of how ecological masculinities can be enacted. There are many more examples in our world and further research might help to highlight and understand them. As Greta Gaard points out they need to be explored through cross-cultural and multicultural perspectives, not to privileging any specific region or ethnicity (Gaard, 2014).

Ecomodern masculinities

These two examples of masculinities are juxtaposed as opposites. I argue there is a third figuration, actually very influential today, coming from a merger of neoliberalism and light green (or reformist environmentalism). In this example I will draw from my dissertation work and exemplify what I call 'ecomodern masculinities' with Arnold Schwarzenegger's way of enacting identity in the late 1990s and early twenty-first century. Ecomodern masculinities has been part of the global shift towards the recognition of environmental issues as an intrinsic part of politics from the 1990s onwards. It can be defined as an asymmetric combination of the determination and hardness of industrial masculinities with appropriate moments of compassion and even sense of care for a vulnerable environment from ecological masculinities. In the end, however, I argue that the combination is a way of dismissing the need for transformational change.

The intense clash between industrial and ecological discourses in the 1970s and 1980s had far-reaching implications for the awareness of global environmental questions and gave way to an ecomodern discourse (Adler *et al.*, 2014). The conflict between ecological and industrial discourses was shifted to the

periphery in the early 1990s by ecomodern discourse. This began to dominate international policies on energy and sustainable development, based on
the premise that economic growth needed to be the basis for a transition of
energy and environmental policies towards a sustainable future (Hultman and
Yaras, 2012). The ecomodern discourse, or ecological modernisation as it has
been called elsewhere, emphasised a continuum from industrial modernisation instead of a hegemonic shift (Adler *et al.*, 2014). The descriptions of
environmental problems changed from being threats to civilisation – which
demanded radical system-wide changes – to being characterised as mostly
under control and soon to be solved (Hultman, 2014a). Conservative politicians and industry actors pushed this discourse that paved the way for market
solutions and a belief that competition would create 'green' jobs. In a concerted move with other neoliberal policies, electricity grids were privatised
and state research money was increasingly directed to private/public environmental research organisations away from basic funding for the universities. In
addition, many environmental organisations participated in this shift by running several campaigns favouring eco-friendly consumption patterns without questioning the total volume of the products consumed. These changes
meant that the focus shifted to finding specific technologies to reduce emissions, describing them in a new way, and creating new coalitions of discourses
(Hultman and Nordlund, 2013).

The ecomodern masculine character demonstrates caring and responsibility for the environment while at the same time, promoting economic growth,
technological expansion and focusing on end-of-pipe solutions. Ecomodern
masculinities demonstrate an in-depth recognition of environmental problems, especially climate change, while at the same time supporting policies
and technologies that conserve the structures of climate-destroying systems
(Hultman, 2013). One example of this masculine configuration is Arnold
Schwarzenegger, discussed below.

Neoliberalism and light green: Schwarzenegger as an example of ecomodern masculinities

In 2003, Arnold Schwarzenegger ran for the governorship of California. He
had been involved closely in Republican Party relationships with Enron and
was the advertising symbol for the energy inefficient Hummer sports utility
vehicle (SUV). The Hummer symbolised a combination of violence and lack of
respect towards nature. As governor, Schwarzenegger thus had quite an image
problem not to be easily combined with his earlier industrial masculinity. The
solution he found was a Hummer with fuel cell and hydrogen technology. In his
political performance as governor, Schwarzenegger aligned himself with the
high expectations for clean technology in the form of fuel cells and hydrogen.
Schwarzenegger's Hummer was marketed as a combination of an aggressive
violent technology, simultaneously caring towards the environment. The
water vapour as the only emission from the pipe made it possible to brand

these giant cars with an image of environmental care/consideration, but at the same time the carbon dioxide emissions from gas powered electricity plants in California that were needed to produce the hydrogen was displaced from the front stage. It is a typical end-of-pipe solution in which the emissions are displaced to another area, thereby hidden from the public.

A similar displacement strategy and therefore an important clue as to how the ecomodern discourse achieved hegemony at the beginning of the 1990s was the argument that economic growth did not destroy the environment; on the contrary, economic growth was claimed to be essential to address various environmental problems. Hence, environmental sociologists have commonly described the large gap between the promises of the ecomodern discourse and its outcome (York *et al.*, 2010; Blühdorn, 2011). Studies testing the ecological modernisation hypothesis demonstrate that the hypothesis of dematerialisation/de-carbonisation is not valid on a global scale (Jorgenson and Clark, 2012). Sweden is one of the countries said to have high economic growth and lower carbon emissions, but the overall carbon (or even more comprehensively, the ecological) footprint continues to rise substantially if air travel and overall consumption are included in the calculation (Hysing, 2014; Berglund, 2011).

In this way, both the ecological modernisation discourse and ecomodern masculinities can be understood as attempts to incorporate and deflect criticism in order to perpetuate hegemony; in other words, to ensure that practices remain in effect industrial modern. Schwarzenegger presented himself as a symbol of ecomodern masculinity, someone who made a serious effort for the environment while not forsaking the epitome of industrial society, the big car (Hultman, 2013).

This displacement practice is visible when Schwarzenegger proposed fuel cells as a silver bullet technology. Schwarzenegger used his image as a science fiction hero and combined that with the image of a responsible man when he transformed himself into governor of California with the nickname 'the Governator'. Ecomodern masculinity, which is presented as a symmetric amalgamation that combines care for the environment and economic growth, may instead be understood as highly asymmetric because it conserves and favours existing system solutions, for example, those in the transportation sector. Here the focus by the ecomodern discourse is on emissions making it possible to create programs that fool the buyer into thinking that the engine is more environmentally friendly than it is (read the emission scandal of VW, which exposed how the whole car industry fakes test results), not on the production of fuel. Established structures of automobility (which are deeply gendered and dominated by industrial masculinities, Carlsson-Kanyama *et al.*, 1999) could be conserved with ecomodern masculinities promoting end-of-pipe technologies.

Towards understanding natures of masculinities

In this chapter I have proposed an analytical framework that conceptualises masculinities in the environmental field. I have discussed how industrial masculinity uses bits and pieces of nature for human ends, proclaims nature as dead, upholds the (hu)man as the chosen dominator, and engineering as the method of creating wealth for mankind. Within the climate change debate this position is no longer tenable, if not denying the whole, or most of, the combined research findings presented by the IPCC. Industrial masculinities tends today to be associated with those denying climate change research. With the ecomodern figuration, environmental problems such as climate change are revised but not completely overhauled. Even though the ecomodern discourse is based on an industrial attitude, such a discourse is still challenging for industrial modern tenants because it opens a debate over, for example, climate change as a societal issue that needs to be addressed by industry, politicians, and the public. With the example of Arnold Schwarzenegger I have discussed how determination, and hardness, is portrayed to go hand in hand with well-chosen moments of compassion, vulnerability and eco-friendly technology in the form of ecomodern masculinities. It appears on the surface to be the ultimate win–win figuration for, for example, Sweden, the US, Canada, Australia and so forth. But looking into it more closely it shows itself more as a cover up to enable these economies to continue down the same modern industrial path that created the problems in the first place. In summary, when industrial masculinites portrays nature as bits and pieces and works with it accordingly, ecomodern masculinities are able to depict nature as alive and in need of care from neoliberal market mechanisms and end-of-pipe technologies. In both cases, however, nature is turned into something external to society, possible to dominate with masculine practices.

In contrast, ecological masculinity is, both historically and in our times, viewed as a different kind of possibility of much needed ideas of how to find paths to a livable earth. Ecological masculinities would be part of remaking economic relationships and facilitating transitions towards a more environmentally benign way of humans being with and not destroying the world. This configuration is indebted to long-standing and inspiring proposals from ecofeminist scholars, being another call for men to take action. Firmly proposing a localisation of economies, use of small-scale technologies, creation of renewable energy, decentralised power structures as well as living with nature are part of everyday practices for ecological masculinities.

I suggest there is a need for more research that seeks to understand the values and practices of men, not least because of the large importance men play in shaping, formulating and deciding about environmental issues globally. This chapter has shown how the figurations of industrial, ecomodern and ecological masculinities are firmly connected to historical discourses. Intra-actions between discourses and actors within the field of environmental politics might be better understood by a closer reading of identities and figurations in the forms of different types of masculinities. This chapter has, from empirically based analyses, outlined an original conceptual way of elaborating upon three configurations of masculinities that can be used to understand historical as well as contemporary case studies of climate change; but it could equally well be applied to mining, agriculture, forestry, extractive industries.

References

Adler, Frank, Beck, Silke, Brand, Karl-Werner, Brand, Ulrich, Graf, Rüdiger, Huff, Tobias and Krüger, Timmo (2014) *Ökologische Modernisierung: zur Geschichte und Gegenwart eines Konzepts in Umweltpolitik und Sozialwissenschaften.* Frankfurt: Campus Verlag.

Alaimo, Stacy. (2009) Insurgent vulnerability and the carbon footprint of gender, *Kvinder Kon Forskning NR*, 3–4: 22–35

Alaimo, Stacy. (2010) *Bodily Natures: Science, Environment and the Material.* Bloomington: Indiana University Press.

Allister, Mark. (ed.), (2004) *Eco-man. New Perspectives on Masculinity and Nature.* Charlottesville: University Virginia Press.

Anshelm, Jonas. (2010) Among demons and wizards: The nuclear energy discourse in Sweden and the re-enchantment of the world, *Bulletin of Science, Technology & Society*, 30(1): 43–53.

Anshelm, Jonas and Hansson, Anders. (2014) The last chance to save the planet? An analysis of the geoengineering advocacy discourse in the public debate, *Environmental Humanities*, 5: 101–123.

Anshelm, Jonas and Hultman, Martin. (2014a) *Discourses of Global Climate Change.* London: Routledge.

Anshelm, Jonas, and Hultman, Martin. (2014b) A green fatwā? Climate change as a threat to the masculinity of industrial modernity, *NORMA: International Journal for Masculinity Studies*, 9(2): 84–96.

Arora-Jonsson, Seema. (2013) *Gender, Development and Environmental Governance: Theorizing Connections.* London: Routledge.

Arora-Jonsson, S. (2014, December) Forty years of gender research and environmental policy: Where do we stand? *Women's Studies International Forum*, 47: 295–308. Oxford: Pergamon.

Barad, K. (2007) *Meeting the Universe Halfway. Quantum Physics and the Entanglement of Matter and Meaning.* Durham, NC: Duke University Press.

Barrett , F.J. (1996) The organizational construction of hegemonic masculinity: The case of the U.S. Navy, *Gender, Work and Organization*, 3(3): 129–142.

Bell, David. (2000) Farm boys and wild men: Rurality, masculinity, and homosexuality, *Rural Sociology*, 65(4): 547–561.

Berglund, M. (2011) *Green Growth? A Consumption Perspective on Swedish Environmental Impact Trends. Using Input–Output Analysis.* Uppsala: Uppsala University, Department of Global Energy Systems.

Bird, S.R. (1996) Welcome to the men's club: Homosociality and the maintenance of hegemonic masculinity, *Gender & Society*, 10(2): 120–132.

Blühdorn, I. (2011) The politics of unsustainability: COP15, post-ecologism, and the ecological paradox, *Organization & Environment*, 24(1), 34–53.

Brandth, Berit. (1995) Rural masculinity in transition: Gender images in tractor advertisements, *Journal of Rural Studies*, 11(2): 123–133.

Buckingham, S. (2004) Ecofeminism in the twenty-first century, *The Geographical Journal*, 170(2): 146–154.

Buckingham, S. (2010) Call in the women, *Nature*, 468(7323): 502.

Campbell, Hugh, and Bell, Michael Mayerfeld. (2000) The question of rural masculinities, *Rural Sociology*, 65(4): 532–546.

Carlsson-Kanyama, Annika, Linden, Anna-Lisa och Thelander, Åsa. (1999) Insights and applications gender differences in environmental impacts from patterns of transportation: A case study from Sweden, *Society & Natural Resources*, 12(4): 355–369.

Christensen, A.D. and Jensen, S.Q. (2014) Combining hegemonic masculinity and intersectionality, *NORMA: International Journal for Masculinity Studies*, 9(1): 60–75.

Cohn, S.M. (1997) *Too Cheap to Meter: An Economic and Philosophical Analysis of the Nuclear Dream*. Albany, NY: SUNY Press.

Connell, Raewyn. W. and James W. Messerschmidt (2005) Hegemonic masculinity: Rethinking the concept, *Gender & Society*, 19: 829–858.

Connell, Raewyn W and Pearse, Rebecca (2014) *Gender: In World Perspective* (3rd edn). Cambridge: Polity Press.

Connell, Raewyn W. and J. Wood. (2005) Globalization and business masculinities, *Men and Masculinities*, 7(4): 347–364.

Connell, Robert W. (1990) A whole new world. Remaking masculinity in the context of the environmental movement, *Gender & Society*, 4, 452–478.

Deitch, R. (2003) *Hemp: American History Revisited: The Plant with a Divided History*. New York: Algora Publishing.

Donald, R.R. (1992). *Masculinity and Machismo in Hollywood's War Films: Men, Masculinity and the Media*. Newbury Park, CA: Sage.

Dymén, Christian, Måns Andersson, and Richard Langlais. (2013) Gendered dimensions of climate change response in Swedish municipalities, *Local Environment*, 18(9): 1066–1078.

Einarsson, Göran, Lars G. Franzén, David Gee, Krister Holmberg, Bo Jönsson, Sten Kaijser, Wibjörn Karlén, Jan-Olov Liljenzin, Torbjörn Norin, Magnus Nydén, Göran Petersson, CG. Ribbing, Anders Stigebrandt, Peter Stilbs, Anders Ulfvarson, Gösta Walin, Tage Andersson, Sven G Gustafsson, Olov Einarsson, Thomas Hellström (2008) LTH 20 toppforskare i unikt upprop: koldioxiden påverkar inte klimatet, *Newsmill*, 17 December.

Emel, Ní Laoire, Catrína. (2001) A matter of life and death? Men, masculinities, and staying 'behind' in rural Ireland, *Sociologia Ruralis*, 41(2): 220–236.

Eriksson, Per-Olof (2008) Jorden går inte under av utsläpp och uppvärmning, *Dagens Industri*, 4/6.

Filteau, M.R. (2015) A localized masculine crisis: Local men's subordination within the Marcellus shale region's masculine structure, *Rural Sociology*, 80(4): 431–455.

Fleming, J. (2007) The climate engineers, *The Wilson Quarterly*, Spring: 46–60.

Gaard, Greta (2014) Towards new ecomasculinities, ecogenders, and ecosexualities, in C.J. Adams and L. Gruen (eds), *Ecofeminism: Feminist Intersections with Other Animals and the Earth*. New York: Bloomsbury Publishing.

Gottzén, L. and Mellström, U. (2014). Changing and globalising masculinity studies. *NORMA: International Journal for Masculinity Studies*, 9(1), 1–4.

Grasswick, H. (2014) Climate change science and responsible trust: A situated approach, *Hypatia*, 29(3): 541–557.

Haraway, Donna. (1988) Situated knowledges: The science question in feminism and the privilege of partial perspective, *Feminist Studies*, 3: 575–599.

Hazlett, Maril. (2004) 'Woman vs. man vs. bugs': Gender and popular ecology in early reactions to silent spring, *Environmental History*, 9(4): 701–729.

Hearn, Jeff, Marie Nordberg, Kjerstin Andersson, Dag Balkmar, Lucas Gottzén, Roger Klinth, Keith Pringle, and Linn Sandberg (2012) Hegemonic masculinity and beyond: 40 years of research in Sweden, *Men and Masculinities*, 15(1): 31–55.

Hearn, Jeff, and Hopcraft, David and Morgan, John (eds) (2014) *Men, Masculinities and Social Theory*. London: Routledge.

Hird, M.J. (2015) Waste, environmental politics and dis/engaged publics, *Theory, Culture & Society*, available online at www.researchgate.net/publication/ 273771646_Waste_Environmental_Politics_and_DisEngaged_Publics (accessed 15 March 2017).

Holt, Douglas B. and Thompson, Craig J. (2004) Man-of-action heroes: The pursuit of heroic masculinity in everyday consumption, *Journal of Consumer Research*, 31(2): 425–440.

Hopkins, J.F. (1951) *A History of the Hemp Industry in Kentucky*. Lexington, KY: University Press of Kentucky.

Hultman, M. (2014a) Transition delayed. The 1980s ecotopia of a decentralized renewable energy systems. In K.Bradley and J. Hedrén (eds), *Green Utopianism: Perspectives, Politics and Micro Practices*. London: Routledge.

Hultman, M. (2014b) "How to meet? Research on Ecopreneurship with Sámi and Māori", paper presented at the international workshop: *Ethics in Indigenous Research – Past Experiences, Future Challenges*, Umeå, March 3–5, 2014.

Hultman, M. (2014c) Ecopreneurship within planetary boundaries: Innovative practice, transitional territorialisation's and green-green values. Paper at *Transitional green entrepreneurs: Re-thinking ecopreneurship for the 21st century* Symposium at Umeå University, June 3–5, 2014.

Hultman, M. (2013) The making of an environmental hero: A history of ecomodern masculinity, fuel cells and Arnold Schwarzenegger, *Environmental Humanities*, 2(1): 79–99.

Hultman, M. (2016) Gröna män? Konceptualisering av industrimodern, ekomodern och ekologisk maskulinitet, i Kulturella Perspektiv vol. 1, Special Issue: *Environmental Humanities* ed. Christer Nordlund, pp. 28–39.

Hultman, Martin and Ali Yaras. (2012) The socio-technological history of hydrogen and fuel cells in Sweden 1978–2005: Mapping the innovation trajectory, *International Journal of Hydrogen Energy*, 37(17): 12043–12053.

Hysing, Erik. (2014) A green star fading? A critical assessment of Swedish environmental policy change, *Environmental Policy and Governance*, 24(4): s. 262–274.

Jorgenson, A.K. and Clark, B. (2012). Are the economy and the environment decoupling? A comparative international study, 1960–2005, *American Journal of Sociology*, 118(1), 1–44.

Laclau, Ernesto, Mouffe, Chantal (1985) *Hegemony and Socialist Strategy*. London: Verso.

MacGregor, S. (2009) A stranger silence still: The need for feminist social research on climate change, *Sociological Review*, 57: 124–140.

Macgregor, S. (2013) Only resist: Feminist ecological citizenship and the post-politics of climate change, *Hypatia*, 29(3): 617–633.

Magnusdottir, Gunnhildur Lily, and Annica Kronsell. (2014) The (in)visibility of gender in Scandinavian climate policy-making, *International Feminist Journal of Politics*, ahead-of-print: 1–19.

McCright, A.M. and Dunlap, R.E. (2003). Defeating Kyoto: The conservative movement's impact on US climate change policy, *Social Problems*, 50(3): 348–373.

McCright, Aron and Dunlap, Riley. (2011) Cool dudes: The denial of climate change among conservative white males in the United States, *Global Environmental Change*, 21(4): 1163–1172.

McCright, A and Dunlap, R. (2015) Bringing ideology in: The conservative white male effect on worry about environmental problems in the USA, *Journal of Risk Research*, 16(2): 211–226.

McNeil, J.R. (2001). *Something New Under the Sun: An Environmental History of the Twentieth-century World (The Global Century Series)*. New York: WW Norton & Company.

McNeil, John. (2003) *Någonting är Nytt Under Solen – Nittonhundratalets Miljöhistoria*. Stockholm: SNS förlag.

Melosi, Martin V. (1987) Lyndon Johnson and Environmental Policy, in Robert A. Divine (ed.), *The Johnson Years: Vietnam, the Environment, and Science*. Lawrence: University Press of Kansas, pp. 119–120.

Merchant, Carolyn (1980) *The Death of Nature: Women, Ecology, and the Scientific Revolution* (1st edn). San Francisco: Harper & Row.

Merchant, Carolyn (1996) *Earthcare: Women and the Environment*. New York: Routledge.

Neumayer, Eric and Thomas Plümper. (2007) The gendered nature of natural disasters: The impact of catastrophic events on the gender gap in life expectancy, 1981–2002, *Annals of the American Association of Geographers*, 97(3) 551–566.

Nordin, Ingemar (2008) Klimatdebatten – en röra, *Östgöta Correspondenten*, 9/6.

Norgaard, K. and York, R. (2005) Gender equality and state environmentalism, *Gender and Society*, 19(4): 506–522.

Nye, D.E. (1996) *American Technological Sublime*. Cambridge, MA: MIT Press.

Öhman, May-Britt. (2007) Taming Exotic Beauties: Swedish Hydro Power Constructions in Tanzania in the Era of Development Assistance, 1960s-1990s, PhD Thesis, KTH, Sweden.

Oreskes, Naomi and Conway, Erik. (2010) *Merchants of Doubt: How a Handful of Scientists Obscured the Truth on Issues from Tobacco Smoke to Global Warming*. New York: Bloomsbury Press.

Parrott, Dominic J., and Amos Zeichner. (2003) Effects of hypermasculinity oh physical aggression against women, *Psychology of Men & Masculinity*, 4(1): 70.

Pulé, Paul. (2013) A declaration of caring: towards ecological masculinism. PhD diss., Murdoch University, Australia.

Radetzki, Marian och Lundgren, Nils. (2009) En grön fatwa har utfärdats, *Ekonomisk Debatt*, 5: s. 58–61.

Räty, Riitta, and A. Carlsson-Kanyama. (2010) Energy consumption by gender in some European countries, *Energy Policy*, 38(1): 646–649.

Rome, Adam. (2003) 'Give Earth a chance': The environmental movement and the sixties, *Journal of American History*, 90 (September): 534–541.

Shiva, Vandana. (2006) *Earth Democracy: Justice, Sustainability and Peace*. London: Zed Books.

Sramek, Joseph. (2006) 'Face him like a Briton': Tiger hunting, imperialism, and British masculinity in colonial India, 1800–1875, *Victorian Studies*, 48(4): 659–680.

Stacey, Judith. (1990) *Brave New Families: Stories of Domestic Upheaval in Late-Twentieth-century America*. Oakland, CA: University of California Press.

White Jr, Lynn. (1967) The historical roots of our ecologic crisis, *Science*, 10 March: 155(3767): 1203–1207.

York, R., Rosa, E.A. and Dietz, T. (2010) Ecological modernization theory: Theoretical and empirical challenges. In M.R. Redclift and G. Woodgate (eds), *The International Handbook of Environmental Sociology*. Cheltenham: Edward Elgar, pp. 77–90.

7 The contribution of feminist perspectives to climate governance

Annica Kronsell

Introduction

Feminist perspectives are a highly useful input to the understanding of climate governance and absolutely crucial if climate governance is to be sustainable. Feminist perspectives build on feminist theory, which has developed considerably through the years, beginning in and forming through the women's movement in civil society, into politics, policymaking and academia. Feminist theory has been particularly concerned with power relations, less frequently addressed in climate governance scholarship. Hence, feminist theory provides useful analytics to shed light on inequities, inequalities and power relations in climate governance. The main contribution of this chapter is to outline how different feminist theories can add depth and critique to the discussion of climate governance.

Feminist analysis focuses on gender relations, thereby it goes beyond a mere focus on women and women's role in climate governance. While the representation of women in politics and governance is of relevance to such studies, feminist theory makes it possible to also seriously discuss and question the relation between women's and men's interests and values in the gender order, and problematize the relation between a critical mass of women and critical acts in policymaking. There are different feminist perspectives, and here I suggest that particular attention be paid to theories with a critical epistemology because they call attention to the gender normative setting of climate governance and climate institutions, and draw attention to intersectional power relations, including how gender relations intersect with other power relations such as class, ethnicity and age. A critical analysis and problematization of climate governance using feminist perspectives is not merely an academic exercise but essential to advance climate governance towards gender sensitivity, equality, inclusion and justice. This chapter outlines the main elements in three different feminist perspectives to illustrate how they are relevant for an analysis of climate governance. Using several empirical examples the chapter demonstrates what feminist insights imply for the prospects of engendering climate governance, regarding representation in policymaking, in scientific discourse on climate and in the local context regarding transport planning. First, it is necessary to determine what climate governance is.

Climate governance

Governance has been extensively studied and theorized in the social sciences (Bäckstrand et al., 2010: 8–12) and this is highly relevant due to the global character of climate change. The Commission on Global Governance (1995) defined governance as 'the sum of the many ways individuals and institutions, public and private, manage their common affairs.' Similarly, Lafferty (2004) refers to governance as the institutions, instruments and mechanisms available to collectively steer a society. Looking at climate governance specifically, the institutions, instruments and mechanisms aim to steer society in the direction of prevention, mitigation and/or adaptations to the risks of climate change (Jagers and Stripple, 2003). As used here, climate governance includes both the efforts to mitigate climate change, for example by reducing carbon emissions, curbing energy or transport use, and adaptation to climate generated risks, for example by setting up policies to prevent as well as manage extreme weather events.

While governance processes can help make schemes more palatable to both governments and corporate interests (e.g. carbon emission trading) they also challenge state authority when policymaking of state governments is pitted against governance forms such as market-based and voluntary instruments (Trieb et al., 2007). State's policies have often been viewed as a problem, obstructing the function and expansion of markets (Rai, 2008: 27). Governance processes also challenge corporate power, for example with Clean Development Mechanisms (CDM), reducing emissions from deforestation and forest degradation (REDD) and climate financing, because they involve an increasing number of non-governmental actors (Okereke et al., 2009). Empirically, we see evidence of complex processes of transnational climate governance (Andonova et al., 2009), emerging from governing institutions as well as societal activities. There is also a fluidity in climate governance across norms, institutions and levels of governance, as politics and decision-making extend from global institutions such as the UNFCCC, to the EU, to national governments and local authorities and across to non-governmental organizations (NGOs), private and civil society actors. The growth in public–private partnerships (Bäckstrand, 2008) is an example of this governance trend, which has witnessed a changing role of states although recent scholarship argues that the state has more influence on climate governance than has previously been assumed in the literature (Bäckstrand and Kronsell, 2015). However, as a general observation: governance 'has had quite distinct imperatives in the North and the South' (Jayal, 2003: 97). In the North it has been about privatization and liberalization, in the South, governance has become the remedy for slow and inefficient development. As there is a complexity of climate governance observable in the world, in making the argument that feminist perspectives can add something significant to the analysis of climate governance, the chapter will focus on a few select examples of climate governance that will hopefully be indicative of future research possibilities.

Research on gender and climate

Gender is a concept developed in feminist theory (Connell, 2002). There are many different feminist theories and frameworks but a commonality is that their view of gender is related to power and a power order (patriarchy). The gender power order is expressed materially in terms of differences in livelihoods, work roles and access to resources, which, in turn, are based on and give rise to gender norms and societal practices (Lorber and Farrell, 1991). Although feminist theory is focused on gender as power relations between men and women, other categories overlap with gender such as ethnicity, class, sexuality, place with implications for power relations (Crenshaw, 1991; Lykke, 2010).

The emerging literature under the broad label of 'gender and climate change' has only just begun to analyse gender as power relations, and gender as intersecting with other categories (cf. Kaijser and Kronsell, 2013). In much of the scholarship on gender and climate change to date, the attention is on women, for example, by demonstrating sex differences in the impacts of climate change, highlighting the vulnerability of women in the South, or emphasizing women's specific role in mitigating climate change. While it is important that women's role in climate governance is studied and the specific effects of climate change on women is an important part of a feminist theorizing of climate governance, there is a concern about *how* 'woman' is conceptualized.

The Intergovernmental Panel on Climate Change (IPCC) has noted how rural women in developing countries are among the groups most vulnerable to climate change and a major emphasis of earlier studies on gender and climate change focused on women in terms of being vulnerable victims, often in the local context and in the South (cf. Resurrección, 2013). Geraldine Terry's edited book *Climate Change and Gender Justice* (2009), one of the first to establish how gender matters in climate issues, studies women as victims in places such as Bangladesh, India and Tanzania (Terry, 2009). Similarly, Irene Dankelman's edited volume *Gender and Climate Change* (2010) is written as an introduction to the field with multiple examples of local practices, mainly in the South. Also in studies of climate adaptation, women in the Global South are cast as vulnerable victims (Aquilar, 2013; Bendlin, 2014: 684; see for example: Denton, 2002; Mearns and Norton, 2010; Oparaocha and Dutta, 2011).

Singh et al., (2010) provide another typical example of some general findings. They studied how the risks of climate change vary for women and men due to their different social, political and economic conditions and show that women are at greater risk, more vulnerable and more likely to become victims of climate change because they do not have the same access to resources, have different living conditions and have more restricted capabilities than many men. Disaster relief and adaptive capacity is often considered in terms of women's specific roles, concerns and livelihoods (Bee et al., 2013; Dominelli, 2013). Margaret Alston (2013a, 2013b) argues that these findings should be taken into account in climate governance. She calls for 'an understanding of significant gendered consequences: that climate change has critical consequences for women and men; and that women are particularly vulnerable to poverty, insecurity and violence during and after climate events' (Alston, 2013: 5). This same tendency, to equate gender with women, is found in adaptation policy practice as Holvoet and Inberg (2014: 272) show in their study of the climate adaptation strategies of 31 countries.

Hence, research in this field has established that women in the rural South are likely to be more vulnerable to climate change. This narrow focus can be questioned also in light of empirical studies that have revealed that gender matters in relation to climate issues also in the richer countries in the North (Carlsson-Kanyama et al., 2010; Nordic Council, 2009; Polk, 2009). Feminist perspectives are critical of the tendency to focus the climate and gender debate mainly on women and women in the local context. It obscures how their predicament is a result of gender power relations. When gender is uncritically equated with women, power relations are made invisible.

The resulting understanding of gender as equal to women is skewed and unfortunate both in light of feminist perspectives that show how gender is about power relations that are intersecting, and that gender is socially constructed. It may also have practical consequences for climate governance as efforts towards climate mitigation and adaptation can be misdirected as a result of restricting the framing of gender in climate governance to one about women in the rural South. However, it should be recognized that, of late, a more nuanced literature is emerging, one that engages with masculinities (Anshelm and Hultman, 2014a, 2014b) with intersectionality (Nagel, 2012) and with the full gamut of gender (MacGregor this volume).

Feminist perspectives

Previous research in the field on climate governance has made scant use of critical feminist work and paid insufficient attention to gender as a power relation and a social construct. Three different feminist perspectives are presented and elaborated on in the following to give a richer and more theorized account of how to understand the relevance of gender in climate governance. What is more, the three feminist perspectives: material, liberal and constructivist are not in agreement on how gender is best theorized. The purpose here is not to grant a winner, that is, the one that delivers the best explanation; my point is that each of these three perspectives contribute specific insights to the understanding of gender in climate governance, on material conditions, on representation and on norms.

Material feminist perspectives

Material feminist perspectives stress that the division between men and women and the resulting gender relations, are material and closely tied to the division between production and reproduction functions in society and in the economy. Sex differences are the base for economic material relations and gender power relations maintained through the continuous devaluation and exploitation of women's reproductive work, and functions. This occurs in a global system where the division of labor into reproductive and productive spheres maps onto sex difference.[1] Accordingly, the gender order is reflected in a climate governance order, which exploits women's reproductive labor as mothers, nurturers and carers. Thus, the difference between men and women in climate change issues can be viewed as the effect of global historic gender power relations. Governments and large corporations are deeply implicated in this and form part of a global power structure (True, 2012). Already in the early 1990s, Ann Tickner (1992: 99) argued that governments are implicated because the nation-state system and states' exploitation of their own resources as well as that of others, not least through neo-colonialism, is connected to the binary gender power order. This order is expressed at multiple levels in relation to resource use in globalization and is tied to decision-making in international relations (Detraz, 2010), for example through agreements and institutions. Gender power is structural but also expressed symbolically and individually. From this perspective, fossil-fuel consumption is not understood as a personal decision. Chris Cuomo (2011) writes that here is a 'widespread disempowerment associated with personal and household efforts' because people's possibility and capability to take complete control over their general energy consumption options – the way they use and access public transportation, local foods or renewable energy – are severely circumscribed by the larger power structures in place in these sectors (Cuomo, 2011: 702). Peter Newell (2005: 73) suggest that the global power order 'feeds on entrenched patterns of social inequality etched along racial, class and gender lines' and that in turn gives rise to 'new patterns of environmental inequality'. Intersectional aspects are expressed at the policy and institutional level of climate governance as well (Nagel, 2012; see also Newell, 2005: 78). Climate policies formulated in the UN, the IPCC and the EU can be expected to reflect the current gender power of those institutions; that is, the concerns of mainly male policymakers and male scientists from the rich North. Gender, class and ethnicity are intersectional aspects that matter here.

The importance of thinking intersectionally is often more concretely illustrated when discussing individuals. In addressing prospects for climate adaptation, Joane Nagel (2012) shows that vulnerability to climate-associated disaster risks is related to intersectional aspects, and speaks about poverty, economic activities and modesty cultures as relevant, rather than just to gender alone. Another example is the devastation of New Orleans by Hurricane Katrina in 2005 where marginalized people were less likely to evacuate, could not afford to live somewhere else and had poorer prospects when displaced (Tuana, 2008: 189). Nicole George (2014: 320) calls this a form of slow violence which affects Pacific Island communities, 'who live on the "front line" of climate change'. For climate issues in the rich North, gender matters, but consumption of transport, energy, food and goods that generate climate emissions and are relevant for climate mitigation is also closely linked to class and economic status. Class matters sometimes more than gender does, as those most affluent consume more and emit most. Consumption and emission also differ depending on ethnicity and age, which calls for these categories to be also considered (Kaijser and Kronsell, 2013). What these studies say is that in taking a material perspective it is necessary to consider intersectional aspects beyond gender. The focus on the vulnerability of women in the South obscures gender differences that exist in other parts of the world. Moreover, there are considerable differences within both the North and the South (Johnsson-Latham, 2007). For example, homeless people in the North hardly produce any carbon emissions, while rich elites in the South may emit more carbon than the average citizen in the North (EIGE, 2012: 21).

Climate change, in the material feminist interpretation, can be seen as the result of a neoliberal masculine order that prioritizes production, growth and expansion over feminine values such as care and nurturing and more solutions generated from women's experiences. From this perspective, women's representation in the politics and policies of climate change is crucial, not just as a matter of democracy and justice but as a source of change because, due to the gender power order, women as individuals and in groups have different experiences to men, and also different values which can become important in generating alternative climate change policies and strategies.

In reviewing studies of gender and climate change, Arora-Jonsson (2011: 745) finds that while women are framed as vulnerable they are also frequently depicted as virtuous heroes. According to Bernadette Resurrección (2013) this framing – casting women as both the problem, 'vulnerable' and the solution, 'heroines' – reflects the persistence in environment and development discourses to view gender as something that only regards women and echoes a more general discourse on gender in global governance (e.g. Leach, 2007) recreated in the contemporary climate change agenda (Resurrección, 2013: 37; Holvoet and Inberg, 2014). Through this dominant framing, both the problem and the solution for gender issues are confined to women and leave little room for an analysis of gender as power relations. Chris Cuomo (2011: 695) suggests that if structural inequalities are framed as differential vulnerability or susceptibility to harm, the result is that attention is drawn to the 'supposed weaknesses or limitations of those who are in harm's way, but says little about whether injustices or other harms have put them in such precarious positions.' If virtuous women are the ones who are called on to solve the problem, it effectively diverts attention from a gender power order.

The celebration of women's agency as 'heroic' and 'virtuous' is an important element in feminist thought, particularly among materialist feminism, such as standpoint theories and ecofeminist thinkers. Standpoint theory suggests that women's positions in society shape ways of knowing, and provides other experiences than those lived, for example, by dominant male elites (Hartsock, 1985; Harding, 1991). Based on mothering and other caring experiences and skills, women can show alternative ways of governance, such as 'earthcare' (Merchant, 1996). Women are viewed as being able to save the world by 'reweaving' it (Diamond and Orenstein, 1990) and by 'healing the wounds' (Plant, 1989).

There are studies that focus on women as a group and show that they collectively contribute less carbon emissions than men as a group do. In this sense, women's behavior could be considered more climate-friendly than men's. The sex difference is most apparent in the South but is also evident in the North (Bendlin, 2014: 684–687; OECD, 2008a, 2008b; Räty and Carlsson-Kanyama, 2010; Schultz and Stiess, 2009). According to various surveys, women in the EU and the US are more concerned about climate issues and more inclined to take climate action (European Commission, 2009, McCright, 2010) and there are gender differences in terms of attitudes to climate change issues (Goldsmith et al., 2013; McCright and Dunlap, 2011). While we should not deny that gender differences may be relevant as a basis for climate agency and for generating alternatives (MacGregor, 2014), this framing does not address the gender and climate problem in terms of power, and does not look beyond women as a homogenous group.

Material feminist perspectives direct attention to the livelihoods of women, and how their role in reproductive labor has led to a structural power imbalance in relation to production and the economic order. In this order, caring and domestic labor, most often done by women, is either not recognized or is exploited and it is an economic order that relies on high fossil fuel consumption. Material feminism focuses on the structural dimension of a world economy in which climate governance takes place and not surprisingly climate governance reflects white elite male rule and steering. The gender structure has effects on multiple levels, leaving women in the poorest countries particularly vulnerable to climate change. At the same time, and due to their less powerful and even marginalized place in the global gender power order, women also have a resource as they, by virtue of their situated knowledge, have the potential to represent an alternative way of living, alternative values and ethics.

Material feminism perspectives are criticized by other feminisms particularly because of the way they seem to treat women as biologically bound, as they argue that women have a privileged insight into the gender order due to their reproductive role (cf. Harding, 2004). This is regarded as problematic in constructivist feminism which considers gender as a constructed category, as something that is made and done in relation to other gendered beings and institutions, and to liberal feminism which has at its core a universal view of humans as basically the same, only limited by discriminating rules and practices.

Liberal feminist perspectives

To be given equal rights to influence climate governance and to receive equal treatment in the climate governance sphere remains a concern for liberal feminist perspectives, although it has been on the agenda of the women's movement since suffrage debates, and devoted to women's possibility to participate in public and political life (Lovenduski, 2005; Phillips, 1995). To that end, the presence of women is believed to assure democratic quality and increase legitimacy (Lovenduski and Norris, 2003; Wängnerud, 2009) in climate institutions as well as enhance women's agency. A strive for gender balance is crucial because there needs to be a critical mass of women for their presence to be felt (Moss Kanter, 1977). Informal gender equality quotas used today aim for gender balance in the 40–60 percent range (Dahlerup, 2006). There is an underlying assumption that women's presence will lead to a different kind of politics as well (e.g. Phillips, 1995).

Women's participation in global climate governance is not gender balanced (Hemmati and Röhr, 2009; Röhr et al., 2008). The United Nations Framework Convention on Climate Change (UNFCCC) was adopted at the UN Conference in Rio de Janeiro in Brazil in 1992 at the same time as a number of other conventions. While the other agreements all officially noted the importance of gender, the UNFCCC did not. Due to this omission, women were not included among the constituencies as observers in the process. There was a major breakthrough at the 13th Conference of the Parties to the Convention (COP 13) taking place in Bali 2007 when the worldwide network Women for Climate Justice (GenderCC) was established and a commitment to gender was made (Hemmati and Röhr, 2009: 25). The report Gender Equality and Climate Change investigated women's participation in climate institutions such as different EU bodies, the EU member states' representation in UNFCCC and in the COPs. According to the report women held on average 39 percent of the seats in national delegations in climate governance and women's representation exhibits a slow but constant increase over the years (EIGE, 2012: 53–58). Equal representation is one of the more important aspects to study, according to liberal feminism, as these perspectives would argue that increased representation in the bodies mentioned above, even a slow one, indicates a move towards gender equality. Women's representation in national climate governance bodies of EU member states shows an average of only 25 percent (EIGE, 2012: 3) and a significant variation between EU member states. Finland and Sweden have more than 50 percent women and hence, have assured gender equality in their climate bodies (EIGE, 2012: 65).

Feminist liberal perspectives are concerned with equal rights, access and opportunities not only as a women's concern but also as an important democratic value that adds legitimacy to political systems. However it does not stop at this; there is also an assumption that something qualitatively different is added through equal representation. When women take part in sufficient numbers in decision-making, their presence is thought to impact governance. Two studies seem to confirm this. Nations with higher proportions of women in parliament are also more prone to ratify environmental treaties (Norgaard and York, 2005). A similar study showed that carbon emissions are lower in nations where women have higher political status (Ergas and York, 2012). This global quantitative comparative study indicated that women's participation in climate governance makes a difference and is in line with what the literature on women's representation assumes, namely that women's contributions to political decisions can be expected to be different from men's contributions. Gender differences suggest that women may have different ideas and opinions on climate politics that could come to the fore with their presence in policymaking. As noted above, the studies of Norgaard and York (2005) and Ergas and York (2012) may confirm this, while other studies do not.

In order to better understand whether the presence of women in climate policymaking had noticeable effects on climate governance and how gender issues were approached in the climate governance institutions, Magnusdottir and Kronsell (2015) studied climate governance in the most gender balanced of the EU member states – the Scandinavian countries – but were unable to establish a link between equal or higher female representation and gender awareness. We looked for evidence through text analysis of climate policy documents and from interviews with climate policymakers but found that known gender differences in material conditions and in attitudes towards climate issues were invisible and excluded from climate policy texts. Both male and female policymakers were largely unaware of the relevance of gender differences and how to consider them in climate policymaking, despite the gender balance of the institutions where climate policy was developed.

A recent study (Kronsell et al., 2015) analysed gender representation in infrastructure decision-making and concluded that gender equality at the macro level of climate governance in Sweden was not reproduced at the micro level. Infrastructure decisions have implications for climate governance as they lock-in certain transport structures for a long period of time. During 2008–2010, long-term infrastructure plans covering national and regional investments in new roads, railway and navigation infrastructure for the period 2010–2021 were developed by the responsible national transport agencies and regional authorities in Sweden. The Network for Women in Transport Policy investigated whether gender balance was achieved in the development of the above long-term infrastructure plans for 2010–2021 by studying the composition of the various groups involved in the infrastructure planning process (Trivector, 2010). The study concluded that the presence of men and women in the process differ according to the nature of tasks. Half of the groups among the national as well as regional working and steering groups, were dominated by men (8 out of 17 studied). In the steering groups where the most important decisions on infrastructure planning were taken, there was no gender balance. In the other groups there was a gender-balanced distribution of men and women but this was in groups that can be assumed to have less influence over the outcome. Women were in the majority only in the working and advisory groups on environmental assessment (Trivector, 2010; Kronsell et al., 2015).

It is important to advance the perspective of liberal feminism in climate governance to assure the values of equality, equal treatment and equal opportunities embedded in democratic systems. However, to assure equal representation whether this is in climate governance bodies at the international, national or local levels, as the above examples show, does not automatically lead to more gender-aware policies. 'Just add women and stir' does not appear to be fully sufficient as a strategy for increased gender awareness in climate governance.

Constructivist feminist perspectives

The basic assumption in constructivist feminism is that gender is socially constructed, and gender power is a normative order that rests on specific constructions of men and women (Locher and Prugl, 2001; Prügl, 2011). Gender is a continuously (re)constructed category and gender construction takes place in the continuous processes and activities of daily lives. Gender subjectivities, or masculinities and femininities, are constructed in relation to one another and to other social categories. It can be argued that climate governance happens in a context in which masculinity is the accepted norm (Hooper, 2001) while this often remains unarticulated and invisible (Hearn and Husu, 2011). The absence of explicit gender symbols or gender recognition is not the same as being gender-neutral. Hence, 'silences' or the lack of conscious and explicit reference to gender within climate institutions or climate strategies does not mean that gender is irrelevant for the issue area. Gender, when understood as normative, implies that climate policies that lack a gender perspective simply re-produce the current gender order. The routine practices of climate institutions will reproduce the existing gender order (Connell, 1995: 212).

The example of infrastructure planning referred to above can illustrate how norms are relevant in gender relations. The study (Trivector, 2010) showed that there was a strong preponderance of male participants in the groups that were responsible for the dialogue between municipalities, businesses and planners in the transport department or county/regional governments. In these groups, interest organizations and other stakeholders are given opportunities to provide feedback on the suggested plans and investments. The opportunity to influence the content of the plans was mostly given to men. Through interviews, it became evident that in connection with the invitations to these meetings the organizations were encouraged to consider gender aspects when appointing their delegates. However, the municipal officials were critical of the appeal to gender aspects since they were inclined to see this as a questioning of their competence in selecting delegates and that the tendency was that the ones that 'are allowed to be involved and think' in these groups should have many years' experience of infrastructure issues. This is a reason why it is often men who are appointed and included in such working groups. The officials explained that men, and often older men, have more experience of previous planning processes and are thus preferred to women delegates (Trivector, 2010). In this context, 'experience' becomes coded as masculine and the call for competence and experience becomes a conserving element, which leads to the reproduction of gendered patterns as women are considered 'less competent', but also to the reproduction of previous norms on infrastructure planning and mobility. This pins decision-making down to a path dependence regarding both who gets to be in charge of decisions but also how transport infrastructure is framed, rather than what might be needed to lower carbon emissions. The study of gender distribution at the micro levels of decision-making suggests that equal participation is not the only aspect that is relevant, since the division of tasks works according to perceived masculine and feminine spheres. Women are involved in groups that have less power in deciding on infrastructure and assigned areas that are coded as less masculine.

Another way to investigate gender power is through an emphasis on normative aspects and on how masculinity shapes climate governance. Studies have discussed the relation between specific masculinities and risk behavior, and found that it is conservative, white, middle-class men who are most likely to deny climate problems and use more risky strategies (Hultman this volume; McCright and Dunlap, 2011). Those who have power and benefit from the current socio-technological system will be more motivated to justify it and less willing to change or promote change (Goldsmith et al., 2013: 168 see also Anshelm and Hultman, 2014b). Climate governance today focuses mainly on incremental improvements and technical fixes in the energy and transport areas. Hemmati and Röhr (2009: 20) write: 'The debate on climate change has been very narrow, focusing on the economic effects of climate change, efficiency, and technological problems.' These are male-dominated sectors and it is mainly men who control and benefit from the research, innovations, and jobs in these sectors. Here male power is material but is also disciplining through the normative structures that have emerged in the fossil-fuel economy. It can be argued, applying the insights of Connell (2002) and Hearn and Husu (2011) that climate governance happens in a context in which masculinity is the accepted norm, where it remains unarticulated and invisible.

Feminist constructivist perspectives are useful because they can disrupt the reproduction of power through a critical ontology that raises epistemological questions. There is a long tradition of feminist critique of science to build upon, for example, when raising questions about the knowledge produced about climate change. The IPCC has become increasingly powerful, as the knowledge is increasingly secured and assured with each new report. According to its webpage, the IPCC's objective is 'to provide the world with a clear scientific view of the current state of knowledge in climate change'. The ambition to 'provide rigorous and balanced scientific information to decision makers' conveys a sense that neutrality in science is possible. The IPCC reaffirms this by saying that its work is 'policy-relevant yet policy-neutral' (IPCC, 2014). The assumption is somehow – because of the consensual process including a global network of scientists – that the IPCC produces *the* knowledge of climate change and that policymakers should act upon it promptly (cf. MacGregor, 2014). It is commendable, and should not be taken lightly that there is such a serious effort of the international community via the IPCC to understand the climate issue. Yet, taking feminist theorizing seriously, it gives reasons for doubt and raises several concerns.

The current body of climate knowledge is centred on global-scale processes, and it has been dominated by a techno-scientific framing that suggests that it is possible to control and dominate nature (Neimanis and Lowen Walker, 2014). This is exemplified in Israel and Sachs' work (2013: 42) who find the IPCC's colourful mapping of future climate conditions to be problematic because through this depiction, the IPCC projects future climates as thoroughly knowable and controllable. Similarly, Joni Seager studied the discourse around the 2°C target and remains sceptical to the way it suggests scientists and politicians can control climate events when they 'identify levels of acceptable danger [i.e., 2 degree target], and hold global warming to that line' as an appropriate global policy (Seager, 2009: 20). She suggests feminist analysis is useful to problematize these illusions of control.

Taking feminist critical perspectives seriously means, as Nancy Tuana (2013) argues, to view scientific knowledge on climate change as situated, for example, within certain scientific traditions such as earth system science, and in this as power-laden. Feminist constructivist perspectives on climate change science takes a step back to argue for the critical inquiry of the foundations of climate governance and are well-fitted to address the epistemological and normative underpinnings of scientific inquiry. They can offer a critical perspective on how the climate problem is scientifically framed and what scientific understandings will inform climate policies and strategies.

Conclusion

Feminist perspectives are necessary as they shed light on inequities, inequalities and power relations in climate governance. Thus, they can provide a richer account of how to understand the relevance of gender in climate governance than previous research that had a tendency to focus on women only, making gender relations less visible. Feminist perspectives are helpful in that they analyse gender as a power relationship and a societal structure.

Material and constructivist feminisms emphasize power relations most succinctly. Material feminist perspectives emphasize how livelihoods of women differ from men's, and in particular how women's reproductive labor has led to a structural power imbalance in relation to production and the economic order. This economic order's fundamentals are fossil fuel-dependent production and growth while caring and domestic labor most often performed by women, is exploited and often not even recognized as labor. Material feminisms call attention to the structural dimension of a world economy in which climate governance is placed and argues it reflects white elite male rule and steering. Climate change is here viewed as the result of a neoliberal masculine order that prioritizes production, growth and expansion over feminine values such as care and nurturing. To fully capture this, it is necessary to consider how women's reproductive labor is differently exploited depending on economic status and location, rendering poor women in deprived areas highly vulnerable to climate change.

Constructivist feminists consider material feminisms' focus on the reproductive labor of women too narrow a definition of gender relations. They problematize the notion of women's specific interests and values, and instead point to the importance of gendered norms that are institutionalized, across societies in general, but also in the context where climate governance takes place. Norms relating to masculinity and femininity impact how climate governance is understood and carried out while the norms are generated from, and have concrete effects on, the material conditions of men and women as they live their daily carbonized lives. Hence, there is a need to combine materialist and constructivist feminist perspectives to fully capture gender power in climate governance, be it in mitigation or adaptation efforts. Liberal feminist perspectives have a place as well, due to the importance it places on the democratic qualities and legitimacy of climate governance. The most interesting liberal feminist research contributions are those that investigate if women's increased representation has the potential to make a qualitative impact on climate policies and institutions and what that might be like. That research definitively challenges the proposal from material feminisms, that knowledge is situated and that women, because of how they are situated in the gender order, have acquired different knowledge that can provide alternatives conducive to a decarbonized and sustainable future.

Note

1 In the category 'material feminism' I include scholars like Rosemary Hennessy (1993) and Maria Mies and Vandana Shiva (1993) and not scholars who write in the (new) materialist tradition such as Alaimo and Hekman (2008).

References

Alaimo, S. and S. Hekman (eds), (2008) *Material Feminism*. Bloomington: Indiana University Press.

Alston, M. (2013) 'Introducing gender and climate change: research, policy and action' in M. Alston and K. Whittenbury (eds), *Research, Action and Policy: Addressing the Gendered Impacts of Climate Change*. Dordrecht: Springer, 3–14.

Andonova, L. B., M. Betsill, and H. Bulkeley (2009) 'Transnational climate governance' *Global Environmental Politics* 9(2): 52–73.

Anshelm, J. and M. Hultman (2014a) *Discourses of Global Climate Change: Apocalyptic Framing and Political Antagonisms*. London: Routledge.

Anshelm, J. and M. Hultman (2014b) 'A green fatwā? Climate change as a threat to the masculinity of industrial modernity' *NORMA: International Journal for Masculinity Studies* 9(2): 84–96.

Aquilar, L. (2013) 'A path to implementation: Gender-responsive climate change strategies' in M. Alston and K. Whittenbury (eds), *Research, Action and Policy: Addressing the Gendered Impacts of Climate Change*. Dordrecht: Springer, 149–157.

Arora-Jonsson, S. (2011) 'Virtue and vulnerability: Discourses on women, gender and climate change' *Global Environmental Change* 21: 744–751.

Bäckstrand, K. and A. Kronsell (eds) (2015) *Rethinking the Green State. Environmental Governance Towards Climate and Sustainability Transitions*. London and New York: Routledge/Earthscan.

Bäckstrand, K., J. Khan, A. Kronsell and E. Lövbrand (2010) *Environmental Politics and Deliberative Democracy. Examining the Promise of New Modes of Governance*. Cheltenham: Edward Elgar.

Bäckstrand, K. (2008) 'Accountability of networked climate governance: The rise of transnational climate partnerships' *Global Environmental Politics* 8(3): 74–102.

Bee, B., M. Biermann and P. Tschakert (2013) 'Gender, development, and rights-based approaches: Lessons for climate change adaptation and adaptive social protection' in M. Alston and K. Whittenbury (eds), *Research, Action and Policy: Addressing the Gendered Impacts of Climate Change*. Dordrecht: Springer, 95–108.

Bendlin, L. (2014) 'Women's human rights in a changing climate: Highlighting the distributive effects of climate policies' *Cambridge Review of International Affairs* 27(4): 680–698.

Carlsson-Kanyama, A., I. Julia Ripa, and U. Röhr (2010) 'Unequal representation of women and men in energy companies boards and management groups: Are there implications for mitigation?' *Energy Policy* 38: 4737–4740.

Commission on Global Governance. (1995) *Our Global Neighbourhood*. Oxford: Oxford University Press.

Connell, R.W. (1995) *Masculinities*. Cambridge: Polity Press.

Connell, R.W. (2002) *Gender*. Cambridge: Polity Press.

Crenshaw, K. (1991) 'Mapping the margins: Intersectionality, identity politics, and violence against women of color' *Stanford Law Review* 43(6): 1241–1299.

Cuomo, C. (2011) 'Climate change, vulnerability, and responsibility' *Hypatia* 26(4): 690–714.

Dahlerup, D. (2006) 'The story of the theory of critical mass' *Politics & Gender* 2(4): 511–522.

Dankelman, I. (ed.) (2010) *Gender and Climate Change: An Introduction*. London and Washington, DC: Earthscan.

Denton, F. (2002) 'Climate change vulnerability, impacts, and adaptation: Why does gender matter?' *Gender and Development* 10(2): 10–20.

Detraz, N. (2010) 'The genders of environmental security' in L. Sjoberg (ed.), *Gender and International Security*. New York and London: Routledge, 103–125.

Diamond, I. and G.F. Orenstein (eds) (1990) *Reweaving the World: The Emergence of Ecofeminism*. San Francisco: Sierra Club Books.

Dominelli, L. (2013) 'Gendering climate change: Implications for debates, policies and practices' in M. Alston, and K. Whittenbury (eds), *Research, Action and Policy: Addressing the Gendered Impacts of Climate Change*. Dordrecht: Springer, 77–93.

EIGE (2012) Review of the Implementation in the EU of Area K of the Beijing Platform for Action: Women and Environment – Gender Equality and Climate Change, European Institute of Gender Equality, Luxembourg: Publications Office of the EU.

Ergas, C. and R. York (2012) 'Women's status and carbon dioxide emissions: A quantitative cross-national analysis' *Social Science Research* 41(4): 965–976.

European Commission (2009) *European Attitude Towards Climate Change*. Brussels: European Commission.

George, N. (2014) 'Promoting women, peace and security in the Pacific Islands: hot conflict/slow violence' *Australian Journal of International Affairs* 68(3): 314–332.

Goldsmith, R.E., I. Feygina and J. Jost (2013) 'The gender gap in environmental attitudes: a system justification perspective' in M. Alston and K. Whittenbury (eds), *Research, Action and Policy: Addressing the Gendered Impacts of Climate Change.* Dordrecht: Springer, 159–171.

Harding, S. (1991) *Whose Science? Whose Knowledge? Thinking from Women's Lives.* Buckingham: Open University Press.

Harding, S. (2004) (ed.), *The Feminist Standpoint Theory Reader. Intellectual and Political Controversies.* New York and London: Routledge.

Hartsock, N. (1985) *Money, Sex and Power: Toward a Feminist Historical Materialism.* Boston, MA: Northeastern University Press.

Hearn, J. and L. Husu (2011) 'Understanding gender: Some implications for science and technology' *Interdisciplinary Science Reviews* 36(2): 103–113.

Hemmati, M. and U. Röhr (2009) 'Engendering the climate-change negotiations: Experiences, challenges, and steps forward' *Gender & Development* 17(1): 19–32.

Hennessy, R. (1993) *Materialist Feminism and the Politics of Discourse.* New York and London: Routledge.

Holvoet, N. and L. Inberg (2014) 'Gender sensitivity of sub-Saharan Africa national adaptation programmes of action: Findings from a desk review of 31 countries' *Climate and Development* 6(3): 266–276.

Hooper, C. (2001) *Manly States: Masculinities, International Relations, and Gender Politics.* New York: Columbia University Press.

IPCC (2014) International Panel on Climate Change. Available online at www.ipcc.ch/organization/organization.shtml (accessed February 5, 2015).

Israel, A. and C. Sachs (2013) 'A climate for feminist intervention: Feminist science studies and climate change' in M. Alston and K. Whittenbury (eds), *Research, Action and Policy: Addressing the Gendered Impacts of Climate Change.* Dordrecht: Springer, 33–51.

Jagers, S.C. and Stripple, J. (2003) 'Climate governance beyond the State' *Global Governance* 9(3): 385–400.

Jayal, N.G. (2003) 'Locating gender in the goverance discourse' in M. Nussbaum, A. Basu, Y. Tambiah, and N.G. Jayal (eds), *Essays on Gender and Governance.* Human Development Research Center, UNDP 96–134

Johnsson-Latham, G. (2007) *A Study on Gender Equality as a Prerequisite for Sustainable Development: What We Know About the Extent to which Women Globally Live in a More Sustainable Way Than Men, Leave a Smaller Ecological Footprint And Cause Less Climate Change.* Stockholm: Ministry of the Environment.

Kaijser, A. and A. Kronsell (2013) 'Climate change through the lens of intersectionality' *Environmental Politics* 23(3): 417–433.

Kronsell, A., Smidfelt Rosqvist, L. and Winslott Hiselius, L. (2015) 'Achieving climate objectives in transport policy by including women and challenging gender norms – the Swedish case' *Journal of Sustainable Transport.* Available online

Lafferty W. E. (ed.), (2004) *Governance for Sustainable Development: The Challenge of Adapting Form to Function.* Cheltenham, UK and Northhampton, MA: Edward Elgar.

Leach, M. (2007) 'Earth Mother myths and other ecofeminist fables: How a strategic notion rose and fell' *Development and Change* 38(1): 67–85

Locher, B. and Prügl, E. (2009) 'Gender and European integration' in A. Wiener and T. Dietz (eds), *European Integration Theory* (2nd edn). Oxford: Oxford University Press.

Lorber, J. and Farrell, S.A. (eds) (1991) *The Social Construction of Gender*. London: Sage.

Lovenduski, J. (ed.), (2005) *State Feminism and Political Representation*. Cambridge: Cambridge University Press.

Lovenduski, J. and Norris, P. (2003) 'Westminster women: The politics of presence' *Political Studies* 51(1): 84–102.

Lykke, N. (2010) *Feminist Studies. A Guide to Intersectional Theory, Methodology and Writing*. New York: Routledge.

MacGregor, S. (2014) 'Only resist: Feminist ecological citizenship and the post-politics of climate change' *Hypatia* 29(3): 617–633.

Magnusdottir, G. and Kronsell, A. (2015) 'The (in)visibility of gender in Scandinavian climate policy-making' *International Feminist Journal of Politics* 17(2): 308–326.

McCright, A. (2010) 'The effects of gender on climate change knowledge and concern in the American public' *Population and Environment* 32: 66–87.

McCright, A. and Dunlap, R. (2011) 'Cool dudes: The denial of climate change among conservative white males in the United States' *Global Environmental Change* 21: 1163–1172.

Mearns, R. and Norton, A. (eds), (2010) *Social Dimensions of Climate Change: Equity and Vulnerability in a Warming World*. Washington, DC: World Bank.

Merchant, C. (1996) *Earthcare: Women and the Environment*. New York: Routledge.

Mies, M. and V. Shiva (1993) *Ecofeminism*. London; New Jersey: Zed Books.

Moss Kanter, R. (1977) *Men and Women of the Corporation*. New York: Basic Books.

Nagel, J. (2012) 'Intersecting identities and global climate change' *Identities: Global Studies in Culture and Power* 19(4): 467–476.

Neimanis, A. and Loewen Walker. R. (2014) 'Weathering: Climate change and the "thick time" of transcorporeality' *Hypatia* 29(3): 558–575.

Newell, P. (2005) 'Race, class and the global politics of environmental inequality' *Global Environmental Politics* 5(3): 70–94.

Nordic Council (2009) *Gender and Climate Changes*. Copenhagen: Nordic Council of Ministers.

Norgaard, K. and York, R. (2005) 'Gender equality and state environmentalism' *Gender & Society* 19(4): 506–522.

OECD (2008a) *Household Behavior and the Environment, Reviewing the Evidence*. Paris: Organisation for Economic Co-operation and Development.

OECD (2008b) *Promoting Sustainable Consumption: Good Practices in OECD Countries*. Paris: Organisation for Economic Co-operation and Development.

Okereke, C., Bulkeley H. and Schroeder, H. (2009) 'Conceptualizing climate governance beyond the international regime' *Global Environmental Politics* 9(1): 58–78.

Oparaocha, S. and Dutta S. (2011) 'Gender and energy for sustainable development' *Current Opinion in Environmental Sustainability* 3: 265–271.

Phillips, A. (1995) *The Politics of Presence*. Oxford: Clarendon Press.

Plant, J. (ed.), (1989) *Healing the Wounds: The Promise of Ecofeminism*. Santa Cruz, CA: New Society.

Polk, M. (2009) 'Gendering climate change through the transport sector' *Women, Gender & Research* 18(3–4): 73–82.

Prügl, E. (2011) *Transforming Masculine Rule. Agriculture and Rural Development in the European Union*. Ann Arbor: University of Michigan Press.

Rai, S. (2008) 'Analysing global governance' in S. Rai and G. Waylen (eds), *Global Governance: Feminist Perspectives*. Basingstoke: Palgrave Macmillan.

Räty, R. and Carlsson-Kanyama, A. (2010) 'Energy consumption by gender in some European countries' *Energy Policy* 38(1): 646–649.

Resurrección, B. (2013) 'Persistent women and environment linkages in climate change and sustainable development agendas' *Women's Studies International Forum* 40: 33–43.

Röhr, U., Spitzner, M., Stiefel E. and von Winterfeld, U. (2008) *Gender Justice as the Basis for Sustainable Climate Policies: A Feminist Background Paper*. Bonn: German NGO Forum on Environment and Development.

Schultz, I. and Stiess I. (2009) *Gender Aspects of Sustainable Consumption Strategies and Instruments*. Frankfurt: Institute for Social-Ecological Research.

Seager, J. (2009) 'Death by degrees: taking a feminist hard look at the 2° climate policy' *Kvinder, Køn & Forskning* 3–4: 11–21.

Singh, A., Svensson, J. and Kalyanpur, A. (2010) 'The state of sex-disaggregated data for assessing the impact of climate change' *Procedia Environmental Sciences* 1: 395–404.

Terry, G. (ed.), (2009) *Climate Change and Gender Justice*. London: Oxfam.

Tickner, J.A. (1992) *Gender in International Relations: Feminist Perspectives on Achieving Global Security*. New York: Colombia University Press.

Trieb, O., Bähr, H. and Falkner, G. (2007) 'Modes of governance: Towards a conceptual clarification' *Journal of European Environmental Policy* 14(1): 1–20.

Trivector (2010) *Jämställdhet i infrastrukturplaneringen – en utvärdering*, Report No. 38.

True, J. (2012) *The Political Economy of Violence Against Women*. New York: Oxford University Press.

Tuana, N. (2008) 'Viscous porosity: Witnessing Katrina' in S. Alaimo and S. Hekman (eds), *Material Feminisms*. Bloomington: Indiana University Press, 188–212.

Tuana, N. (2013) 'Gendering climate knowledge for justice: Catalyzing a new research agenda' in M. Alston, and K. Whittenbury (eds), *Research, Action and Policy: Addressing the Gendered Impacts of Climate Change*. Dordrecht: Springer, 17–31.

Wängnerud, L. (2009) 'Women in parliaments: Descriptive and substantive representation' *Annual Review of Political Science* 12: 51–69.

Part II

Case studies

8 Gender, climate change and energy access in developing countries

State of the art

Javier Mazorra, Julio Lumbreras,
Luz Fernández and Candela de la Sota

Introduction

The relationship between gender, climate change and energy access is complex and varies between developed and developing countries. The way these linkages are related to the international policy context had particular importance in 2015 as that year was critical to determine the international agenda for the following 15 years.

The Rio+20 outcome document, *The future we want* (United Nations and General Assembly, 2012), inter alia, set out a mandate to establish an Open Working Group (OWG) to develop a set of Sustainable Development Goals (SDGs) for consideration and appropriate action by the General Assembly at its 68th session. The Rio+20 outcome gave the mandate that the SDGs should be coherent with and integrated into the United Nations (UN) development agenda beyond 2015.

After more than four years of work, the OWG established a set of 17 goals, with respective targets and indicators, which were approved by 193 member states of the UN, during the UN Sustainable Development Summit in New York in September 2015.

Three of these goals are related to the aim of this chapter (United Nations and General Assembly, 2015): "SDG 5: Achieve gender equality and empower all women and girls", "SDG 7: Ensure access to affordable, reliable, sustainable, and modern energy for all" and "SDG 13: Take urgent action to combat climate change and its impacts". In addition, the Conference of Parties (COP) 21 to the United Nations Framework Convention on Climate Change (UNFCCC) had the aim "to reach, for the first time, a universal, legally binding agreement that will enable us to combat climate change effectively and boost the transition towards resilient, low-carbon societies and economies" (UNFCCC, 2015c).

In December 2015, 195 countries adopted the first universal climate agreement (UNFCCC, 2015b), the Paris Agreement. The agreement aims "to strengthen the global response to the threat of climate change, in the context of sustainable development and efforts to eradicate poverty" (UNFCCC,

2015a). To achieve this, parties agreed to i) hold the increase in the global average temperature to well below 2°C while pursuing efforts to limit it to 1.5°C above pre-industrial levels; ii) increase the ability to adapt to climate change, foster climate resilience and lower greenhouse gas (GHG) emissions, without threatening food production; and iii) make finance flows consistent with the previous statements. The agreement also recognizes the need to support developing countries (UNFCCC, 2015a).

As in other areas, the lack of gender equality is important in relation to climate change and energy access. The literature on gender and climate change grew rapidly during the first decade of 2000, but gender issues around climate change, as well as the response to gender-specific impacts, particularly the political response, are still not clear (Röhr, 2007). Many women in their role as household energy managers and traditionally responsible for providing energy to the household, are affected in most cases by the lack of access to energy (Lambrou and Piana, 2006a), although this varies across contexts (see Gonda, Chapter 11 in this volume). Therefore, access to energy in the near future will have much relevance for gender equality. Also, how people without access to energy at present gain it in the future will have a great impact on climate change, positive or negative.

In this context, the main objective of this chapter is to establish the relationships between gender equality, climate change and energy access in developing countries based on an extensive literature review.[1]

The first part of the chapter summarizes current knowledge of the gender dimension of climate change, making a distinction between adaptation and mitigation. Climate change and development are inextricably linked, as are development and gender equality. On the one hand, the economic development of societies has contributed to the unsustainable increase of greenhouse gas emissions, which are destabilizing the global climate system while fostering an unequal distribution of people's ability to cope with these changes. On the other hand, there is now a consensus that climate change directly impacts so-called sustainable development. There is a growing agreement that climate change will substantially threaten societies' capacities to eradicate poverty in the medium to long term (Fernández, 2014; Harrold et al., 2002; Shepherd et al., 2013).

The relationship between gender and energy access is presented in the second part of the chapter. It is estimated that in 2013, 1.4 billion people (20% of the global population) lacked access to electricity, and that 2.7 billion people (40% of the global population) relied on the traditional use of biomass for cooking (IEA, 2015). The traditional use of biomass refers to the basic technology used, such as three-stone fires or inefficient cookstoves, rather than the resource itself. This data provides an idea of the importance of energy access in developing countries.

Access to affordable, reliable, sustainable and modern energy is critical, as defined in SDG 7 and in how this concept is understood in this chapter, for sustainable development and to eradicate poverty due to its capacity to improve productivity, to make easier new income generation activities and to ensure the provision of basic services. The UN Secretary-General Ban Ki-moon recognized that "energy is the "golden thread" that connects economic growth, increasing social equity, and an environment that allows the planet to thrive" (UN Meeting Coverages and Press Releases, 2014).

In the conclusion, the linkages between gender, climate change and energy access in developing countries and the research gaps detected during the literature review are discussed.

Gender and climate change

The literature available about the role of gender in climate change is large (Otzelberger, 2011; Skinner, 2011; Dankelman, 2010; Brody et al., 2008; Demetriades and Esplen, 2008; Dankelman, 2002; Denton, 2002) and has many points of agreement. First of all, women's and men's contribution to climate change differs (Röhr, 2007). Second, in their respective social roles and responsibilities, women and men are differently affected by the effects of climate change and by climate protection and adaptation instruments. Third, they differ in their respective perception of, and reactions to, climate change (Röhr, 2007). Lastly, as the male perspective dominates at all levels, climate protection and climate adaptation measures often fail to take into account the practical and strategic needs of large parts of the population (Röhr, 2007).

The COP to the UNFCCC has adopted important decisions regarding gender and climate change, raising the profile of gender. Decision 36/CP.7 in COP 7 (2001) (UNFCCC, 2001) and Decision 23/CP.18 in COP 18 (2012) (UNFCCC, 2012b) include the promotion of gender balance and the improvement in the participation of women in bodies established under the UNFCCC or the Kyoto Protocol (UNFCCC, 2001; UNFCCC, 2012b). The Lima Work Programme on Gender adopted at the COP 20 (2014) (UNFCCC, 2014) aims to "advance gender balance, promote gender sensitivity in developing and implementing climate policy and achieve gender-responsive climate policy in all relevant activities under the Convention" (UNFCCC, 2014).

Regarding COP 21, the "Women and Gender Constituency" (WGC), established in its position paper which agreements needed to be taken with special attention to gender aspects (WGC, 2015). The Paris Agreement recognizes in its preamble, the need to respect, promote and consider gender equality and empowerment of women, when taking action to address climate change. Also, it points out that adaptation and capacity-building actions should be gender-responsive. However, the Paris Agreement does not include most of the demands of the WGC and there is not a single mention of gender on the "mitigation", "finance", "loss and damage" or "the technology development and transfer" sections (UNFCCC, 2015a).

Gender and adaptation

Adaptation to climate change is defined as "the adjustment in natural or human systems in response to actual or expected climatic stimuli or their effects, which moderates harm or exploits beneficial opportunities" (IPCC, 2001).

Gender aspects of adaptation to climate change are better documented than gender aspects of mitigation. This draws on the generalized assumption that women are more vulnerable to impacts of climate change (Arora-Jonsson, 2011) and that gender inequalities will be exacerbated by climate change (Skinner and Brody, 2011; Brody et al., 2008). Another common agreement is that climate change impacts will affect poor countries more, in particular the poorest in developing countries (Lambrou and Piana, 2006b).

Despite many small-scale studies with evidence of differentiated climate change impacts on gender (Goh, 2012), it is not an area that attracts funding for large-scale transnational projects. Also, there is no systematic collection of gender-disaggregated data. Moreover, most case studies are focused on limited sectors, like heatlh, water, food security or agiculture, where adaptation responses have addressed the specific vulnerability of women (Otzelberger, 2011; Brody et al., 2008). This makes it difficult to scale up from case studies, and can explain why some poorly verified statistics continue to be used (Arora-Jonsson, 2011). Despite this lack of scientific evidence, women's organizations and feminists have used some of these materials to introduce gender issues to the climate change agenda and negotiations (Resurrección, 2013).

In agriculture production, impacts are various. The increased climate variability lowers agriculture production in most developing countries with different impacts on women and men depending on the context. In the case of Mexico, different studies (Buechler, 2009; Biskup and Boellstorff, 1995) have shown that income losses from climate impact on agriculture produced serious difficulties for women to cover their basic needs. These kinds of impacts also increase the workload of both women and men, mainly as they diversify their crop production, but as women are mainly responsible for other household tasks, like childcare, impacts tend to affect them more severely (Goh, 2012). In this context, access to information and technologies is crucial in coping with climate impacts (Goh, 2012). Several studies have shown that men have more access to information and technology than women, like in South Africa (Archer, 2003), Burkina Faso (Roncoli et al., 2009) and Jamaica (Vassell, 2009).

Some specific impacts on women that related to scarcity of natural resources, mainly water and fuel shortages, had also been evidenced in different geographic contexts, due to their traditional roles as resource collectors for the household. The distances that women have to walk for collecting water have increased after several droughts in Vietnam (Shaw et al., 2007), Ethiopia (Asheber, 2010) and Senegal (Dankelman et al., 2008).

All of these climate change impacts could provoke migrations, or may even precipitate conflicts. Men are more likely to migrate than women, who normally stay in the affected areas to care for the family and household, producing specific positive or negative impacts (Goh, 2012). In Nigeria, temporary migration of men induced the development by women of new livelihoods strategies and the increase of women's workload (Agwu and Okhimamhe, 2009). On the contrary, seasonal migration of men in South Africa (Nelson and Stathers, 2009) and in Tanzania (Babugura, 2010) has increased the prevalence of HIV/AIDS among this group due to unprotected sex outside marriage with the subsequent infection of their wives.

Summarizing, gender-sensitive interventions to support adaptation to the direct and indirect impacts of climate change aim to address the different vulnerabilities of men and women. However, such a gender perspective is still lacking in sectors related to energy, transport and infrastructures (Otzelberger, 2011; Brody et al., 2008).

Gender and mitigation

Mitigation of climate change is defined as "an anthropogenic intervention to reduce the sources or enhance the sinks of greenhouse gases" (IPCC, 2001). Regarding gender and mitigation, there are few policy briefings, NGO reports or scientific literature of how mitigation is treated in developing countries (Marshall et al., n.d.).

As has been discussed previously, that gender has been inserted into the climate change agenda through women's specific vulnerability has obscured the contribution that women can make to mitigation actions, and the effects that mitigation actions could have on them. To untap this potential, it is important to determine how emissions characteristics (amount and patterns) of men and women differ (Lambrou and Piana, 2006b).

The information related to gender on mitigation is very sparse, mainly because mitigation actions are normally seen from a technological or economic view (Brody et al., 2008), even when it is recognized that climate change is not just a technological problem and needs to take into account political and socio-economic aspects (Dankelman, 2002).

Before COP 21, countries were asked to communicate their Intended Nationally Determined Contributions (INDCs), in which they outlined their national post-2020 climate action commitments. This INDC will become the first Nationally Determined Contribution (NDC) when a country ratifies the agreement, unless a new NDC is submitted at the same time. The inclusion of gender aspects in the NDC will be fundamental for equitable and sustainable development outcomes, but only 37.5 percent of INDCs submitted included gender as an aspect to consider (IUCN, 2016). To really achieve gender-responsive policies to climate change, the development of national gender and climate change action plans together with a deep monitoring of the implementation of INDCs regarding gender aspects needs to be promoted (Aguilar et al., 2015). Also, it is important to mention that many INDCs include references to countries' Clean Development Mechanism (CDM) strategy; Nationally Appropriate Mitigation Actions (NAMAs); and Low-Emission Development Strategies (LEDS) (IUCN, 2016).

Regarding CDMs or any future market mechanism developed after the Paris Agreement, several studies recognized their potential to promote positive gender impacts but there remains a lack of knowledge and specific tools to achieve it (FORMIN, 2010; Lambrou and Piana, 2006; UNFCCC, 2012a). Also, at the national level, NAMAs could be an opportunity for developing countries to secure the support of developed countries to reduce or limit their emissions. Moreover, NAMAs could be an opportunity for the integration of more gender-aware indicators, although there are no clear guidelines or gender indicators for them (Skinner, 2011).

Gender issues, and especially opportunities for women's empowerment through mitigation actions, are not commonly addressed, with some exceptions where this is the main focus (EIGE, 2012; WHO, 2014). One notable exception is the policies promoted by the Reducing Emissions from Deforestation and Forest Degradation (REDD), where gender dimensions have been largely overlooked (Skinner, 2011; UN-REDD, Programme 2011; Nhantumbo and Chiwona-Karltun, 2012). (Although see Bee, Chapter 12 in this volume for further discussion.)

Overall, the role of women in mitigation measures should not be underestimated, as developing countries have the potential to reduce or store greenhouse gases, particularly in areas in which women are already active, such as emission reduction through the implementation of clean cookstoves, lighting or electrical appliances (Bailis et al., 2015; FORMIN, 2010).

Gender and energy access in developing countries

In order to establish the relationship between gender and energy access in developing countries, it is useful to understand how "energy access" has been treated in the international agenda since the beginning of the twenty-first century.

When the Millennium Development Goals (MDGs) were established in the Millennium Summit of the UN in 2000, no explicit mention to energy was formulated in any of the eight global goals, even though a debate was already established around the role of energy for human development (Brew-Hammond, 2012). It is later, in 2005, that the relationship between access to modern energy sources, both electricity and improved cooking systems, and human development, and their importance in achieving the MDGs, was clearly established (Takada et al., 2007; UNDP, 2005).

Since then, energy access has been increasingly gaining attention in the international agenda and it is set as a critical factor for human sustainable development. The year 2012 was declared by the United Nations as the "International Year of Sustainable Energy for All" and 2014–2024 the "Decade of Sustainable Energy for All". Accordingly, research and development moved from energy for sustainability towards sustainable energy (Brew-Hammond, 2012; Halsnæs and Garg, 2011; Bazilian, Sagar et al., 2010). Promising energy access initiatives have also been launched such as Sustainable Energy for All (SE4all) by the UN and the World Bank (WB), and the Global Alliance for Clean Cookstoves (GACC) by UN Foundation or Lighting Africa by WB.

Finally, a specific goal for energy was include in the SDGs. The SDG 7 is an ambitious goal, especially target 7.1 stating "by 2030 ensure universal access to affordable, reliable, and modern energy services" (Sustainable Development Knowledge Platform, 2015). One study developed just after the approval of the SDGs shows that to achieve SDG 7 the progress should be between three to four times faster than the one projected at that time (Nicolai et al., 2015).

Despite the increased recognition of energy access in the international agenda, different stakeholders working on the topic have identified significant gaps that should be covered for better commitment in energy interventions. Consensus has not yet been reached on what is understood by 'energy access'. While a universal definition has not emerged due to its intrinsic complexity (Pachauri, 2011), this has not deterred some attempts to try to frame energy access by finding common aspects in all of them.

The International Energy Agency (IEA) highlights four significant elements that commonly appear in most definitions of energy access: (a) minimum level of electricity; (b) safe and sustainable cooking and heating fuels and stoves; (c) energy for productive activities; and (d) energy for public services (IEA, 2014). The UN Secretary-General's Advisory Group on Energy and Climate Change (AGECC) (United Nations, 2010) defines energy access as "access to clean, reliable and affordable energy services for cooking and heating, lighting, communications and productive uses".

This has lead to several approaches for measuring energy access based on multi-tier methodologies. Before these methodologies emerged, a binary measurement – household with access to electricity and modern technologies for cooking or without them – was used. Compared to this, multi-tier frameworks are a significant advance, as they are based on the performance of households in different attributes of its energy sources like capacity, duration and availability, reliability, quality, affordability, legality, health and safety. These attributes are measured by tiers, from Tier 0 (no access) to Tier 5 (the highest level of access). Some examples are the "Energy Supply Index" (Practical Action, 2010), the "Incremental levels of access to energy services" (United Nations, 2010), the "Multi-tier standards for cookstoves" from the GACC and the Partnership for Clean Indoor Air (PCIA) (ISO, 2012) and the "Multi-Tier approach to measuring energy access" from SE4all (ESMAP et al., 2013).

Gender and energy access in the development context

Compared to climate change, gender considerations are fully mainstreamed in energy access in developing countries at policy level (Danielsen, 2012; Lambrou and Piana, 2006a; Dutta, 2003; Cecelski and Unit, 2000), at project level (Cecelski and Dutta, 2011; UNDP and ENERGIA, 2004) and in research (Pachauri and Rao, 2013; Sovacool et al., 2013; Batliwala and Reddy, 2003; Clancy et al., 2002).

Energy poverty in developing countries is experienced differently and more severely by women than by men (UNIDO and UNWOMEN, 2013). Specific impacts of the lack of access to modern energy services suffered by women, particularly in rural areas, are large and varied. Women and girls spend hours a day gathering fuelwood and carrying water (UNDP, 2005; Cordes and GACC, 2011). During these tasks, women and girls have to walk long distances with a heavy load with its consequent health impacts. In addition, these walks, together with the lack of street lighting, increase the risk of them suffering gender-based violence (Cordes and GACC, 2011), especially in conflict situations and in refugee camps (UNHCR, 2014). Indoor air pollution produced by the use of traditional biomass for cooking is the fourth largest cause of morbidity and premature mortality worldwide, with women more affected than men (IEA et al., 2010; WHO, 2006). As a result of this, the importance of energy access to women's development and empowerment is clear and has been widely established, as Table 8.1 illustrates.

Gender mainstreaming in the energy sector has received wide attention. In this sense, the required components (i. Integration of women's and men's concerns throughout the energy development process; and ii. specific activities within energy intervention aimed at reducing gender disparities and empowering women) (Oparaocha and Dutta, 2011), the capacity-building needs as set out in Table 8.2, and how to translate them to project level (Skutsch, 2005) to effectively mainstream gender in the energy sector, have been defined by different authors, international institutions and NGOs.

As previously stated, a non-binary definition based on energy quality has reached wider acceptance but a unique framework has not yet been standardized, so the process of measuring and monitoring energy access is still undefined. Multi-tier methodologies have only general indicators regarding gender or treat gender issues superficially, as a review of research on this area has shown (Hailu, 2012; Nussbaumer et al., 2012; Pachauri and Spreng, 2011; Bazilian, Nussbaumer et al., 2010). Consequently, although the possibilities to improve the position of women through developing access to energy is large, there is a need to use specific gender indicators and disaggregate data on energy access by gender in order to improve and promote gender-sensitive energy programmes.

Despite previous statements, there is a recognition of the importance of introducing gender issues and gender-specific indicators that measure improved energy access and its productive uses (UNDP, 2015; UNIDO and UNWOMEN, 2013). Also, several considerations to not exclude women

Table 8.1 Possibilities for improving the position of women through energy

Energy Form	Impacts for women of access to energy		
	Practical	*Productive*	*Strategic*
Electricity	– Pumping water: reducing need to haul and carry – Mills for grinding – Lighting improves working conditions at home	– Increase possibility of activities during evening hours – Provide refrigeration for food production and sale – Power for specialized enterprises	– Make streets safer: allowing participation in other activities (e.g. evening classes and women's group meetings) – Improved access to information through ICT
Cooking technologies and fuels	– Improved health through reduction on indoor air pollution and reduction on carrying heavy loads of firewood-Less time and effort in gathering and carrying firewood – Less time and effort for cooking	– More time for productive activities – Lower cost of process heat for income-generating activities – Participation in related business	– Can allow for improved monitoring and control of natural forests in community forestry management frameworks – Can allow for gained recognition in the case of women's participation in related business – More time for education and income generation activities (specially for young women and girls)
Mechanical	– Milling and grinding – Transport and portering of water and crops	– Increases variety of enterprises	– Transport: allowing access to commercial and social/political opportunities

Source: adapted from Yianna Lambrou and Piana (2006a); Clancy et al., (2002)

and marginalized groups from energy access initiatives have to be taken into account. In the feasibility and initial planning, questions about specific energy needs, access to information, ability to participate in the decision-making process of different groups, and tasks and works done by each social group have to be asked. During the programme design and implementation phases, specific training and payment schemes for each group, and ensuring their capacity to use energy productively and to own and control the operation of the energy service, must be assured. Finally, different indicators related to practical and strategic benefits for each group should be used during monitoring and evaluation.

Table 8.2 Capacity-building needs for mainstreaming gender in energy

Target Group	Capacity-Building needs	Means
National policymakers	Sensitization towards openness to try out new methods and tools Willingness to make space and strengthen women staff in organization's set up	Advocacy through media and print messages Well-structured and focused interaction with researchers and NGOs
Implementers of energy programmes	Sensitization towards gender issues Practical tools and techniques to incorporate women's role in planning	Field-level workshops in local language Exchange visits and interaction with local organizations working on gender issues
Village communities	For men, sensitization and assurance that women can meaningfully participate in programmes while respecting their traditionally accepted space and roles Willingness to participate in social empowerment process of women	Exposure visits Focus groups discussions
NGOs	Tools and techniques to incorporate women's role in planning orientation towards new methodologies	Local-level workshops Interaction with researchers and policymakers

Source: UNDP and ENERGIA (2004)

In conclusion, the importance that energy access could have for women's development and empowerment and for mainstreaming gender in energy access policies, programmes and projects is well established and receives wide attention. However, with the new approaches adopted to measure energy access and the importance of this topic in the SDGs, future work and research on this area are still important.

Discussion: the linkages between gender, climate change and energy access

During the previous sections, the linkages between gender and adaptation, mitigation to climate change and energy access in developing countries have been discussed separately. Summarizing, gender and adaptation and energy access have been widely studied while mitigation and gender is not well covered. Some important linkages have been identified when looking at the intersection between gender, climate change and access to energy, and they are discussed in this section.

Some significant knowledge gaps still remain. There is a lack of documentation on linkages between gender, climate change mitigation and energy access. The linkages between energy access and climate change mitigation are clear and the importance to mainstream gender in mitigation is widely recognized. However, the literature is rather thin on the evidence on the key role that women could have on mitigation actions through access to energy projects, and on how to mainstream gender on these actions. Also, while linkages between gender and climate change adaptation are clearly established and the literature recognizes the importance of integrating gender in adaptation actions, energy access has not received wide attention as impacts of climate change on this are not clear.

Regarding the linkages, first of all, and importantly, if we compare which gender dimensions have been studied or used when analysing climate change and energy access separately (see Roehr, 2007; Skutsch, 2005; Table 8.1 and 8.2) we can see that most of them are the same or very similar.

There are a lot of well-known possibilities for improving the position of women through energy (Table 8.1) but, as stated in the previous section, there are several considerations to be taken into account so as not to exclude women and marginalized groups from energy access initiatives. At the same time, the ways in which access to energy is achieved will determine whether the impact is positive or negative to mitigate climate change, as energy is at the core of climate change mitigation discussions (Aguilar et al., 2015).

However, in the case of access to electricity, some considerations regarding the previous statement have to be taken into account. On one hand, there is a risk that the higher costs, especially in the short term, of renewable energy investments, the main option for decarbonizing electricity in developing countries, are passed on to consumers, with resulting gender-specific impacts (Skinner, 2011). On the other hand, women's engagement in these initiatives could be an effective way for them to gain access to new jobs or to improve their income generation activities (Aguilar et al., 2015).

Regarding access to modern fuels and technologies for cooking, the linkages between women's development and both mitigation and adaptation to climate change are large. However, although this is a great example on how to simultaneously address these three issues at the same time, the knowledge and lessons learned from these kinds of projects have not received wide attention in the general debate about gender and climate change. First of all, the main beneficiaries of initiatives like improved cookstoves and clean fuel projects are women, and the positive impacts are various, including financial, time-savings and health improvements. Second, improved cookstoves and clean fuel projects have great potential to mitigate climate change (GACC, n.d.) as they reduce the consumption of solid fuels and consequently the emissions of GHG during combustion. It is estimated that GHG emissions from solid fuels are 1.0–1.2 Gt CO_2e/yr. (1.9–2.3% of global emissions) and that successful deployment and utilization of 100 million improved stoves could reduce this by 11–17 percent (Bailis et al., 2015). Also, during the combustion of solid fuels black carbon (BC) and other short-lived climate pollutants are emitted. Recent studies show that BC may be responsible for close to 20 percent of the Earth's warming, and that residential solid fuel burning produces up to 25 percent of global BC emissions (GACC, n.d.). Finally, the availability of traditional fuel sources, like solid fuels, is affected by climate change and energy for cooking becomes more commoditized, scarce and expensive (Skinner, 2011). In that way, improved cookstoves and clean fuel projects could be an adaptation strategy to traditional fuel scarcity, especially in semiarid regions.

A recent study conducted in Senegal, The Gambia and Guinea Bissau by the authors of this chapter exemplifies the potential of improved cookstoves projects to link energy access, climate change and gender equality. The results of the study show that there are several social benefits including the reduction of indoor air pollution, the reduction in collecting and cooking time, and some economic savings where woodfuel was paid for by the family. At the same time, there was a GHG emissions reduction (Mazorra et al., 2015; Sota et al., 2014). In addition, the profitability of these actions is confirmed by the results, which showed that the installation of the ICS generates benefits that outweigh the associated costs both at household and project level (Mazorra et al., 2015).

As climate change impacts could reduce agricultural production, the availability of sustainable and efficient energy can play a key role to reduce these impacts (GTZ, 2010). In these ways, energy access projects can play an important role in climate change adaptation.

The foregoing argues that women must play a key role in determining energy access in developing countries, both in access to electricity and in modern cooking technologies. If this is done in a sustainable way (e.g. renewable energy in the case of access to electricity, and really clean and sustainable cookstoves and clean fuels for the first step towards modern cooking technologies,) developing countries can contribute widely to the mitigation of climate change, adaptation to its impacts and, at the same time, to women's empowerment and development as household energy managers (GTZ, 2010; UNFCCC, 2012a; FORMIN, 2010).

Conclusions

Although there is some literature on linkages between gender, climate change and access to energy, more evidence on how they interplay at all scales is needed. These scales range from the international level where many decisions are taken, to the local level where specific projects are implemented and where information is a key issue.

Specifically, wider scientific evidence of women's vulnerabilities and differentiated gendered impacts of climate change is needed. Not all women are vulnerable to climate change and they are not vulnerable in the same way. Their vulnerability varies between countries, and within different communities in the same country. In this sense, the emphasis on women's vulnerability to climate change impacts as the main reason for the work on gender and adaptation has hidden some important questions, such as: what are the gender inequalities underlying this vulnerability and, how can men and women work together even with different roles, preferences, needs, knowledge and capacities?

Regarding the mitigation of climate change, the potential contribution of women is often underestimated and there is a need for research on how to include gender issues in mitigation actions to better address the views and needs of often ignored social groups. In the first place, it is needed to determine the differences in the emissions characteristics (amount and patterns) between men and women (Lambrou and Piana, 2006b). In addition, there is low knowledge about the nexus between mitigation and adaptation, and a gendered understanding of this is even lower. Focusing on the third issue of this chapter, some energy projects with great potential to contribute to mitigation and gender equality could fail if the impacts of climate change on energy access are not addressed to determine the best gender sensitive adaptation strategies in each case. Cooking technologies are a good example for this situation (Jerneck and Olsson, 2013; Skinner, 2011).

Nevertheless, the approval of the Post 2015 Agenda constitutes great progress. For the first time, the environmental agenda, especially climate change, and the development agenda have converged through the SDG. Given that in the development agenda it is widely accepted that gender considerations must be taken into account, and there are specific goals for gender equality, access to energy and climate change, this convergence could be a unique opportunity to really embed gender issues in the climate change agenda (Women's Major Group, 2013).

Note

1 The need for this literature review was identified during a "Short Term Scientific Mission" of the GenderSTE Cost Action done in the Centre of Human Geography of Brunel University. Also, this chapter is a central part of the PhD thesis of Javier Mazorra where these linkages are studied through a case study of the improved cookstoves project in Senegal, The Gambia and Guinea-Bissau.

References

Aguilar, L., Granat, M. and Owren, C. (2015) *Roots for the future: The landscape and way forward on gender and climate*, Washington, DC: IUCN & GGCA.

Agwu, J. and Okhimamhe, A. (2009) *Gender and climate change in Nigeria: a study of four communities in north-central and south-eastern Nigeria*. Nigeria: Heinrich Böll Stiftung.

Archer, E.R.M. (2003) Identifying underserved end-user groups in the provision of climate information. *Bulletin of the American Meteorological Society*, 84(11): 1525–1532.

Arora-Jonsson, S. (2011) Virtue and vulnerability: Discourses on women, gender and climate change. *Global Environmental Change*, 21(2): 744–751.

Asheber, S.A. (2010) *Mitigating drought: policy impact evaluation. A case of Tigray Region, Ethiopia.* Twente, Netherland: University of Twente, Faculty of Geo-Information and Earth Observation ITC, Enschede.

Babugura, A. (2010) *Gender and climate change: South Africa case study.* Nigeria: Heinrich Böll Stiftung.

Bailis, R. Drigo, R., Ghilardi, A. and Masera, O. (2015) The carbon footprint of traditional woodfuels. *Nature Climate Change* (January): 1–7.

Batliwala, S. and Reddy, A.K.N. (2003) Energy for women and women for energy (engendering energy and empowering women). *Energy for Sustainable Development,* 7(3): 33–43.

Bazilian, M., Nussbaumer, P., Cabraal, A., Centurelli, R., Detchon, R., Gielen, D. and Rogner, H. (2010) *Measuring energy access: Supporting a global target.* New York: Earth Institute, Columbia University.

Bazilian, M., Sagar, A., Detchon, R. and Yumkella, K. (2010) More heat and light. *Energy Policy,* 38(10): 5409–5412.

Biskup, J.L. and Boellstorff, D.L. (1995) The effects of a long-term drought on the economic roles of hacendado and ejidatario women in a Mexican ejido. *Nebraska Anthropologist,* Paper 80.

Brew-Hammond, A. (2012) Energy: The missing millennium development goal. *Energy for Development.* Netherlands: Springer, 35–43.

Brody, A., Demetriades, J. and Esplen, E. (2008) *Gender and climate change: Mapping the linkages A scoping study on knowledge and gaps.* Brighton, UK: BRIDGE, Institute of Development Studies, University of Sussex.

Buechler, S. (2009) Gender, water, and climate change in Sonora, Mexico: Implications for policies and programmes on agricultural income-generation. *Gender & Development,* 17(1): 51–66.

Cecelski, E. and Dutta, S. (2011) *Mainstreaming gender in energy projects: A practical handbook.* Salem, OR: ENERGIA.

Cecelski, E. and Unit, A.A.E. (2000) Enabling equitable access to rural electrification: Current thinking and major activities in energy, poverty and gender. *World Development Report,* 1: 2–3.

Clancy, J.S., Skutsch, M. and Batchelor, S. (2002) *The Gender-Energy-Poverty nexus: Finding the energy to address gender concerns in development,* Project Report CNTR998521, Department for International Development, London, UK.

Cordes, L. and GACC (2011) *Igniting change: A strategy for universal adoption of clean cookstoves and fuels.* Global Alliance for Clean Cookstoves.

Danielsen, K. (2012) *Gender equality, women's rights and access to energy services: An inspiration paper in the run-up to Rio+20.* Denmark: Danish Ministry of Foreign Affairs.

Dankelman, I. (2002) Climate change: Learning from gender analysis and women's experiences of organising for sustainable development. *Gender & Development,* 10(2): 21–29.

Dankelman, I. (2010) *Gender and climate change: An introduction.* New York: Routledge.

Dankelman, I., Alam, K., Ahmed, W.B., Gueye, Y.D., Fatema, N. and Mensah-Kutin, R. (2008) *Gender, climate change and human security: Lessons from Bangladesh, Ghana and Senegal,* WEDO, ABANTU for Development in Ghana, ActionAid Bangladesh and ENDA Senegal.

Demetriades, J. and Esplen, E. (2008) The gender dimensions of poverty and climate change adaptation. *IDS Bulletin,* 39(4): 24.

Denton, F. (2002) Climate change vulnerability, impacts, and adaptation: Why does gender matter? *Gender & Development*, 10(2): 10–20.

Dutta, S. (2003) *Mainstreaming Gender in Energy Planning and Policies*, Background paper for Expert Group Meeting, UNESCAP Project on "Capacity Building on Integration of Energy and Rural Development Planning", Bangkok, Thailand.

EIGE (2012) Review of the Implementation in the EU of area K of the Beijing Platform for Action: Women and the Environment Gender Equality and Climate Change, European Institute of Gender Equality, European Union.

ESMAP, World Bank and IEA (2013) SE4All Global Tracking Framework 2013, Energy Sector Mangement Assistance Program, World Bank and International Energy Agency.

Fernández, L. (2014) *Evaluación de los co-beneficios sobre el desarrollo sostenible y la reducción de la pobreza de proyectos de mitigación del cambio climático en Brasil*. School of Industrial Engineering, Technical University of Madrid (UPM), Madrid, Spain.

FORMIN (2010) Gender and the Clean Development Mechanism (CDM): Opportunities for CDM to Promote Local Positive Gender Impacts, Ministry for Foreign Affairs of Finland.

GACC, Clean Cookstoves and Climate Change, Global Alliance for Clean Cookstoves.

Goh, A.H.X. (2012) A Literature Review of the Gender-differentiated Impacts of Climate Change On Women's and Men's Assets and Well-being in Developing Countries, CAPRi Working Paper No. 106. Washington, DC: International Food Policy Research Institute.

GTZ (2010) *Climate change and gender: economic empowerment of women through climate mitigation and adaptation?* Eschborn, Germany: The Governance Cluster, Deutsche Gesellschaft für Technische Zusammenarbeit (GTZ) GmbH.

Hailu, Y.G. (2012) Measuring and monitoring energy access: Decision-support tools for policymakers in Africa. *Energy Policy*, 47(SUPPL.1): 56–63.

Halsnæs, K. and Garg, A., 2011. Assessing the role of energy in development and climate policies: Conceptual approach and key indicators. *World Development*, 39(6): 987–1001.

Harrold, M. et al., (2002) *Poverty and climate change: Reducing the vulnerability of the poor through adaptation*. AfDB, ADB, DFID (UK), European Comission, Federal Ministry for Economic Cooperation and Development, Germany Ministry of Foreign Affairs – Development Cooperation, The Netherlands, OECD, UNDP, UNEP, WB.

IEA (2014) World Energy Outlook – Methodology for Energy Access Analysis, Organisation for Economic Co-operation and Development, International Energy Agency.

IEA (2015) World Energy Outlook 2015, Organisation for Economic Co-operation and Development, International Energy Agency.

IEA UNDP and UNIDO (2010) Energy Poverty: How to make modern energy access universal? Special early excerpt of the World Energy Outlook 2010 for the UN General Assembly on the Millennium Development Goals, International Energy Agency.

Intergovernmental Panel on Climate Change (IPCC) (2001) *Climate Change 2001: Synthesis Report. A Contribution of Working Groups I, II, III to the Third Assessment Report of the Intergovernmental Panel on Climate Change*. Cambridge and New York: Cambridge University Press.

ISO (2012) IWA 11:2012 – Guidelines for evaluating cookstove performance. Geneva Switzerland: International Standard Organisation.

IUCN (2016) Gender in Mitigation Actions. (November).

Jerneck, A. and Olsson, L. (2013) A smoke-free kitchen: Initiating community based co-production for cleaner cooking and cuts in carbon emissions. *Journal of Cleaner Production*, 60: 208–215.

Lambrou, Y. and Piana, G. (2006) *Energy and gender in rural sustainable development*, Rome: United Nations, pp. 1–41.

Lambrou, Y. and Piana, G. (2006a) *Energy and gender issues in rural sustainable development*. Rome: Food and Agriculture Organization of the United Nations (FAO).

Lambrou, Y. and Piana, G. (2006b) *Gender: The missing component of the response to climate change*. Rome: Food and Agriculture Organization of the United Nations (FAO).

Marshall, M., Ockwell, D. and Byrne, R. (n.d.) Sustainable energy for all, or sustainable energy for men? Gender and the construction of identity within climate technology entrepreneurship in Kenya In press – Accepted for publication in Progress in Development Studies, 8. *Progress in Development Studies*.

Mazorra, J. et al., (2015) Cost benefit analyses of improved cookstoves project: The case of the cassamance natural subregion, Western Africa. *Healthy Polis Kushan*.

Nelson, V. and Stathers, T. (2009) Resilience, power, culture, and climate: A case study from semi-arid Tanzania, and new research directions. *Gender & Development*, 17(1): 81–94.

Nhantumbo, I. and Chiwona-Karltun (2012) His REDD+, her REDD+: how integrating gender can improve readiness, International Institute for Environment and Development Briefing.

Nicolai, S., Hoy, C. Berliner, T. and Aedy, T. (2015) *Projecting progress: Reaching the SDGs by 2030. Flagship Report*. London: Overseas Development Institute.

Nussbaumer, P., Bazilian, M. and Modi, V. (2012) Measuring energy poverty: Focusing on what matters. *Renewable and Sustainable Energy Reviews*, 16(1): 231–243.

Oparaocha, S. and Dutta, S. (2011) Gender and energy for sustainable development. *Current Opinion in Environmental Sustainability*, 3(4): 265–271.

Otzelberger, A. (2011) *Gender-responsive strategies on climate change: recent progress and ways forward for donors*. Brighton, UK: BRIDGE, Institute of Development Studies, University of Sussex.

Pachauri, S. (2011) Reaching an international consensus on defining modern energy access. *Current Opinion in Environmental Sustainability*, 3(4): 235–240.

Pachauri, S. and Rao, N.D. (2013) Gender impacts and determinants of energy poverty: Are we asking the right questions? *Current Opinion in Environmental Sustainability*, 5(2): 205–215.

Pachauri, S. and Spreng, D. (2011) Measuring and monitoring energy poverty. *Energy Policy*, 39(12): 7497–7504.

Practical Action (2010) *Poor people's energy outlook 2010*, Rugby, UK: Practical Action Publishing.

Resurrección, B.P. (2013) Persistent women and environment linkages in climate change and sustainable development agendas. *Women's Studies International Forum*, 40: 33–43.

Röhr, U. (2007) *Gender, climate change and adaptation. Introduction to the gender dimensions*, Background paper prepared for Both Ends briefing paper Adapting to climate change: How local experiences can shape the debate. Berlin, Genanet, August 2007.

Roncoli, C., Jost, C., Kirshen, P., Sanon, M., Ingram, K.T., Woodin, M., Somé, L., Ouattara, F., Sanfo, B.J., Sia, C., Yaka, P. and Hoogenboom, G. (2009) From accessing to assessing forecasts: An end-to-end study of participatory climate forecast dissemination in Burkina Faso (West Africa). *Climatic Change*, 92: 433–460.

Shaw, R., Prabhakar, S.V.R.K, Nguyen, H. and Price-Thomas, S. (2007) *Drought management considerations for climate-change adaptation: Focus on the Mekong Region*, Oxfam Vietnam and International Environment and Disaster Management (IEDM), Graduate School of Global Environmental Studies of Kyoto University, Japan.

Shepherd, A., Mitchell, T., Lewis, K., Lenhardt, A., Jones, L., Scott, L. and Muir-Wood, R. (2013) *The geography of poverty, disasters and climate extremes in 2030*. London, UK: ODI, MET Office Hadley Centre, RMS.

Skinner, E. (2011) *Gender and Climate Change Overview Report*. Brighton, UK: BRIDGE, Institute of Development Studies, University of Sussex.

Skinner, E. and Brody, A. (2011) *Gender and climate change*. BRIDGE.

Skutsch, M.M. (2005) Gender analysis for energy projects and programmes. *Energy for Sustainable Development*, 9(1): 37–52.

Sota, C. de la, et al., (2014) Effectiveness of improved cookstoves to reduce indoor air pollution in developing countries: The case of the cassamance natural subregion, Western Africa. *Journal of Geoscience and Environment Protection*, 02(01): 1–5. Available online at www.scirp.org/journal/PaperInformation.aspx?PaperID=42023&#abstract (accessed May 23, 2015).

Sovacool, B.K., Clarke, S., Johnson, K., Crafton, M., Eidsness, J. and Zoppo, D. (2013) The energy-enterprise-gender nexus: Lessons from the Multifunctional Platform (MFP) in Mali. *Renewable Energy*, 50: 115–125.

Sustainable Development Knowledge Platform (2015) Proposal for Sustainable Development Goals. Sustainable Development Knowledge Platform. Available online at https://sustainabledevelopment.un.org/focussdgs.html (accessed August 12, 2015).

Takada, M., Rijal, K. and Morris, E. (2007) Energizing the Least Developed Countries to Achieve the Millenium Development Goals: The Challenges and Opportunities of Globalization. Ministerial Conference "Making Globalization Work for the LDCs" Istanbul, 9–11 July 2007. Istanbul: United Nations Development Programme Issues Paper.

UNDP (2005) Energizing the Millennium Development Goals: A Guide to Energy's Role in Reducing Poverty. New York: United Nations Development Programme.

UNDP (2015) Energyplus Guidelines: Planning for Improved Energy Access and Productive Uses of Energy. New York: United Nations Development Programme.

UNDP and ENERGIA (2004) Gender and Energy for Sustainable Development: A Toolkit and Resource Guide. New York: United Nations Development Programme, ENERGIA.

UNFCCC (2015a) Adoption of the Paris Agreement, FCCC/CP/2015/L9 United Nations Framework Convention on Climate Change.

UNFCCC (2012a) CDM and women, United Nations Framework Convention on Climate Change.

UNFCCC (2015b) COP21 | 195 countries adopt the first universal climate agreement. Available online at www.cop21.gouv.fr/en/195-countries-adopt-the-first-universal-climate-agreement/ (accessed April 1, 2016).

UNFCCC (2015c) COP21 main issues | COP21 – United Nations Conference on Climate Change. Available online at www.cop21.gouv.fr/en/cop21-cmp11/cop21-main-issues (accessed August 16, 2015).

UNFCCC (2012b) Decision 23/CP.18 Promoting gender balance and improving the participation of women in UNFCCC negotiations and in the representation of Parties in bodies established pursuant to the Convention or the Kyoto Protocol, FCCC/CP/2012/8/Add.3 United Nations Framework Convention on Climate Change.

UNFCCC (2001) Decision 36/CP.7: Improving the participation of women in the representation of Parties in bodies established under the United Nations Framework Convention on Climate Change or the Kyoto Protocol, FCCC/CP/2001/13/Add.4 United Nations Framework Convention on Climate Change.

UNFCCC (2014) Lima work programme on gender, United Nations Framework Convention on Climate Change.

UNHCR (2014) Global Strategy for Safe Access to Fuel and Energy (SAFE): A UNHCR Strategy 2014–2018, Division of Programme Support and Management, United Nations High Commissioner for Refugees, Geneva, Switzerland.

UNIDO and UNWOMEN (2013) Sustainable Energy for All: The Gender Dimensions, United Nations Industrial Development Organization, United Nations Entity for Gender Equality and the Empowerment of Women.

United Nations (2010) Energy for a Sustainable Future: The Secretary-General's Advisory Group on Energy and Climate Change (AGECC): Summary Report and Recommendations, New York, US.

United Nations and General Assembly (2012) The Future We Want: Outcome document adopted at Rio+20 Conference on Sustainable Development, A/RES/66/288 (July 27, 2012).

United Nations and General Assembly (2015) Transforming our world: the 2030 Agenda for Sustainable Development, A/RES/70/1 (September 25, 2015).

UN Meeting Coverages and Press Releases (2014) Sustainable Energy "Golden Thread" Connecting Economic Growth, Increased Social Equity, Secretary-General Tells Ministerial Meeting. SG/SM/15839-EN/28, May 12, 2014. Available online at www.un.org/press/en/2014/sgsm15839.doc.htm (accessed August 12, 2015).

UN-REDD Programme (2011) The Business Case for Mainstreaming Gender in REDD+, The United Nations Collaborative Programme on Reducing Emissions from Deforestation and Forest Degradation in Developing Countries.

Vassell, L. (2009) *Gender climate change and disaster risk management: Case study of Jamaica.* Barbados: UNDP Caribbean Risk Management Initiative.

WGC (2015) Women and Gender Constituency: Position Paper on the 2015 New Climate Agreement Overview, Women and Gender Constituency.

WHO (2006) Fuel for life: household energy and health. Geneva: World Health Organisation.

WHO (2014) Gender, Climate change and health. Geneva: World Health Organisation.

Women's Major Group (2013) Gender Equality, Women's Rights and Women's Priorities: Recommendation for the SDG and Post-2015 Development Agenda, WECF International.

9 Everyday life in rural Bangladesh

Understanding gender relations in the context of climate change

Alex Haynes

Introduction

The United Nations and the World Bank have heralded the South Asian country of Bangladesh a development success story. In reporting on progress towards the Millennium Development Goals, the Bangladesh Government (2015) cites impressive poverty reduction from 48.9 percent in 2000 to 24.8 percent in 2015. The report also notes significant progress on gender parity in primary and secondary school enrolment and a significant decline in both fertility among married women and maternal mortality. Despite this progress, many of the 160 million (BBS, 2016) Bangladeshis still face high levels of poverty and lack sufficient access to food, drinking water and basic services. Patriarchal attitudes and practices, like the continuation of child marriage, dowry and related violence, persist despite legislative reform, advocacy and pressure from international agencies. The combination of poverty and discriminatory social structures can increase people's vulnerabilities and exposure to climate change, and undermine their capacities to adapt.

As one of the largest river deltas in the world Bangladesh is particularly vulnerable to sea level rise, tidal surges and flooding from its 700 rivers and tributaries. The Climate Change Vulnerability Index (2015), which evaluates the sensitivity of populations, the physical exposure of countries, and governmental capacity to adapt to climate change over the next 30 years, ranks Bangladesh first and most at risk. This exposure to changing weather patterns, climate variability and intensifying climate events is likely to contribute to increasing inequality and instability.

Uneven progress towards gender equality and the gendered impacts of climate change in Bangladesh demand we look beyond what we know and continue to ask provocative and unsettling questions to unveil assumptions in policies, program design, institutions and unquestioned processes. Gender inequality and climate change are proving less amenable to change than is sometimes suggested by the reductionist demands of positivist approaches or linear thinking. By focusing on the dynamic nature of people's lives and paying more attention to interactions between everyday life and elements of social, political, economic and ecosystems, we can build our understanding of how gender relations are being, or could be, reshaped.

This chapter explores climate change and gender relations through the lens of everyday life as it was captured in a research study exploring the gendered impacts of climate variability in rural Bangladesh. An overview of the research is followed by a moment in time constructed from the data that illustrate the dynamic and gendered nature of life. Some specific research findings are discussed in the context of current efforts to address gender inequality and initiatives to manage the impacts of climate change.

Research overview

Oxfam Australia, Oxfam in Bangladesh and Monash University in Australia undertook a three-year study to explore the gendered impacts of climate variability in three rural districts of Bangladesh shown in Figure 9.1.

• Gaibandha, in the north of Bangladesh, has the Jamuna river, one of the largest and most dynamic of Bangladesh's rivers, as its eastern border. Gaibandha is prone to lengthy drought periods, significant flooding and riverbank erosion.
• Satkhira, in the south west of Bangladesh, borders India on the Bay of Bengal and includes the internationally significant Sundarbans mangrove forest. Temperatures are rising and salinity from tidal surges and inundation of productive land is an increasing problem. The district was recovering from Cyclone Aila (2009).
• Barguna, a low-lying coastal area in the south, is experiencing an increase in tidal surges and saltwater intrusion. The district was hard hit by Cyclone Sidr (2007).

The first qualitative stage, completed in 2012, involved 33 interviews and 29 focus group discussions with women and men in nine villages across the three districts. It focused on capturing daily life, identifying changes to weather and seasons, the impacts of these changes and the responses and strategies of women and men. The discussions explored the sense women and men are making of these changes. The first stage also involved twelve interviews with activists, academics and policymakers in the capital Dhaka. The second stage, completed in 2014, involved conducting a face-to-face quantitative survey in the same sites with 298 women and 319 men to ascertain the significance and prevalence of the themes identified in the first stage.

A moment in time

The flood event described below took place along the Jamuna River in Gaibandha in 2011 during the first stage of the research. It is recorded in

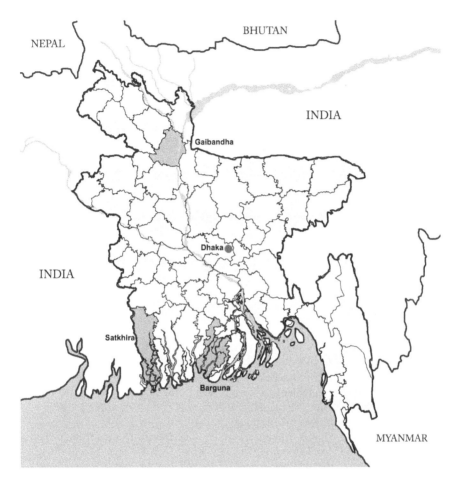

Figure 9.1 Map of Bangladesh showing three rural districts where research was undertaken and the capital city, Dhaka.

interviews, focus groups and field notes but, drawing from Gardner's (1997)
approach, is told here in my language and through my lens of assumptions
and unfamiliarity. The women and men involved would no doubt tell things
in quite a different way or would tell about something else that has happened
since this flood, as surely much has.

Presenting the data in the context of everyday life brings the varied and gendered experiences of rural Bangladeshi women and men into focus, both in
their interactions and their responses. It begins to make the interplay between
social, cultural and natural processes in the human situation of Bangladesh
more visible.

The annual flood comes earlier than expected and it is more intense causing significant river erosion. It washes away a large section of the char,[1] swallowing farmland, houses, livestock and possessions. Luckily everyone survives this time. An older man points out into the river to where his house was, a mile from where we are standing. He had four acres but all is lost now. The local NGOs have tried to help but they have limited capacity. He says the government does very little for them. The local official who is in charge of emergency food relief sells the rice rations supplied by the national government in the market and provides them one-tenth of the ration in poor quality old rice.

The char village he points to and describes is devastated. People take what little they have left and try to find a place to settle albeit temporarily. Some men try to find laboring jobs nearby, some head to the nearest city or Dhaka to find work, often leaving wives and children in makeshift housing. Past events indicate that some of these men will return or come back to take their wives and children to the city with them, some will not come back and may take other wives and a few will never be heard from again. One woman says she and her husband may go to Dhaka to work in the garment factories to make money so that they can buy land to build a house and start their life again. That's what her husband has said. She has never been to Dhaka and is worried about how she will live there but she told her husband she would go wherever he takes her. Her husband has said she needs to bring twenty thousand taka[2] from her parent's house but she knows they don't have that much. As if warning of the dangers of Dhaka, a mother shares the story of her daughter whose husband went to Dhaka and started rickshaw pulling but was having a problem living alone, he needed someone to cook for him. Her daughter was not interested but the mother told her daughter that 'even if your husband keeps you under a tree you should be with him'. So her daughter went to Dhaka and within a year her husband left her for a younger woman.

The riverbank erosion separates families. There were five brothers living together on the char. Three lost their land from river erosion, two of these moved away as they could not rebuild in the village. One brother stayed and started working for a small shop owner. In addition to losing the support of his brothers he feels people treat him differently, he has lost his social status and is largely ignored. He explains that 'once no one would have dared point fingers at our family but now, now we have nothing, no money, no muscles, we cannot defend our daughters against rumours because of our weakness ... and you know how rumours are, they spread like wildfire'. When they were five brothers together they had strength and status.

The disruption is palpable. One man is quick to tell that his char village didn't practice dowry but the village on the riverbank where many of them have moved did, so quite soon after relocating, some char dwellers with sons to marry adopt the practice. This intensifies pressure on families with girls. One mother tells how an 'uncle' took her youngest daughter to the city to work as a housemaid, and her earnings contribute to her elder sister's dowry. In the telling she is visibly upset and says she made a mistake and would not let her daughter go again. Another woman says defiantly that she was not married with dowry and would not let her daughters marry with dowry.

The girls and boys miss months of schooling, as the school was washed away. I talk with the school teacher whose English is good. We are next to a group of girls and young women in a focus group. They are talking about marriage and one says her life was over when she had to get married at fifteen. She could not go to school or see her friends and was seldom allowed to visit her family. They talk about girls as young as 9 and 10 preparing for marriage. This prompts the teacher to talk about how she actively intervenes when she hears about a child marriage being planned. She visits the house sometimes several times and encourages the family to think again about their decision. She explains it is illegal but this is not the basis for her argument. She has had some success, and is optimistic that as more girls continue their education and get jobs, the value of women will change and will help parents see another future for their girls. She also wants to say that these children's dreams and aspirations are not yet limited by their situation. They want to grow up to be police officers, teachers, and one in her class, an astronaut.

In stark contrast to the children's dreams one woman grasps her saree and explains this is all she has. The river took everything else and so she has to go to the river to wash every morning before dawn so she is not seen, but she is scared of the river. She wants the government to build a wall to keep the river back. Large pieces of the newly exposed riverbank continue to fall into the river as we talk. River bank erosion is an active and ongoing process.

And for those who didn't lose their land or house it is only temporary relief. They know the river will come for them eventually. One older man explains that if the people from the south side of the char want to leave, they can do so by selling their land. But if he wants to leave he won't be able to sell his land because no one wants to buy land that is so vulnerable to river erosion. And if he stays, the amount of land left won't grow a quarter of the food his family needs. He ends by saying that only the Almighty knows what the future holds for them.

For those who remain on the char and continue to farm, the unpredictability of the weather makes decisions about what and when to plant much more difficult. One older man suggests the knowledge of their forefathers from whom they learned agriculture is no longer much use, because they don't really know when to cultivate. This year the flood came very early so the cultivated paddy (rice) was washed away before it could be harvested. Crop fields were free for cultivation two months earlier than normal, too early to plant rice so they had to plant corn in the field for paddy. There are new types of pests destroying the crops and there is little money to buy insecticides. One man says even if he had the money he wouldn't buy the chemicals as he thinks they are killing the fish in the river.

An older man shares his explanation of what is happening. In the ancient time the river was very deep, but at present rivers are being silted. Now the rivers are not so deep, the water moves faster and increases the river erosion. He finds the present situation very alarming even though he has experienced river erosion thirteen times since the Liberation War in 1971. 'Just think if one family faces this 13 times then what would be left for them.'

Summary of findings

The findings are reported in detail elsewhere (see Alston, 2014) but in brief the research found that changing weather patterns are negatively impacting food production, increasing poverty and internal migration and disrupting social stability – factors that have gendered impacts. Women and men reported that the changing weather conditions, combined with both rapid and slow onset climate events, have resulted in the loss of life, land, homes, livestock and assets, extreme hardships including permanent injury and poor health, displacement and the need to negotiate temporary settlement rights (for example to build a house on another's land). There was also greater food insecurity, reduced self-sufficiency and the need to change livelihood strategies and agricultural practices. Many men and women reported a reliance on local NGOs to a significant extent for information, support and resources.

Increasing poverty and uncertainty led some parents to take their daughters and sons out of school to work or to marry. Both men and women described feeling under pressure to arrange marriages quickly for their daughters in the wake of a natural disaster, or in the anticipation of one, particularly families who faced losing their home and land through the destruction caused by river erosion. This is consistent with the Human Rights Watch Report (2015) and PLAN survey (2013) about child marriage.

Men reported that they are more likely now to temporarily move or relocate for seasonal or casual work or to consider internal migration to secure permanent work. They spoke of the difficult jobs and long hours they were required to work away from home and their, often precarious, living arrangements. Girls and young women are also moving to cities for education and work. With this comes a shift in roles and responsibilities for men, women and children, particularly with roles performed outside the house, like collecting water and fuel. Both men and women commented on a breakdown in community structure and networks as people move in and out of the area and a perceived increase in freedom to act outside norms.

The research reconfirmed the visible signs of gender inequality, including violence against women and girls, child and forced marriage, the impact of dowry on girls, young women and their families, increased incidence of divorce and polygamy and the exclusion of women from decision-making forums, even where their participation is legislated. Gender inequality appears to be prevailing regardless of government and NGO efforts to change cultural norms and, in some instances, is being exacerbated by a changing climate.

The research findings add to existing evidence of how men and women are experiencing the impacts of climate variability and climate change, but simply restating that climate variability exacerbates gender inequalities will not enable the change required to move more rapidly towards gender equality, in Bangladesh, or anywhere else. The following section discusses specific research findings relevant to our understanding of gender relations, social practices, efforts to address gender inequality and initiatives to manage the impacts of climate change. It attempts to highlight where and how gender roles are changing.

Decision-making, paid work and household dynamics

One of the areas of everyday life with the most striking gender division of roles was household decision-making. Seventy percent of women reported that their husband makes all the decisions, including those about the women's health, fertility and the conditions under which women can leave the house, and 72 percent of men reported that they themselves make all the decisions. Only 5 percent of women surveyed reported that they make all or most decisions together with their husband. Another 25 percent of women reported that they make some decisions and while some of these women were working outside the house, there were also examples of women who earned income or took loans to benefit the family but did not participate in household decision-making. A more nuanced understanding of relationships, influence and individual agency is needed to understand the intersection between earning and household decision-making. Others who reported making decisions were in female-headed households or not married. Female-headed households are on the rise due to divorce, abandonment and death of husbands, or husbands working elsewhere. Women in this position talked about experiencing social discrimination, feeling less safe living alone or needing to take on other roles while men were away. For example in Gaibandha a woman described how she needed to sell the livestock and prepare for the flood as her husband was often away working in the lean times before the flood came, and if the flood came early he was not there to help.

In addition to taking on more livestock and agricultural work there is a significant increase in women seeking and securing paid work. Rao (2012) suggests that this is becoming more accepted due to poverty and family survival. A woman's life-stage is critical for the status implications of her participating in the workforce – married women were more likely to prioritize the need to support their family and children than their personal reputation. Kabeer (2000) suggests that women's work participation in Bangladesh has always had contradictory meanings partly because local notions of femininity emphasize domesticity so work outside the home can be seen as a threat to family honour and status and can be a source of conflict in marital relations (see White, 1992, Koenig et al., 2003). One imam in Satkhira described the changes to weather and climate as a punishment, a direct consequence of Bangladeshi women going out of the house to work, but many men in the focus group discussions explained how it was now vital for women to earn.

> It is quite impossible to maintain our life without the job of husband and wife ... now even though both men and women are working together, the earning is not enough to meet the high price of daily necessities ... earlier we grew our own food such as milk, fruit, vegetables but now with the salinity nothing grows, we have to buy everything.
>
> (Male, Satkhira)

The most common reasons for girls and young women to move from rural to urban areas was to work in urban households to earn money for the family, or to work in the garment industry. The garment manufacturing industry in Bangladesh is the largest source of foreign earnings and female industrial labour. Feldman (1992) cautions against glorifying this contribution to women's economic development. It is a relatively high-risk occupation for women who often have poor housing conditions, low wages, and are at risk of sexual assault or physical violence. She argues that the government prioritizes the 'interests of the multinational garment firms in their search for cheap sources of labour' (1992: 113) above the welfare of Bangladeshi garment workers, the majority of them women.

As women become more visible through increased workforce participation and the activities of NGOs, they are gaining a stronger sense of their rights but Bates et al., (2004) suggest that in some cases this has been associated with a backlash – 'a rising incidence of public forms of violence against women including rape, assault, acid throwing and the punitive use of fatwas to enforce women's moral conduct' (Kabeer 2011: 519). Anxieties about changing gender relations feed directly into debates on Islamism (Basu, 1998; Feldman; 1998; Gardner, 1998; Shaheed, 2010) with Feldman (1992) suggesting that Islamic movements are in part a response to women's increased economic independence in Bangladesh. Religious politics create opportunities for women's activism while simultaneously undermining women's autonomy (Shaheed 1998) highlighting the contradictory nature of the structures of domination and the possibilities of resistance.

In addition to the political and structural backlash, women's increased control over assets or access to financial resources can upset household dynamics as established power relations are challenged. Women reported that earning money through paid work outside the home has the potential to lead to physical aggression, verbal abuse or community isolation and gains from their improved income can be compromised by husbands renegotiating their own contributions (for example, refusing to pay for food or children's education expenses). Despite the risks, most women in the study had a strong desire to earn income.

Across our field sites income earning and microcredit initiatives have been a primary focus for many of the NGOs opening up new opportunities for women and new avenues for mobilizing women. Often implied in these programs are presumptions that earning, saving and access to credit directly contribute to individual economic empowerment, which will convert into the general well-being of women and their families and somehow enable women to renegotiate gender relations within the household. This logic doesn't allow for other factors that can impede women, or the agency of women and others they interact with. Koenig et al's., (2003) comparison of two different settings in rural Bangladesh showed how increased female empowerment challenged long-established gender roles and led to conflict and domestic violence in a more conservative setting, but in a less culturally conservative area women's increased autonomy was not associated with an increased risk of violence.

This suggests the effects of women's earning and status on their experience of violence is highly context-specific.

The social instability created through climate and disaster-related displacement, migration and greater participation of women in the workforce is creating both risks and opportunities for improving gender relations. Men and women in the study suggested that individual families are largely in control of gender norms, and this was evident in the variation in attitudes and actions across many households in the same local areas. Since 33 percent of men and 44 percent of women in the survey sample had no education at all, there are opportunities for changes in gender relations to come through education and awareness at the household level.

Variation, dynamism and history of social practices

The research captured significant variations in custom, norms, practices and responses between households in one village, from one village to the next and across the three districts, and changes within households over time. For example there was wide variation in dowry demands made by the groom's family of the bride's family across the field sites, even between adjacent villages. There were men and women who opposed dowry, villages in Gaibandha that did not practice dowry, brides' families incapable of paying the demanded dowry, and additional demands being made soon after the marriage had taken place. In these cases the girl or young woman often faced humiliation, verbal and physical abuse, and/or deprivation. Women shared several cases of young women being killed or maimed by their husband or in-laws, and other cases of newly married daughters returning or being sent back to their parents after less than three months. The variation in practice and rate of change reinforces the importance of local context.

The practice of dowry is an example of relatively rapid change in social practice in Bangladesh (Shahnaz, 2006). The demand for dowry and the violence associated with such demands was a relatively uncommon phenomenon until the 1970s (Ahmed and Naher, 1987; Amin and Mead, 1997). Lindebaum (1981) argues the rise of dowry needs to be contextualized in the historical shifts of the region's political economy and social relations. Declining land fertility in the 1970s and rising urban unemployment suggests a salaried worker is a more desirable groom than a land owner. Similarly Rao (2012) suggests that the prestige system has shifted from being land-based to one centred on accumulation of money. In highlighting the new forms of dowry, like the provision of visa papers, consumer goods or capital for an enterprise, White (1992) argues that, rather than signifying increased subordination, dowries indicate the use of marriage for upward social mobility.

If reducing the prevalence of dowry, particularly the violence associated with it, is to be achieved then contextualized and pre-emptive responses are needed. In many fields much value is placed on identifying and disseminating 'best practice' but Anderson et al., (2005) suggest that there is likely to be more than one way to be successful – that is more than one successful process, structure, organization or configuration within any system, particularly a system with so many people living in such dynamic environments.

Law reform and its limits

One of the Bangladeshi government's responses to dowry and dowry-related violence has been law reform to address domestic violence against women. In cases of physical or psychological dowry-related abuse, women can now take action through the Domestic Violence (Prevention and Protection) Act, 2010. Despite this initiative Dr Nahid Ferdousi, Associate Professor of Law at Bangladesh Open University (2014) suggests there has been no noticeable change in the dowry practice since the enactment of the law.

NGO workers and advocates interviewed suggest that the inadequacy of legal protection and lack of enforcement of existing laws, the influence of Shariah law, the absence of adequate direct services and lack of familial support means that many women do not assert their legal rights. Women in the study suggested there is often a lack of cooperation from family members and some victims do not want to, or cannot afford to, continue the legal battle against their husbands for fear of reprisal or further isolation. In the southern districts of Satkhira and Barguna women suggested there was more emphasis on non-formal decision-making processes and traditional adjudication forums that may involve social, cultural and religious barriers as well as gender bias.

In contrast to law reform efforts related to reducing the prevalence of dowry and dowry-related violence the reduction in child marriage appears to be suffering from a lack of political commitment. This is despite Bangladesh having the highest rate of marriage involving girls under fifteen in the world (UNFPA, 2012). Child marriage has been illegal in Bangladesh since 1929, and the minimum age of marriage set at 18 for women since the 1980s. Despite this, Bangladesh's Prime Minister Sheikh Hasina, with the support of her cabinet, attempted to lower the legal age of marriage for girls from 18 to 16 years old, raising serious doubts about her public commitment in 2014 to end child marriage in Bangladesh by 2041. The End Child Marriage advocacy network argues that it is essential that the government focus on how to change the mindset of people rather than looking for easy ways to 'improve' the statistics. While government and NGO documents often state the intention to change cultural attitudes the subsequent interventions rarely go beyond gender sensitivity and awareness raising, which are unlikely to change underlying attitudes, gender relations or the way households operate.

Cumulative effects, uncertain causes and unintended consequences

As discussed above there are multiple and intersecting factors and influences on gender relations (for example social, cultural, familial, religious, political, legal) making it difficult to claim direct causal relationships. This complexity and potential for compounding impacts relates to climate change and climate events as much as social norms and practices.

The research did not attempt to separate the influences of rapid onset events (cyclones, tidal surges, flooding and riverbank erosion, for example) from slow onset change (sea level rise, increasing temperatures, salinization, land degradation and loss of biodiversity, for example) but recognized there are some important relationships. For example, drought can be an extreme weather event, while also being closely linked to slow onset, incremental climatic change (IPCC, 2012; UNFCCC, 2012). The synergistic interactions between rapid onset and slow onset events can increase the risk of loss and damage (Siegele, 2012). In Gaibandha people were experiencing drought-induced food shortages causing many to sell their livestock and assets, spend their savings on fuel for irrigation, or seek paid work to survive. This was difficult for char dwellers, as they couldn't navigate the river to access markets or temporary jobs, as it was particularly low. When the annual flood came early and quickly, it caused riverbank erosion and the flooding of crops about to be harvested. Many of the men were away working and households didn't have sufficient resources to re-establish their farming as they had sold their assets and depleted their savings to survive the drought.

Human factors such as land-use patterns, water management and control of river flow upstream can compound climate-related hazards. The river referred to earlier in the narrative is the highly dynamic Brahmaputra (Jamuna) River that rises in the Himalayas in Tibet before flowing through India and into Bangladesh. Its annual flooding and silt deposits are critical in maintaining soil fertility and sustaining fisheries. In recent years there has been more erosion and less river island (char) building, and this trend is accelerating (BWDB, 2012). According to IPCC (2014) it could be linked to increased river flow as glaciers melt in the Himalayas and the Tibetan Plateau, due to global warming, and/or the construction of hydropower projects upstream in India, which hold back essential silt.

Adaptation interventions have also been shown to contribute to the intensity of disasters and cause further displacement. Since the mid-1960s there has been a steady growth of flood control projects in Bangladesh, involving the construction of embankments, drainage channels, sluices and regulators, with Rahman and Salehin (2013) estimating the coverage area of interventions is 36 percent of Bangladesh's total area. They point to increasing population, ill-planned infrastructural development and massive flood control interventions as reasons for flood disasters becoming larger and more frequent. They argue that the lack of consideration of the interdependence of land, water, ecosystems and socio-economic development means most projects are designed to solve immediate problems or single issues without giving adequate attention to the potential, undesirable and difficult to predict long-term consequences of disrupting the hydrological functioning of the floodplains.

While climate challenges and the international attention they bring with them, particularly to Bangladesh, are highlighting gender inequalities, gender 'sensitive' disaster planning and recovery and climate adaptation is not enough to address them, and can have unintended consequences. One woman described a disaster response targeting women to receive food aid that required them to go to the food distribution centre. Several women in her area did not go, as they were not allowed to leave the house or feared leaving. Another woman sold what she considered surplus to her families needs on the riverbank where men had gathered but when she got home her husband was angry as she had sold them too cheaply. Another woman told of the local NGO giving her a cow for breeding as part of a flood recovery and livelihood program. Pirates came along the river with guns and demanded the cow or her daughter. She felt more vulnerable than ever before. A father talked about the corrugated iron he received from an NGO to reroof his house damaged in the cyclone. He was pleased he was able to sell the iron in the market to contribute to his daughter's dowry. These individual accounts illustrate that government and NGO inputs do not necessarily deliver the expected outputs, partly because the power structures that maintain inequalities are dynamic and can adapt to re-establish control and capitalize on any new or redistributed resources.

Discussion

These findings point to the significant variation in norms, practices and impacts within villages and across sites, highlighting the importance of understanding both the ecological and social context and the variation in women and men's roles and status within the local community. Overall the intersections, the uncertainty and the unintended consequences of dealing with changing social and ecological systems and environments requires a flexible way of anticipating, responding and relating across and within systems.

The findings confirm the importance of acknowledging power dynamics, particularly in households, and the need to transform gendered relations by being more explicit about power and recognizing that different power dynamics may be operating in different aspects of a woman's life (Wee and Shaheed, 2008). Shaheed (2009) warns that

> ignoring the power structures operative in women's lives, development initiatives imply that improving a particular right (legislation on gender based violence), or access to some specific resource (microcredit or education), or service (health) or instituting affirmative action (political representation quotas) is sufficient to change gender relations.
>
> (2009: 5)

She suggests that while these can create an environment conducive to change they may not shift thinking and may interrupt or destabilize existing power relations.

Public policy and programs that present as gender neutral or misunderstand the power dynamics between men and women in rural communities risk being less effective and, more disturbingly, have the potential to perpetuate or exacerbate existing gender inequality. For example, what does a climate change policy that is gender-neutral on its surface, say or assume about women and men? More specifically, what attributes, characteristics or roles does it ascribe to women and men? And in what way does it disadvantage women or reinforce men's privilege and dominance? Does this apply to all women or men, or does it affect different men or different women differently? We need to uncover the gender implications of public policy and program logic that might otherwise appear to be neutral or objective. We might consider how institutions and structures are gendered and how this might affect the implementation of a research study, public policy, programs and advocacy. We can be more rigorous in interrogating assumptions and expanding our frameworks. Even where there is a strong commitment to gender 'sensitive' policy or programs it is often focused on targeting women as a group separate from men, providing opportunities to participate in decision-making or delivering services and access to women and girls as an under-resourced or under-represented category (Connell, 2012).

That gender issues are being addressed is an important gain but raises the question of whether the understanding of gender embedded in research proposals, development programs and policies that aim to address climate variability is adequate to address underlying attitudes, relationships and norms and how these are changing. Connell (2012) argues that we need

> a way of conceptualising the dynamics of gender: that is, the historical processes in gender itself, the way gender orders are created and challenged ... we cannot rely any longer on categorical thinking if we are to come to terms with the actual gender processes that affect [equality], the complex social terrain on which they emerge, and the urgency of these issues.
>
> (2012: 1676, 1681)

Considering the limits of the law and the lack of political will discussed above there is a need to find ways to change cultural attitudes, perhaps by considering men's and women's lives as central to change rather than particular issues. This chapter presents a partial view of how climate change and gender relations intersect and relate to everyday life in three rural districts in Bangladesh. The obvious signs of continuous change and uncertainty suggest that it is not helpful to maintain a world view of nature and society as static, predictable or independent of each other. Government and development structures, policies and programs tend to be sector specific without the capacity or mechanisms to view people's lives as a whole. They often struggle to respond effectively to unique actors, political situations and random events that interfere with implementation or replication. A dynamic view of everyday life, relations and systems may be a prerequisite for better understanding these complex social and ecological systems, and with that understanding more targeted and integrated steps towards the familial, social, cultural, legal, economic and political change that is necessary to improve gender relations and respond effectively to climate challenges.

Notes

1 In Bangladesh a 'char' is a river island formed from sedimentation and people who live on a river island are often referred to as char dwellers.
2 Taka is Bangladeshi currency (20,000BDT was equivalent to US$250).

Reference

Ahmed, R. and Naher, M.S. (1987) *Brides and the Demand System in Bangladesh.* Dhaka: Centre for Social Studies.
Alston, M. (2014) *Women and Climate Change in Bangladesh.* London: Routledge.
Amin, S. and Mead, C. (1997) The rise in dowry in Bangladesh, in G.W. Jones, R.M. Douglas, J.C. Caldwell and R.M. D'Souza (eds), *The Continuing Demographic Transition* (pp. 290–306). Oxford: Clarendon Press.
Anderson, R., Crabtree, B., Steele, D., and McDaniel, R. (2005) Case study research: The view from complexity science, *Qualitative Health Research* May, 15(5): 669–685.
Bangladesh Government (2015) Millennium Development Goals: Progress Report 2015, prepared by the General Economics Division, Bangladesh Planning Commission.
Basu, A. (1998) Appropriating gender, in Jeffery, P and Basu, A (eds), *Appropriating Gender: Women's Activism and Politicised Religion in South Asia.* New York: Routledge.
Bates, L., Schuler, S., Islam, F. and Islam K. (2004) Socioeconomic factors and processes associated with domestic violence in rural Bangladesh, *International Family Planning Perspectives* Dec 30(4): 190–199.
BBS (2016) Bangladesh Bureau of Statistics. Available online at www.bbs.gov.bd/Home.aspx (accessed April 10, 2016).
Connell, R. (2012) Gender, health and theory: Conceptualising the issue, in local and world perspective, *Social Science and Medicine* 74(11): 1675–1683.
Feldman, S. (1992) Crisis, Islam and gender in Bangladesh: The social construction of a female labour force, in L. Beneria and S. Feldman (eds), *Unequal Burden: Economic Crises, Persistent Poverty and Women's Work.* Boulder, CA: Westview, pp. 105–130.
Feldman, S. (1998) (Re)presenting Islam: manipulating gender, shifting state practices and class frustrations in Bangladesh, in P. Jeffery and A. Basu (eds), *Appropriating Gender: Women's Activism and Politicised Religion in South Asia.* New York: Routledge, pp. 33–51.
Ferdousi, N. (2014) Improving the standard of human rights, Daily Star, December 9, 2014. Available online at www.thedailystar.net/improving-the-standard-of-human-rights-54237 (accessed March 10, 2015).
Gardner, K. (1997) *Songs at the River's Edge: Stories from a Bangladeshi Village.* London: Pluto Press.

Gardner, K. (1998) Women and Islamic revivalism in a Bangladeshi community, in P. Jeffery and A. Basu (eds), *Appropriating Gender: Women's activism and politicised religion in South Asia*. New York: Routledge.

Human Rights Watch (2015) Marry Before Your House is Swept Away: Child Marriage in Bangladesh. Available online at www.hrw.org/report/2015/06/09/marry-your-house-swept-away/child-marriage-bangladesh (accessed August 10, 2015).

IPCC (2012) Managing the risks of extreme events and disasters to advance climate change adaptation, Special report of the Intergovernmental Panel on Climate Change. New York: Cambridge University Press.

IPCC (2014) Managing the risks of extreme events and disasters to advance climate change adaptation, Special report of the Intergovernmental Panel on Climate Change. New York: Cambridge University Press.

Kabeer, N. (2000) *The Power to Choose: Bangladeshi Women and Labour Market Decisions in London and Dhaka*. London and New York: Verso.

Kabeer, N. (2011) Between affiliation and autonomy: navigating pathways of women's empowerment and gender justice in Bangladesh, *Development and Change*, 42(2): 499–528.

Koenig, M., Ahmed, S., Hossain, M., Khorshed, A. and Mozumder, A. (2003) Women's status and domestic violence in rural Bangladesh: Individual and community level effects, *Demography*, May, 40(2): 269–288.

PLAN (2013) Child Marriage in Bangladesh, Findings from a National Survey,

Rahman, R and Salehin, M. (2013) Flood risk and reduction approaches in Bangladesh, In R. Shaw, F. Mallick and A. Isla (eds), *Disaster Risk Reduction Approaches in Bangladesh*. Tokyo and New York: Springer, pp. 65–90.

Rao, N. (2012) Breadwinners and homemakers: Migration and changing conjugal expectations in rural Bangladesh, *The Journal of Development Studies*, 48(1): 26–40.

Shaheed, F. (1998) The other side of the discourse: Women's experience of identity, religion and activism in Pakistan, in P. Jeffery and A. Basu (eds), *Appropriating Gender: Women's Activism and Politicised Religion in South Asia*. New York: Routledge, pp. 143–166.

Shaheed, F. (2009) Empowerment and Development Planning: A forced South Asian marriage?, Research paper from the Research Programme Consortium on Women's Empowerment in Muslim Contexts (WEMC) funded by UKaid from the UK Department for International Development (DFID). Available online at http://r4d.dfid.gov.uk/PDF/Outputs/WomenEmpMus/Shaheed_Emp_and_Development_in_South_Asia.pdf (accessed September 12, 2015).

Shaheed, F. (2010) Contested Identities: Gendered Politics, Gendered Religion in Pakistan for the special edition of *Third World Quarterly*, 31(6): 851–867.

Shahnaz H. (2006) Dowry in Bangladesh: Compromising women's rights, *South Asia Research*, November 26(3): 249–268.

Siegele, L. (2012) *Loss and Damage: The Theme of Slow Onset Impact*. Bonn, Germany: Germanwatch.

UNFCCC (2012) Background Paper to the Expert Meeting on: a Range of Approaches to Address Loss and Damage Associated with the Adverse Effects of Climate Change, Including Impacts Related to Extreme Weather Events and Slow Onset Processes. Available online at http://unfccc.int/files/adaptation/cancun_adaptation_framework/loss_and_damage/application/pdf/literature_review_barbados.pdf (accessed August 5, 2015).

UNFPA (2012) Marrying too young: End child marriage, United Nations Population Fund

Wee, V. and Shaheed, F. (2008) Women empowering themselves: a framework that interrogates and transforms, by Research Programme Consortium on Women's Empowerment in Muslim Contexts: gender, poverty and democratisation from the inside out, Southeast Asia Research Centre (SEARC)

White, S.C. (1992) *Arguing with the Crocodile: Gender and Class in Bangladesh*. London: Zed Books.

10 Investigating the gender inequality and climate change nexus in China

Angela Moriggi[1]

Introduction

In November 2014, the Intergovernmental Panel on Climate Change (IPCC, 2014) released its 5th Assessment Report, providing a gloomy picture of the various phenomena that are and will be affecting China, its environment and its people. With regards to Asia, scientists have observed an increase of 0.4–2.5°C, that will surge to 2–4°C over the mid-term (2046–2065) and to 4–6°C over the long term (2081–2100). Rising temperatures have resulted in different impacts for such a vast country like China, with drastically different geographies ranging from mountains, grasslands and deserts to tropical forests and low-lying coastal areas. Increase in temperatures and decrease in annual mean soil moisture – essential determinants of plant growth – are adversely affecting rice and other crop yields. Changing climate is contributing to the altering of ecosystems and the loss of biodiversity. Warming temperatures are also expanding the geographic range of infectious diseases such as dengue fever and schistosomiasis to the northern part of the country. Rapid change in temperatures is also directly linked to glaciers melting and sea-level rise. The latter puts China's eastern regions under the risk of coastal inundation (Gupta, 2014). According to the China Meteorological Administration, the extraordinary frequency and intensity of extreme weather events in the last few years, such as floods, droughts and storms, is also to be attributed to climate change (Blanchard, 2007). These phenomena not only translate into great economic losses and adverse human health impacts, but also further exacerbate the environmental crisis resulting from 20 years of dramatic economic development (Brombal et al., 2015).

Over the past ten years, several United Nations agencies have advocated for the need to couple mitigation efforts with adaptation ones, with a particular focus on the social dimensions of climate change (United Nations Task Team on Social Dimensions of Climate Change, 2011). A renewed attention to the "human face" of climate change has brought awareness to the specific climate-induced struggles experienced by women, particularly in underdeveloped rural areas. It is now an established fact that conditions of gender inequality may cause greater vulnerability for women as a result of climate change impacts, while greatly undermining their coping and adaptation capacities (Lambrou and Piana, 2006; Haigh and Vallely, 2010). On the other hand, various authors have highlighted the need for more context-specific studies, to avoid simplified categorizations of the vulnerability-resilience dualism (Arora-Jonsson, 2011). The 4th Assessment Report of the IPCC (Parry et al., 2007) demanded a gender-based approach to be applied in the drawing-up of development, human rights and climate change policies, in order to a) avoid perpetuating and exacerbating existing gender vulnerability and inequalities; and, b) guarantee an effective and concretely applicable adaptation. Development cooperation organizations have devoted a great amount of resources to document and tackle gender vulnerabilities to climate change all over the world (Aguilar, 2009; United Nations Task Team on Social Dimensions of Climate Change, 2011; Jost et al., 2014). Against this background, case studies focusing on China rarely feature in such international studies and programs, while domestically the institutional debate on climate change has been almost blind to gender-sensitive perspectives. Existing studies focusing on countries in Sub-Saharan Africa, South America and Southern Asia, mostly employ four recurring variables to document conditions of gender inequality that might lead to greater vulnerability in the face of climate change: division of labor, access to resources, access to decision-making and gender norms. No similar comprehensive study can be found with relation to China. Drawing from an extensive literature review and to a lesser extent primary research, this chapter aims at bridging this gap. Its objective is to draw attention to the long-lasting conditions of inequality affecting women in China, with regards to the abovementioned variables, laying the basis for more context-specific studies on the ground that might unveil the complex web of interrelated socio-economic and natural factors affecting gender vulnerability to climate change. In the discussion, the chapter touches upon the degree of institutional awareness over the climate change and gender nexus, to understand what socio-cultural and political factors might hinder a stronger commitment on the topic on the part of Chinese policymakers. Finally, potential and limits for further research are presented.

Gender vulnerability to climate change and its relevance for China

For a long time, research attention has been primarily focused on investigating factors of biophysical vulnerability in China. Only recently, a handful of authors have started to explore the profile of regional social vulnerability, with respect to different phenomena directly or indirectly affected by climate change. Zhou et al., (2014) provide evidence of the spatiotemporal patterns in overall vulnerability to natural hazards in 31 Chinese provinces between 2000 and 2010. Using a social vulnerability index composed of 30 indicators, the authors find that economic status, employment in the rural sector, urbanization, percentage of children and population growth played the most important roles in socio-economic vulnerability variability among provinces in 2010, followed by population density and medical services. Gender, population change (natural population growth) and percentage of elderly served as the third group of dominant factors, with unemployment as the least influential component.

Another study, aimed at evaluating the vulnerability of the livelihoods of grazing households in the grasslands and steppes of Northern China, reveals that people's gender, grassland area, livestock numbers and net incomes had significant effects on the vulnerability of grazer households. Families with female householders were found to own less grassland, smaller houses, fewer or no vehicles, fewer young livestock and fewer numbers of livestock slaughtered annually (Ding et al., 2014). A joint research collaboration between Lanzhou University and UN Women carried out in the drought-prone northwestern province of Gansu, found that, as a result of climate-induced extreme weather events, women's workload increased by 2.65 hours a day (Ding et al., 2012). Data also show that women's disease rate was 2.87 percent higher than men. According to the study, women in general had a poorer knowledge of climate change than their male counterparts and participated to a lesser extent to local decision-making (Ding et al., 2012). Liu (2014) laments that gender differences are not taken into consideration in times of disaster relief, when facilities are established and materials are allocated. The author also affirms that disaster-prevention trainings offered at community level are usually mostly attended by males. This is because women have less free time than men due to the great housework burden, and because their lower education levels hinder their capability to engage in the trainings. Liu also claims that traditional gender norms might impair women's survival during extreme weather events, as happened during the Wenchuan earthquake in 2008, when many women were reportedly unable to flee their collapsing houses in time, because they tried to get dressed first (Liu, 2014).

While providing only a very partial account, the above-mentioned literature indicates that, in China, as in many other climate-stressed regions of the world, climate change might affect women in different ways than men, due to a number of different factors related to access to resources, access to decision-making and traditional gender norms. Studies also testify that lack of sex-disaggregated data has so far prevented researchers from providing more substantial quantitative analysis of the existing situation across the country (Liu, 2014; Zhou et al., 2014). The same claims have also been gathered during fieldwork research carried out by the author in Beijing between July and December 2014. Semi-structured interviews were administered to several NGOs involved in the fields of climate change and/or gender advocacy.[2] The few organizations familiar with the gender and climate change nexus affirmed that adaptation projects increasingly targeted rural women. Through past experience on the ground with poverty alleviation programs, NGOs had learnt that women in poor rural Chinese regions were most vulnerable to climate change.[3] Yet, because data were only anecdotal, advocacy on the issue was only at the initial phase. Moreover, people interviewed could only partially explain the reasons that might have contributed to such vulnerability, failing to recognize the structural roots of gender inequalities. Of the

12 interviews held among relevant NGOs, only in two cases did people demonstrate awareness of the link between gender inequality and climate change vulnerability. As mentioned in the introduction, in many parts of the world, gender vulnerability to climate change has resulted as a consequence of conditions of gender inequality manifested in unfavorable divisions of labor, unequal access to resources, unequal access to decision-making and cultural patriarchal norms (Lambrou and Piana, 2006; Haigh and Vallely, 2010). The following section will explore these variables with reference to China.

An analysis of key factors in gender inequality and their link to climate change in China

Feminization of agricultural labor

It is important to understand gender relations in the context of China's extraordinary socio-economic changes, brought about by three decades of continuous economic growth, following the "opening reforms" of 1978. Some of the most significant transformations have been triggered by dramatic patterns of migration and an unprecedented rate in urban growth. According to a number of studies, one of the most remarkable consequences of massive out-migration has been the feminization of agricultural labor (*nongye nvxinghua*), in Chinese also referred to as "men work and women plough" (*nan gong nv gang*) (Chang et al., 2011; Judd, 2007; Zuo, 2004). With men seeking off-farm employment and joining the "floating population" (*liudong renkou*) of 230 million rural migrant workers (ILO, n.d.) many women have been left in the countryside to care for the farm and the household, with a resulting pattern often described as a concentration of "women, children and the elderly" (*sanba liuyi jiujiu*) (Judd, 2007). In fact, although the difference in the likelihood of migration between men and women has decreased over time, migration is still primarily male, mainly because of cultural and consuetudinary reasons (Chang et al., 2011). A report released by The State Council Information Office of the People's Republic of China in September 2015 states that women living in rural areas account for about 70 percent of the total agricultural labor force (SCIO, 2015). Feminization of agriculture has not occurred evenly across all regions nor across different groups of women (Chang et al., 2011). Women seem to be particularly active in planting and stockbreeding, activities which are time-consuming, arduous and generate low economic returns (UNDP, 2003). Moreover, surveys indicate that it is particularly middle-aged women who have been increasingly participating in farming more than men (Mu and van de Walle, 2011). State rhetoric and literature now commonly refer to them as the "left-behind women" (*liushou funv*), often portrayed as a passive, helpless "vulnerable group" (*ruoshi qunti*) (Jacka, 2012). Reality hardly reflects such a monolithic picture. However, a few studies, attempting to reveal the impact of male outmigration on women's well-being, have documented some of the factors contributing to such vulnerability. As a result of a quantitative and qualitative analysis carried out in three villages of Guangxi province, Zuo (2004) finds that an increased burden of agricultural responsibilities has widened the gap in cash income between women and their husbands, the latter enjoying a better salary through off-farm employment. Through an analysis of data retrieved from the China Health and Nutrition Survey related to the period 1991–2006, Chang et al., (2011) find that, in rural areas, women are increasingly participating in both farm and off-farm work, but also continue to provide for the majority of unpaid domestic work. This not only includes the care of children and the elderly, but also fetching water, gathering fuel, collecting and preparing food which translates into greater exposure to indoor pollution and water and soil contamination, with negative effects on their health (Wang, 2011). Because of physiological reasons, women are also particularly exposed to certain kinds of contaminants: a research conducted in the southern province of Guangzhou revealed that 95 percent of victims of cadmium contamination were women (Wang, 2011). Reliance on the primary sector also means that women are directly impacted by the consequences of climate change on natural resources, crops, livestock and fishing production. For instance, water shortages might further increase the burden on women and girls responsible for fetching water, a back-breaking and time-consuming task. Deforestation greatly impacts the capabilities of women to secure firewood, honey, mushrooms and medicinal plants for them and their families. Depletion of fish stocks or crops failure might further exacerbate women's income insecurity and inequality (Otzelberger, 2014). On the other hand, traditional gender norms might greatly undermine their coping and adaptation capabilities because of unequal access to material and immaterial resources and because of lack of decision-making leverage at family and community level.

Access to education and health

At the legislative level, China has made noteworthy efforts to guarantee equal rights to men and women, through the enactment of various laws beginning from the mid-1990s, following the Fourth World Conference on Women held in Beijing in 1995. The conference provided a unique capacity-building opportunity for the emerging Chinese women's NGO community and women' studies network, getting in touch with the international women's movement for the first time (Kaufman 2012). Despite progress made in the last few decades, traditional ideas still play a major role in influencing gender inequality. The Sixth National Census revealed a further deterioration of the long denounced sex ratio. In 2010, 118.0 boys were recorded for every 100 girls, in the under-15 age group (Attané, 2012). Discrimination at birth is only one of the consequences of patriarchal traditions that favor the masculine line, and impair women's access to education, healthcare and property. All three are particularly relevant in relation to climate change vulnerability and adaptation capacities.

Education achievements have increased substantially in the past four decades, and while the gender gap in education has greatly narrowed, it is still quite significant. From 1980 to 2000, illiteracy rates decreased from 49 percent to 14 percent for women and from 21 percent to 5 percent for men. However, still in 2000, 1 person out of 12 was illiterate, and 3 out of 4 illiterate people were women (WB and ADB, 2006). The 2014 Gender Inequality Index reveals that, among people aged 25 years and more, with reference to the period 2005–2014, 58.7 percent of women received at least some secondary education, in comparison to 71.9 percent of men (World Economic Forum, 2015). Financial difficulties are the main cause of both boys and girls leaving school early. However, most parents, especially in rural areas, consider education unnecessary for girls much more than boys (Attané, 2012). This has led certain poor Chinese provinces, such as the north-western province of Ningxia, to enact affirmative regulations that demand priority for girls in rural areas in high school enrolment, and in getting tuition and accommodation fee waivers. Most disadvantaged in terms of education is the current generation of elderly women, that was born and raised during times of poverty and social instability, and thus received much less education than subsequent generations (Zhang et al., 2015).

Insufficient education might have several consequences when it comes to coping and adapting to climate change. First, it might hinder people's capacity to respond to early warning systems in times of natural disasters, and to participate in trainings for disaster-risk prevention (Aguilar, 2009). Second, education has an important social impact on mental and physical health. Education helps to promote and sustain healthy lifestyles and reduce the possibility to engage in risky behaviors. Zhang et al., (2015) examine comprehensive longitudinal data from a multi-country WHO survey on the health and well-being of adult populations in six countries, including China. The research confirms that poor levels of education still make a substantial contribution to the health gender gap in the country. The disadvantage is particularly problematic for elderly women, who, as mentioned above, received little education in their early years, because of traditional gender norms. But education does not act in isolation to other factors. Chinese elderly women in rural areas might also have low socio-economic status, little or no effective pension or healthcare arrangements, and if widowed, they might lack access to support and resources altogether (Zhang et al., 2015). Third, lack of formal education might also limit people's capacity to receive appropriate training during adaptation programs, and thus hinder their possibility to access credit, technology and extension services to address the agricultural impacts of climate change (Aguilar, 2009).

As for health, studies on the ground also reveal that there still is discrimination against girls in getting health treatment, as demonstrated by a survey administered to 1,428 rural households of 48 villages in eight Chinese provinces (Gao and Yao, 2006). Poor mental and physical health negatively impact people's vulnerability and undermine their coping and adaptation capacities. Gender-sensitive sanitary conditions are also often missing during post-disaster relief, creating insalubrious and socially awkward environments for women, especially when they happen to be pregnant or breast-feeding (Liu, 2014). Poor health conditions are also connected to dietary deficiencies. Context-specific studies aimed at investigating maternal nutrient intake among pregnant and lactating Chinese women living in rural areas, have revealed that the majority of the sample had low intakes of nutrients essential for pregnancy (Cheng et al., 2009; Ma et al., 2008) and that nutritional inadequacy in both rural and urban women was strongly linked to dietary precautions prescribed by traditional Chinese culture (Gao et al., 2013).

Access to resources

Along with unequal access to education and health, restricted access to material and financial resources such as land, income, and credit, also greatly impacts the capacity of people to cope and adapt to climate change (Jost et al., 2014). According to a report by the Asian Development Bank, as a result of low-profit activities and unpaid care work, women have lower incomes and are thus particularly vulnerable to poverty, especially in China's rural areas (WB and ADB, 2006). Marital breakdown through divorce or the premature death of a partner in many cases also contributes to poverty and is exacerbated by it (Judd, 2007). Poverty also means inability to contract insurance coverage, which is often the case for low-income groups (Otzelberger, 2014). Many women are also still discriminated against in the entitlement to family legacy and inheritance and thus denied possible additional economic and material resources and support (Attané, 2012).

Equity in land tenure is also a long-standing issue in China, since access to land has been systematically differentiated by gender lines. A new land-use rights regime has come into force since the mid-1990s, authorizing an adjustment in land allocation, which is then frozen for 30 years, the so-called "no change for thirty years" (*sanshi nian bu bian*). This means that the majority of women, who marry or remarry patrilocally, are de facto separated from land previously allocated to them. Newly arrived spouses and their children are thus excluded from local land-use rights, which results in unevenness in household land resources (Judd, 2007). Lack of control and property over natural resources strongly limits the capacity for people to implement adaptation strategies to face the consequences of climate change (Jost et al., 2014). This situation is further aggravated by other factors: the results of a large-scale survey covering 464 households of 27 villages in eight Chinese provinces, reveal that rural women face difficulties in accessing technology, credit and market information due to their limited social and human capital (Song et al., 2009; UNDP, 2003). Doss et al., (2011) report that in large contract-farming schemes involving many thousands of farmers, contracts are exclusively stipulated with men, although women perform the bulk of the work related to contract farming.

Access to decision-making and traditional gender norms

Not surprisingly, limited access to resources, coupled with the persistence of traditional ideas about gender roles, strongly impacts women's possibility to influence decisions. Many authors have denounced the lack of equal representation of men and women in Chinese decision-making arenas, with the result that women's knowledge, experience and needs are often ignored or underestimated (Hu, n.d.; UNDP, 2003; Wang, 2011). Despite the great human advancements achieved in the last few decades, the vast majority of people still adhere to the old idea of "men belong outside, women belong inside" (*nan zhu wai, nv zhu nei*), implying that men are the breadwinners turning towards society, while women are the homemakers devoted to the family. Data show that the notion is shared by most men and women alike, proving the very deep-seated internalization of masculine domination in Chinese society. This is not only true in rural areas, where traditional gender roles might be harder to tackle (Zuo, 2004), but also in urban areas (Attané, 2012). Social pressure (*shehui yali*) is harder on men, demanding that they buy a house and pursue a successful career, while the vast majority of people believe that for women "a good marriage is better than a career" (*gan de hao buru jia de hao*). This creates a vicious circle that might be one of the main causes of women's unemployment rates, 50 percent higher than for men (Attané, 2012).

Unequal access to decision-making also results from insufficient political participation. Several laws have been enacted to push for stronger female representation at all levels of the governance machine. However, their nature is rather formalistic and symbolic, and has failed to exercise any concrete policy pressure on the party and government organizations. The Gender Inequality Index 2014 shows that women only account for 23.6 percent of representatives to the National People's Congress (NPC), China's nearest equivalent to a parliament (World Economic Forum, 2015). In 1988, the figure was 21.3 per cent. Only 21 percent of all party members and 23 percent of all national-level civil servants are women. Further down the system, little more than 1 percent of village committee chairs are female (Howell, 2014). Not much has changed since the early '90s, when women's political participation was summarized with the saying "one low and three small": low in overall proportion of women participating in politics, and small in the number of women taking positions at higher levels, taking head positions and taking positions in key sectors (Zeng, 2014). This means that affirmative action has had only a superficial impact and no substantial improvement has been brought to women's inclusion in power structures both at high and grassroots levels (Guo and Zheng, 2008).

Potential and limits for a gender-based approach to adaptation strategies at the institutional level

As mentioned in the introduction, the 4th Assessment Report of the IPCC demanded the application of a gender-based approach in the drawing-up of development, human rights and climate change policies, in order to a) avoid perpetuating and exacerbating existing gender vulnerability and inequalities; and b) guarantee an effective and concretely applicable adaptation (Parry et al., 2007). When implementing adaptation programs, it is important that all stakeholders are taken into consideration, and that their input is viewed as a valuable resource of local knowledge and indigenous expertise (Moriggi, 2016b; Aguilar, 2009). Studies have also demonstrated that coping and adaptation measures are not gender-neutral. A recent survey carried out in Anhui province, Yongqiao district, the first Chinese district to ever issue wheat drought weather index insurance, reveals gender differences in adopting climate change adaptation measures: male respondents were more likely to adopt new technologies and undertake risky business investments than female respondents (Jin et al., 2015). Research on adaptation and resilience to water-related hazards carried out in the south-western province of Yunnan, found that women and men attributed the incidence of drought to different causes in significantly different percentages: female interviewees believed that anthropogenic activities were mainly responsible for droughts (50.8%), followed by government policies (25.4%), environmental changes (20.3%) and weak leadership (3.4%); male respondents indicated an incidence of 37.9 percent for both anthropogenic activities and environmental changes, 24.1 percent for government policies, and no incidence at all for weak leadership (Pradhan et al., n.d.).

Such gender-specific differences must be taken into account in the design and implementation of adaptation strategies. The structural conditions of gender inequality affecting China suggest that the country still has a long way to go in order to pursue such a goal. Professor Tamara Jacka, renowned expert in gender relations and social change in contemporary China, affirms that "most senior Chinese policymakers, social activists and scholars do not recognize gender inequality as a social problem, or regard the problem as trivial" (Jacka, 2013: 985). Official documents appear to confirm such a claim. An element worth investigating is that there seems to be a contradiction between the rhetoric employed at institutional level and the facts lamented by many scholars, as reported in the previous section of this chapter. In September 2015, the Information Office of the State Council issued a white paper on "Gender Equality and Women's Development in China", stating that "it is obvious to all that, in tandem with rapid economic and social development, great progress has been achieved in the promotion of gender equality and women's development in China over the past two decades" (SCIO, 2015). According to the white paper, gender-disaggregated indicators have been standardized and improved, providing routine and statistical surveys of women's health, well-being and development. Moreover, overall consideration has been given to the impacts of urbanization, aging, climate change and other social and market factors on women's poverty. A particular mention is made to women employed in the agricultural sector and to the beneficial effects of the so-called "Sunshine Project", which is claimed to have trained nearly 200 million women in new agricultural technologies and new crop species.

The document also affirms that, thanks to specific laws and policies, China now boasts a higher level of female participation in politics and a greater role of women in decision-making and management of state and social affairs. It is clear how these statements are in conflict with the data reported in the previous section. Particularly noteworthy in the white paper is a special session dedicated to "Women and the Environment", which denotes how the gender and environment/climate change nexus has been elaborated in a rather problematic way. No mention is made of gender-specific environmental and climate change vulnerabilities. The word "environment" is meant to indicate the social, cultural and natural environment, and women's development is always associated with their central role in the family sphere. The section recites:

> China attaches great importance to creating a social and cultural environment conducive to boosting gender equality; through building a healthy and safe natural environment and fostering equal and harmonious family traditions, sound conditions have been created for women' s development. Women are playing an increasingly prominent unique role in the fostering of social culture, protection of the ecological environment and family management.
>
> (SCIO, 2015)

Only a small part is dedicated to issues related to sanitary conditions and environmental protection: women are said to be increasingly participating in ecological protection and conservation, also at high decision-making levels (SCIO, 2015). The same understanding of "environment" is also evident in the "National Program on the Development of Women (2011–2020)", released by the State Council in 2011. The outline sets major objectives to promote women's development, covering six areas: 1) women and the economy, 2) women in decision-making and management, 3) education of women, 4) women and health, 5) women and law, 6) women and the environment. The document is intended as a guideline for more than 30 government and nongovernment institutions responsible for the implementation and monitoring of measures aimed at achieving the above-mentioned goals. In the "women and the environment" section, a reference is made to the need for undertaking gender risk assessment for air and water pollution, and in taking women's needs and capabilities into consideration when designing disaster risk reduction measures. The remaining part of the section explains how women should foster education and good practices in the family, and pursue their own intellectual and personal advancement in society. They should also be "organized and mobilized" to participate in ecological construction and protection and raise their "ecological civilization awareness" (SCIO, 2011). The employment of such discourses clearly resonates with the government main rhetorical strategies, centred around three main concepts: "ecological civilization" (*shengtai wenming*), "harmonious society" (*hexie shehui*) and "raise the quality" discourse (*tigao suzhi*). The latter in particular has been strongly internalized by the All China Women Federation (ACWF), the main organization promoting women's development in China since its establishment in 1949. The ACWF is active at all administrative levels, working with national and international donors in development cooperation projects on the ground. Numerous campaigns launched by ACWF have focused on enhancing women' self-esteem and improving their "quality", understood as the innate and nurtured physical, psychological, intellectual, moral and ideological qualities of people (Jacka, 2009; Howell, 2014). The employment of the "raise the quality" discourse, coupled with the one of "left-behind women" might cast the spotlight on women as a problem and a solution, thus further contributing to their stigmatization as a "vulnerable group" on the one hand, while aggravating their burden on the other. In this way, the vulnerability discourse becomes an instrumental tool to mask the institutional, political and socio-economic factors that cause gender inequality (Otis, 2015; Jacka, 2012; Howell, 2014).

A gender-based approach to adaptation strategies might also be hindered by a problematic understanding of the issue of adaptation in China. Xu and Liu (2014) lament that China has yet to undertake targeted work to address adaptation, and that coordinated action is still missing. This is not surprising if we consider that China's first National Strategy for Climate Change Adaptation was published only in 2013, by the National Development and Reform Commission (NDRC), the country's ministry for macroeconomic management. Other studies undertaken by the Ministry of Water Resources and the Ministry of Health and Family Planning Commission have investigated the impacts of climate change on water resources and human health (Nachmany et al., 2015). China faces a great challenge in adapting to climate change, considering its variety of climate types, geomorphologies and ecosystems. Great resources must be channeled to address awareness gaps and needs especially at local levels. A survey conducted among 85 administrative and management personnel from government departments responsible for climate change adaptation planning in five provinces, has reported a lack of scientific knowledge, inexperience in vulnerability and risk assessment and cost-benefit analysis. Respondents, mainly male (81.7%), indicated that water resources, agriculture and ecosystems were more heavily affected by climate change than human health and urban construction. About 60 percent of the staff surveyed were also unaware of measures such as mental health education and post-disaster psychological trauma. A lack of training was identified as the main problem in developing professional knowledge on climate change and adaptation. With regards to implementation, several constraints were reported: lack of sufficient funding and policy support, weak public awareness, lack of departmental cooperation, human resources shortage, weak awareness of climate change, and inadequate scientific evidence and research (Liu et al., 2016).

Adding to this, the national debate on climate change often fails to consider the "human face" of climate change. A focus on social aspects of climate change, and on the small-scale manifestations of the issue, remains largely marginalized in favor of grand technological solutions and financial mechanisms. Moreover, top-down approaches still dominate decision-making processes, failing to incorporate bottom-up stakeholder participation in the planning and implementation procedures (Moriggi, 2016a).

Failure to recognize gender inequalities, insufficient understanding of adaptation, and little attention to issues of social vulnerability, all contribute to explain how the vast majority of policymakers, scholars and practitioners may lose sight of the gender nexus with climate change, and contribute to the subordination and poor implementation of gender equity goals (Guo and Zheng, 2008).

Conclusions

The debate on climate change in China has so far been predominantly gender-blind. Only a handful of researchers and commentators have warned about the risks of such an approach, while providing case study accounts of the gendered impacts of climate change. Meanwhile, a national rhetoric reiterating the idea of women as a "vulnerable group" has failed to highlight the structural causes of such vulnerability, while further exacerbating gender discrimination at many levels. This study is a first attempt to provide a comprehensive picture of the structural conditions of gender inequality affecting many women in China today, shedding light over their close link to social vulnerability to climate change. By looking at feminization of labor, access to resources, access to decision-making and traditional gender norms, this chapter has demonstrated how several factors may hinder Chinese women's capacity to cope and adapt to climate change, while exposing them to the risk of being disproportionally harmed by its effects. In order to effectively bridge the gap in the literature of gender and climate change in China, more context-specific studies are needed. Studies should concentrate not only in rural areas, but also in urban areas. Research should not only be supported by sex-disaggregated data, but also by feminist-informed sociological approaches to understand deeply the ways in which climate change is gendered. Intersectionality might be endorsed as an analytical tool to reveal how structures of power emerge and interact, in relation to gender, class, race, and other categories of difference in individual lives, social practices, institutional arrangements and cultural ideologies (Kaijser and Kronsell, 2014). Most importantly, scientific evidence and political will are required to eradicate the structural conditions of inequality that prevent those pointed at as vulnerable groups proving themselves to be important resources in the fight against climate change.

Acknowledgement

The research leading to these results has been funded by a research fellowship granted by University Ca' Foscari Venice (Department of Environmental Sciences, Informatics and Statistics and Department of Asian and North

African Studies) and the Euro-Mediterranean Center on Climate Change (CMCC). Funds for fieldwork were provided by the EU FP-7 project GLOCOM (Global Partners in Contaminated Land Management), Grant agreement no. 26923.

Notes

1 Angela Moriggi is currently a Researcher at Natural Resources Institute Finland (Luke), in the Society and Economics Unit and EU Marie Curie ITN Fellow of the SUSPLACE program. She contributed to this book in her capacity as Research Fellow at University Ca' Foscari Venice.
2 This fieldwork was part of a research work developed in the framework of "GLOCOM", an EU FP7 Marie Curie IRSES Project, coordinated by University Ca' Foscari Venice, and involving the Chinese Research Academy of Environmental Sciences, Beijing Normal University and Umeå University. Fieldwork aimed at investigating the relevance of the gender perspective in climate change programmes implemented by both local and international NGOs operating in China, either in the field of climate change adaptation and mitigation or in the field of gender advocacy. Results of the study will be published in the course of 2017.
3 An indicative example is provided by the activities carried out by the NGO Oxfam, one of the major international organizations operating in China. See www.oxfam. org/en/countries/china.

References

Aguilar, L. (2009) *Training manual on gender and climate change.* San José, Costa Rica: GGCA (Global Gender and Climate Alliance), pp. 1–278. Available online at https://portals.iucn.org/library/efiles/documents/2009–012.pdf (accessed July 2, 2015).

Arora-Jonsson, S. (2011) Virtue and vulnerability: Discourses on women, gender and climate change. *Global Environmental Change,* 21(2): 744–751.

Attané, I. (2012) Being a woman in China today: A demography of gender. *China Perspectives,* (4): 5–15.

Blanchard, B. (2007) China blames climate change for extreme weather. *Reuters.* Available online at www.reuters.com/article/environment-china-floods-dc-idUSPEK35318820070802 (accessed June 12, 2015).

Brombal, D., Wang, H., Pizzol, L., Critto, A., Giubilato, E. and Guo, G. (2015) Soil environmental management systems for contaminated sites in China and the EU. Common challenges and perspectives for lesson drawing. *Land Use Policy,* 48: 286–298.

Chang, H., MacPhail, F. and Dong, X. (2011) The feminization of labor and the time-use gender gap in rural China. *Feminist Economics,* 17(4): 93–124.

Cheng, Y., Dibley, M.J., Zhang, X., Zeng, L. and Yan, H. (2009) Assessment of dietary intake among pregnant women in a rural area of western China. *BMC Public Health,* 9(222): 1–9.

Ding, W., Ren, W., Li, P., Hou, X., Sun, X., Li, X., Xie, J. and Ding, Y. (2014) Evaluation of the livelihood vulnerability of pastoral households in Northern China to natural disasters and climate change. *The Rangeland Journal,* 36: 535–543.

Ding, W.G., Wei, Y. and Xian, Y. (2012) "Qihou bianhua yingdui zhengce xu guanzhu shehui xingbie mingan" [Climate change policy response needs to give attention to social gender sensitiveness). *Zhongguo nv wang [China women network)*. Available online at www.clady.cn/2012/nxgz_0509/20293.shtml (accessed February 20, 2014).

Doss, C. and the SOFA Team. (2011) *The role of women in agriculture*. FAO (Food and Agriculture Organization of the United Nations). Available online at www.fao.org/docrep/013/am307e/am307e00.pdf (accessed June 2, 2015).

Gao, H., Stiller, C.K., Scherbaum, V., Biesalski, H.K., Wang, Q., Hormann, E. and Bellows, A.C. (2013) Dietary intake and food habits of pregnant women residing in urban and rural areas of Deyang City, Sichuan Province, China. *Nutrients*, 5(8): 2933–2954.

Gao, M. and Yao, Y. (2006) Gender gaps in access to health care in rural China. *Economic Development and Cultural Change*, 55(1): 87–107.

Guo, X. and Zheng, Y. (2008) Women's political participation in China. *China Policy Institute, Briefing Series*, (34): 1–13.

Gupta J. (2014) How climate change will impact China – latest IPCC report. *China Dialogue*. Available online at www.chinadialogue.net/blog/7458-How-climate-change-will-impact-China-latest-IPCC-report/en (accessed June 2, 2016).

Haigh, C. and Vallely, B. (2010) *Gender and the climate change agenda. The impacts of climate change on women and public policy*. Women's Environmental Network. Available online at www.gdnonline.org/resources/Gender and the climate change agenda 21.pdf (accessed March 3, 2016).

Howell, J. (2014) Where are all the women in China's political system? *East Asia Forum*. Available online at www.eastasiaforum.org/2014/10/15/where-are-all-the-women-in-chinas-political-system/ (accessed June 3, 2016).

International Labour Organisation (ILO). (n.d.) *Labour migration in China and Mongolia*. ILO. Available online at www.ilo.org/beijing/areas-of-work/labour-migration/lang–en/index.htm (accessed June 27, 2016).

Intergovernmental Panel on Climate Change (IPCC). (2014) *Climate Change 2014: Synthesis Report. Contribution of Working Groups I, II and III to the Fifth Assessment Report of the Intergovernmental Panel on Climate Change*. [Core Writing Team, R.K. Pachauri and L.A. Meyer (eds)). Geneva, Switzerland: IPCC, 151 pp.

Jacka, T. (2009) Cultivating citizens: Suzhi (quality) discourse in the PRC. *Positions*, 17(3): 523–535.

Jacka, T. (2012) Migration, householding and the well-being of left-behind women in rural Ningxia. *The China Journal*, 67: 1–22.

Jacka, T. (2013) Chinese discourses on rurality, gender and development: A feminist critique. *The Journal of Peasant Studies*, 40(6): 983–1007.

Jin, J., Wang, X. and Gao, Y. (2015) Gender differences in farmers' responses to climate change adaptation in Yongqiao District, China. *Science of The Total Environment*, 538: 942–948.

Jost, C., Ferdous, N. and Spicer, T.D. (2014) *Gender and inclusion toolbox: Participatory research in climate change and agriculture*. CGIAR Research Program on Climate Change, Agriculture and Food Security (CCAFS), CARE International and the World Agroforestry Centre (ICRAF): 1–200. Available online at www.ccafs.cgiar.org/ (accessed May 2, 2016).

Judd, E.R. (2007) No change for thirty years: The renewed question of women's land rights in rural China. *Development and Change*, 38(4): 689–710.

Kaijser, A. and Kronsell, A. (2014) Climate change through the lens of intersectionality. *Environmental Politics*, 23(3): 417–433.

Lambrou, Y. and Piana, G. (2006) *Gender: The missing component of the response to climate change*. FAO, pp. 1–58. Available online at www.fao.org/docrep/010/i0170e/i0170e00.htm (accessed January 2, 2016).

Liu, B. (2014) Gender equality and climate change. *Women of China*. Available online at www.womenofchina.cn/womenofchina/html1/columnists/17/4240-1.htm (accessed January 16, 2017).

Liu, T., Ma, Z., Huffman, T., Ma, L., Jiang, H. and Xie, H. (2016) Gaps in provincial decision-maker's perception and knowledge of climate change adaptation in China. *Environmental Science & Policy*, 58: 41–51.

Ma, G., Jin, Y., Li, Y., Zhai, F., Kok, F.J., Jacobsen, E., Yang, X. (2008) Iron and zinc deficiencies in China: What is a feasible and cost-effective strategy? *Public Health Nutrition*, 11(06): 632–638.

Moriggi, A. (2016a) Climate change activism "with Chinese characteristics": The role of domestic and foreign NGOs. *China Policy Institute Blog*. Available online at http://blogs.nottingham.ac.uk/chinapolicyinstitute/2016/05/11/climate-change-activism-with-chinese-characteristics-the-role-of-domestic-and-foreign-ngos/ (accessed May 12, 2016).

Moriggi, A. (2016b) Una prospettiva di genere sui cambiamenti climatici. Vulnerabilita' e adattamento, discorso internazionale e gender mainstreaming. *DEP*, 30: 38–57.

Mu, R. and Van De Walle, D. (2011) Left behind to farm? Women's labor re-allocation in rural China. *Labour Economics*, 18: S83–S97.

Nachmany, M., Fankhauser, S., Davidová, J., Kingsmill, N., Landesman, T., Roppongi, H. et al., (2015) Climate change legislation in China. An excerpt from the 2015 Global Climate Legislation Study. *Grantham Research Institute on Climate Change and Environment*. Available online at www.lse.ac.uk/GranthamInstitute/wp-content/uploads/2015/05/CHINA.pdf (accessed June 27, 2016).

Otis, E. (2015) Inequality in China and the impact on women's rights. *The Conversation*. Available online at http://theconversation.com/inequality-in-china-and-the-impact-on-womens-rights-38744 (accessed April 2, 2016).

Otzelberger, A. (2014) *Tackling the double injustice of climate change and gender inequality*. CARE International. Available online at www.careclimatechange.org/files/Double_Injustice.pdf (accessed May 2, 2016).

Parry, M.L., Canziani, O.F., Palutikof, J.P., van der Linden,P.J., Hanson, C.E. (eds), (2007). Climate Change 2007: impacts, adaptation and vulnerability. Contribution of Working Group II to the Fourth Assessment Report of the Intergovernmental Panel on Climate Change. Cambridge, UK: Cambridge University Press, pp. 1–976.

Pradhan, N.S., Bisht, S. and Zu, Y. (n.d.) *A case study on adaptation and resilience to water related hazards: Analyzing gendered responses to climate change in Yunnan province, China*. International Centre for Integrated Mountain Development (ICIMOD). Available online at www.icimod.org/resource/9799 (accessed June 12, 2016).

SCIO (State Council Information Office of the PRC). (2015) Gender equality and women's development in China. *english.gov.cn*, pp. 1–12. Available online at http://english.gov.cn/archive/white_paper/2015/09/22/content_281475195668448.htm (accessed February 12, 2016).

Song, Y., Zhang, L. and Jiggins, J. (2009) *Feminization of agriculture in rapid changing rural China: Policy implication and alternatives for an equitable growth and sustainable development*. Working paper presented at the FAO-IFAD-ILO Workshop on Gaps, trends and current research in gender dimensions of agricultural and rural employment: differentiated pathways out of poverty. Available online at www. fao-ilo.org/fileadmin/user_upload/fao_ilo/pdf/Papers/24_March/Song_et_al._-_formatted.pdf (accessed June 1, 2016).

UNDP (United Nations Development Program). (2003) *China's accession to WTO: Challenges for women in the agricultural and industrial sectors. Overall report.* Available online at www.cn.undp.org/content/china/en/home/library/democratic_governance/china-s-ascension-to-wto--challenges-for-women-in-the-agricultur. html (accessed June 1, 2016).

United Nations Task Team on Social Dimensions of Climate Change. (2011) *The social dimensions of climate change.* Available online at www.who.int/globalchange/mediacentre/events/2011/social-dimensions-of-climate-change.pdf (accessed September 10, 2016).

Wang, H. (2011) Nvxing zai huanjing zhili zhong de "quexi" yu "zaichang" [Women in environmental governance: To be absent or to be present). *Huazhong kejidaxue xuebao (shehuikexue ban). [Journal of Huazhong University of Science and Technology (Social Science Edition)]*, 2: 63–69.

WB (World Bank) and ADB (Asian Development Bank). (2006) *China: Research report on gender gaps and poverty reduction.* Available online at http://documents. worldbank.org/curated/en/2006/10/9334773/china-research-report-gender-gaps-poverty-reduction (accessed May 1, 2016).

World Economic Forum. (2015) *Global gender gap report 2015.* Available online at http://reports.weforum.org/global-gender-gap-report-2015/economies/#economy=CHN (accessed February 20, 2016).

Xu, N. and Liu, J. (2014) China faces a unique challenge in adapting to climate change. *China dialogue.* Available online at www.chinadialogue.net/article/show/single/en/6809-China-faces-a-unique-challenge-in-adapting-to-climate-change (accessed February 2, 2016).

Zeng, B. (2014) Women's political participation in China: Improved or not? *Journal of International Women's Studies*, 15(1): 136–150.

Zhang, H., Bago d'Uva, T. and van Doorslaer, E. (2015) The gender health gap in China: A decomposition analysis. *Economics & Human Biology*, 18: 13–26.

Zhou, Y., Li, N., Wu, W. and Wu, J. (2014) Assessment of provincial social vulnerability to natural disasters in China. *Natural Hazards*, 71(3): 2165–2186.

Zuo, J. (2004) Feminization of agriculture, relational exchange, and perceived fairness in China: A case in Guangxi province. *Rural Sociology*, 69(4): 510–531.

11 Revealing the patriarchal sides of climate change adaptation through intersectionality

A case study from Nicaragua

Noémi Gonda

Introduction

Nicaragua is the third most climate change-affected country in the world (Harmeling and Eickstein, 2012) and its government identifies climate change adaptation as one of its key priorities (Campos Cubas et al., 2012; IPCC, 2007). Since the early 2010s, this national priority is translated into measures that support rural populations to adapt to climate change impacts. The introduction of climate change as a national priority has occurred in a context in which the Nicaraguan Sandinista government (ruling since 2007) promotes a discourse on the environment that denounces the destructive character of neoliberalism, and calls for a post-neoliberal[1] era in which humans live in harmony with "Mother Earth" (Nicaraguan Government, 2010, 2012; Houtard, 2011).

This stance, reflected in Nicaraguan policy documents (e.g. Nicaraguan Government, 2010, 2012), as well as in official declarations on climate change made by governmental institution workers, or in the written media (e.g. Wilder Pérez, 2012; Martínez and Rodríguez, 2015), has two main gendered features. First, it feminizes Nature by attributing to the latter traits that are recognized as feminine, such as vulnerability and the capacity to give life. Second, it naturalizes women. Indeed, most environmental and climate change policies and measures in post-neoliberal Nicaragua encourage the participation of women in environmental management and give them the priority to contribute to environmental protection-related actions. This prioritization is justified with the belief that women have a natural connection to nature, and therefore that they are especially apt to fight environmental degradation and climate change.

The National Environmental and Climate Change Strategy for the 2010–2015 period (Nicaraguan Government, 2010) is illustrative of the feminization of Nature and the naturalization of women in the Nicaraguan policy discourse. The strategy considers that the earth is to be "loved, respected, protected as our own mother" (Nicaraguan Government, 2010: 3). The word 'mother' is mentioned 21 times in the 27 page-long document, most often as "Mother Earth" with initial capital letters. Women are referred to in the strategy through the roles to which they are traditionally attributed in Nicaraguan society: (environmental) education, water management, fuelwood provision and the use of medicinal plants. The link between the necessary environmental education and women becomes evident as the strategy explains that the goal of environmental education is "life" itself, as if women were not only giving birth to children but also environmental consciousness.

In addition, gender-related measures and policies implemented since the beginning of the post-neoliberal regime place Nicaragua sixth in terms of the Global Gender Gap Index and fourth concerning the political participation of women, out of 142 countries (World Economic Forum, 2014). Since the beginning of the post-neoliberal era in 2007, development, environmental and climate change policy discourses and interventions give special visibility to women. Programs targeted at rural and urban women living in poverty have been implemented since then: access to micro-credit and agricultural assistance to cite just two of them.

However, together with feminist scholars working in Nicaragua (e.g. Cupples 2004; Bradshaw et al., 2008; Kampwirth, 2008; Lacombe, 2014), I suggest that these policies and measures, while they integrate gender considerations, do not reflect a feminist perspective. The prestigious ranking of Nicaragua in terms of gender equality hides a context in which women's participation in decision-making spaces is not always the result of their empowerment, nor does it contribute to empowerment in the majority of the cases.

In this chapter I argue that despite the discursive inclusion of gender dimensions and the promotion of the participation of women, Nicaraguan climate change politics contribute to reproducing patriarchy as well as other types of oppressions. They do so by essentializing women, reinforcing hegemonic masculinities, consigning climate change adaptation among the bulk of women's reproductive roles, blaming smallholder farmers for climate change, and excluding indigenous knowledge on environmental changes.

Feminist scholars who have engaged with climate change have seldom paid attention to climate change politics and interventions that already integrate concerns for gender.[2] They have mostly called for attention to gender in the predominantly gender-blind climate change debate (e.g. Nelson et al., 2002; Lambrou and Piana, 2006), or to the analysis of the gendered exclusions climate change may aggravate (e.g. Ahmed and Fajber, 2009; Alston, 2011; Buechler, 2009; Denton, 2002; Cannon, 2002; Tschakert, 2012). Some have drawn attention to the reproduction of hegemonic gender stereotypes through masculinist and science-oriented climate change discourses (e.g. Arora-Jonsson, 2011; MacGregor, 2010), or have written about how hegemonic gender identities are reinforced in post-disaster reconstruction work (e.g. Bradshaw, 2002). I argue that the insufficient focus by feminist scholars and practitioners on climate change politics that already integrate gender concerns is a deficiency that needs to be urgently addressed.

In this chapter, I first discuss the necessity for a systematic engagement of feminism with gendered climate change politics to unveil the reproduction of patriarchy and other oppressive forces. In the two sections that follow, I illustrate through the mobilization of an intersectional lens how gendered, class, ethnicity and age-based exclusions are reproduced through climate change politics in Nicaragua. In the conclusion, I strengthen my call for feminist scholars and practitioners to engage more systematically with gendered climate change politics.

Feminists need to engage with gendered climate change politics

Feminism seldom talks to environmentalism and vice-versa (MacGregor, 2010), an argument already sufficient *per se* to call for more feminist studies of climate change. The feminist perspective unveils the obstacles that impede transforming unequal power relations. In the Nicaraguan climate change discourse, women end up bearing most responsibilities in the fight against climate change, thus making this gendering an oppressive process that constructs women with particular traits, considered as immutable. A feminist approach to climate change invites a systematic analysis of this type of pernicious oppressive processes. My endeavour is a feminist critique that serves the interests of both gender justice and climate justice by unveiling some of the subtle forms of patriarchy and other oppressions inherent in climate change politics that encourage the participation of women without being conceived from a feminist perspective.

Such an endeavour does not only mean 'tracking' patriarchy, understood as a particular form of gender relations in which masculine dominance is constantly reinforced (Ford and Gregson, 1986). Following post-structural feminist political ecologists (e.g. Ge et al., 2011; Nightingale, 2011; Tschakert, 2012; Mollett and Faria, 2013), I argue that in order to understand how gendered climate change politics reproduce gendered and other types of oppressions, the differences and diversity among the ways in which people experience and react to environmental changes is crucial. This is a claim for an intersectional perspective: indeed, "people are not just men and women with culturally defined roles, but inhabit multiple and fragmented identities that intersect with class, race, ethnicity, sexuality, etc." (Elmhirst, 2011; in Tschakert, 2012: 149). These identities all contribute to shaping the way they experience climate change, their capacities to adapt or resist the imposed adaptation measures. The differences emerge and are produced out of everyday practices (Nightingale, 2011) such as the implementation of adaptation measures in farming, the attendance of training workshops on climate change, as well as the discourses conveyed by climate change projects, the media and local leaders, among other things. They are both symbolic (for example when women are discursively constructed as saviours or when young people are decreed community leaders in climate change adaptation) and material (like the droughts from which rural inhabitants of specific territories suffer). As feminist political ecologist Andrea J. Nightingale stresses, the operation of intersectional power is continuous with, sometimes unexpected, consequences on the creation of new subjectivities and bodies. It occurs in multiple dimensions under the interaction of a multiplicity of axes of privilege and oppressions (2011) such as gender, class, age, and ethnicity.

In this chapter, I am interested in understanding how intersectional identities of gender, class and ethnicity created through climate change politics produce new subjectivities and sometimes reinforce unequal power relations. The empirical data I use as background information was collected between 2013 and 2014 through participant observation and interviewing rural women and men in two rural communities of Nicaragua: El Nancite, in the municipality of Telpaneca, department of Madriz in the dry region, and El Pijibay, in the municipality of El Rama, in the Autonomous Region of the Southern Caribbean Coast in the humid region of the country.[3] Figure 11.1 shows the approximate geographical location of the two research sites.

A total of 72 rural women and men were interviewed about the changes they experienced in their lives. Interviews were also held with 18 climate change and development experts, as well as with ten feminist activists. I also analysed available climate change project and policy documents.

In the following sections, I first illustrate how Nicaraguan climate change politics reinforce 'traditional' gender roles and imaginaries by including

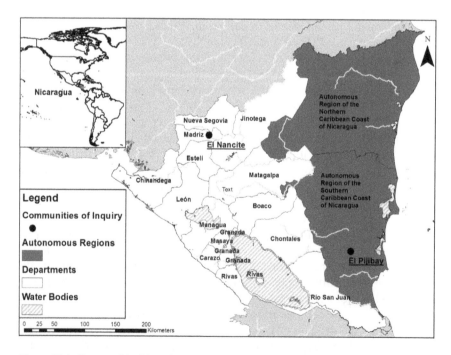

Figure 11.1 Geographical location of the research sites

climate change adaptation among the reproductive roles of women. Second, I show how class, ethnicity and age-based oppressions mark the politics of knowledge creation in the field of climate change in Nicaragua.

Gendered exclusions in gendered climate change politics

Women are given core responsibilities but unequal gender relations are not challenged

The Nicaraguan government makes significant efforts to integrate formerly excluded women into productive and environmental activities, both related to climate change adaptation. By giving them access to production means, this type of initiative reveals its conceptual reliance on the belief that women are more efficient in executing environmental adaptation. In post-neoliberal Nicaragua, this belief manifests itself in that the same women who under the neoliberal era (1990–2007) were encouraged to only implement 'secondary' activities through, for example, home garden production (Larracoechea Bohigas, 2011), become the direct beneficiaries of programs and are now called upon to implement activities considered as 'core'. An example of such a program is 'Hambre Cero' (Hunger Zero). It promotes inclusive rural development through directly and exclusively supporting rural women's access to means of agricultural production.

Doña Liliana[4] from El Nancite in the 'Dry Corridor' of the country, is 52 years-old, married, and the mother of five children, including a 26-year-old disabled son. She receives a cow and some chickens through the program, and explains:

DOÑA LILIANA: – This [program] is only for women, it is women, so we women have to go to attend the training, meetings and issues like this because it is for women, including the signature [of the contract], and everything is for women.

ME: – And what is your opinion about that? That it is for women only?

DOÑA LILIANA: – Ummm, well... I don't know, as they say it benefits women but as (...) the [project] technician said, well, the women receive [the benefit], but the husband is the one who has to look after it ...

ME: – He has to look after it?

DOÑA LILIANA: –Yes, because the men have to look after the cow. At least [in my case], I only went to receive it [from the project], but my poor son went to bring the cow [to the community], he is the one who looks after it, there, in the pastures.

(Interview, El Nancite, April 24, 2014)

Both Doña Liliana and the project technician think that at the end of the day it is men who benefit from the project, as they are the ones who are taking care of the cow. The fact that the cows rapidly pass on to be the responsibility of men is also due to the fact that it is not a gift, but a loan at very favorable rates that women have to pay back even though they seldom engage in cash generating activities themselves.

While the strategies to take care of the cow and pay back the debt are different from one woman to another, they often involve men and it is common for women to give a lot of importance to the cow, but not to talk about it as if it were their own animal. An illustrative example is a situation that happened when together with Doña Rosibel, a 48-year-old married woman, I crossed her family plot where four cows were grazing. I asked about the owners of the animals. Doña Rosibel answered that three of them were her husband's and one belonged to 'Hambre Cero'. This shows how these women, who are direct beneficiaries of these projects intended to serve both climate change adaptation and their empowerment, do not feel the benefits as their own. Women have difficulties assuming their new leadership roles in climate change adaptation projects because they are given responsibilities that continue to be considered and reproduced as typically masculine. Indeed, in the Nicaraguan social imaginary, cattle-ranching generates social differentiation, social status and power among men (Flores and Torres, 2012), something that is exceptionally visible in cattle-ranching communities, such as El Pijibay. For example, despite increasing NGO efforts to convince smallholder cattle-ranchers from El Pijibay to convert into cocoa producers as part of a climate change adaptation strategy, cattle-ranchers resist because cocoa production is seen as a far less masculine activity than cattle-ranching.

A feminist activist who mourns that there is no coherent opposition to the measures of the government she qualifies as "pseudo-feminist", explains that 'Hambre Cero' is illustrative of the governmental approach to gender in climate change adaptation policies: "women matter a lot to this government but [the government] takes much care not to disrupt the power relations between men and women" (Interview, Managua, December 14, 2014). Indeed, through governmental programs like 'Hambre Cero', the government institutionalizes the (pseudo-) feminist discourse and takes it to the climate change policy level while NGO efforts do not manage to challenge unequal gender relations either.

The post-feminist discourse increases the burdens of both poor rural women and men

Instead of helping progress towards gender equality, the leadership role given to women both in the discourse and in the programs enables a post-feminist discourse that renders feminism useless (Lazar, 2007). Indeed, post-feminist discourses usually build on achievements that relate to equality indicators, such as the ones of the Global Gender Gap Report (World Economic Forum, 2014), or the fact that women are positioned as having better aptitudes and are more keen on fighting environmental changes than men. Michelle M. Lazar, a feminist discourse analyst, warns about the dangers of the post-feminist discourse:

The discourse of popular postfeminism requires urgent need of critique, for it lulls one into thinking that struggles over the social transformation of the gender order have become defunct. The discourse is partly a reactionary masculinist backlash against the whittling away of the patriarchal dividend.

(2007: 154)

Such a critique is urgent in Nicaraguan society. As feminist scholar Karen Kampwirth warns: "the return to the left in Nicaragua does not look very left-wing, at least not from a feminist perspective" (Kampwirth, 2008: 132).

By institutionalizing the (pseudo-)feminist discourse and taking it to the climate change policy level, the impression that the government wants to give is that the structural conditions are created for both climate change adaptation and gender equality to be reached. When addressing climate change, women are called on to participate both in decision-making and climate change adaptation because they are seen as the ones who have the best understanding of what should be done. This relates to a typically problematic assumption of the post-feminist discourse: that women, because they are given the possibilities to participate, only need to try hard enough (Lazar, 2007) to achieve in this case, both climate change adaptation and gender equality. This assumption that they can fully exercise their personal freedom results in the fact that it "obscures the social and material constraints faced by different groups of women" (Lazar, 2007: 154) in the face of climate change. This discourse is one of the reasons why for feminism in Nicaragua, the panorama is dark.

The post-feminist discourse is also contributing to increasing particular men's vulnerability when the needs of smallholder male farmers are also excluded on the basis of gender. This is best illustrated with the case of Don Leandro from El Nancite, a 56-year-old widower who is raising his three grandsons alone, the oldest of whom is 14 years old. Don Leandro, despite his important needs, cannot be part of programs like 'Hambre Cero' for the simple reason that he is not a woman. This illustrates how the promotion of the empowerment of women, in the way it is conceived by the government, can contribute to the gendered exclusion of this poor, old, widowed man who fulfils many of the gender roles typically assigned to women.

Rather than contributing to transforming unequal gender hierarchies and tackling hegemonic masculinities, 'traditionally' masculine activities (like raising a cow) offered to women are easily appropriated by men for the simple reason that they mostly serve the interests of men. Meanwhile, the social structures and the symbolic power that maintain hegemonic masculinities remain unchallenged. The discursive invisibilization of climate vulnerable men who implement activities that are traditionally constructed as feminine, also mirrors the non-engagement with the necessary transformation of patriarchy.

Reinforcement of traditional gender roles

In addition to their 'spiritual connection' to 'Mother Earth', another reason why women are considered the most apt to fight climate change relates to their material relations with the environment through fuel-wood and water fetching, two chores considered part of their reproductive roles. It is assumed that with increased fuel-wood and water scarcity linked with climate change, not only are they likely to suffer more, but they will also be more eager than men to implement measures alleviating the execution of these activities (Soares, 2006). While this argument may be true in some cases, it reveals an essentialist view of women tied to their traditional gender roles considered as immutable.

One 'concrete' solution for climate change adaptation with a gender perspective that climate change adaptation project practitioners I interviewed would mention is the construction of wood-saving cooking stoves. Indeed, wood is used for cooking in most rural communities in Nicaragua. With the justification that firewood fetching is part of women's 'traditional' gender roles, together with a concern for women's respiratory health and deforestation, the climate change adaptation project selected 26 women to be the direct beneficiaries of wood-saving cooking stoves in El Nancite. Interestingly, the discussions I had on this topic through my interviews with 12 women and 8 men of 16 different households showed that women are not predominantly in charge of fetching fuel-wood and that this role is changing. Table 11.1 shows a classification of the answers I got about who fetches wood for household needs as well as the perceived advantages of the wood-saving cooking stoves for the households that benefited from it.[5]

In all households but two, men were involved in fetching firewood. In 11 of the 16 households, men were predominantly in charge of this chore. The only household where a woman alone was responsible for gathering wood was that of Doña Leonor who lived on her own. One household that bought or exchanged fuel-wood for goods such as eggs or maize consisted of an elderly couple with limited mobility, two adult children with mental disabilities and a five-year-old granddaughter. Of the 14 households in which men were involved in firewood gathering, 12 received wood-saving stoves. This was seen as a clear benefit in all but one household consisting of a young couple and their two daughters. Before they were provided with a new cooking stove, they had built one from local materials such as clay and stones and their stove was as energy efficient, according to them, as the cement stove received from the project. But the project staff told the couple that in order to receive a project stove, they had to destroy the former stove, something that both husband and wife told me they regretted.

In the 11 cases in which men were predominantly involved in fuel-wood gathering, the stove benefited them. This situation is contrary to the project aim of reducing the time women spent on gathering wood. In this situation, the women would have benefited more had they been given a stove that

Table 11.1 Responsibility for fetching wood and use of improved cooking stoves in households of the community

Responsible for fetching wood in the household	Number of households	Number of households that received improved cooking stove	Number of households where reduction in use of wood is observed with new stove	Number of households where reduction in smoke is observed with new stove
Men only	7	4	3	2
Usually men, occasionally women and children (when men not available)	4	4	4	1
Men and women with children (alternating upon availability and needs)	3	3	3	1
Women only	1	0	0	0
Do not fetch wood: household members buy or exchange it for goods with people external to the household	1	1	1	1
Total	16	12	11	5

(Source: Individual interviews in El Nancite, January–December 2014)

emitted less smoke. Indeed, my participant observation showed that mostly women were in charge of cooking (with some exceptions among the younger generation, which are discussed further). However, in seven cases no smoke reduction was observed. For example, the plastic roof of the house (related to the extreme poverty of the households) was not suitable to install a chimney. In one case, the increase in the quantity of smoke related to the fact that the household decided to keep both their traditional stove and their new stove against the advice of the project.

While this illustration may not be quantitatively representative of the 42 households[6] of El Nancite, it shows that the construction of wood fetching as an exclusively female chore by the climate change projects, reinforces 'traditional' gender roles that are 'traditional' only in the view of the project. Interestingly, the drivers for the transformation of this 'traditional' role are linked with two apparently contradictory factors: first, they relate to the maintenance of the 'traditions', that is to the fact that local society generally attributes to women tasks around the household. While women stay at home, it is increasingly men who bring back wood, to be found further and further away, after a day of agricultural work. Concerning water, for similar scarcity-related reasons, there is an increased use of donkeys and horses to fetch water at the sources to be found further and further away compared to ten years ago. The fact that in the local society animals are generally the responsibility of men explains why men become increasingly involved in water fetching, something I also observed. These examples show that while gender roles are changing due to decreasing water and fuel-wood availability, the direction in which they are transforming is still influenced by hegemonic gender identities, which confine women in the house while they give men freedom and the responsibility for handling the animals. The second factor to influence how water and fuel-wood fetching are less the responsibility of women than they used to be relates to an observable change in the distribution of roles within young, heterosexual couples of El Nancite. The difference between the habits of older couples in which women are mostly in charge of fetching water and fuel-wood and young couples in which the woman and the man share this responsibility illustrates this tendency.

Class, age and ethnicity-based exclusions in the politics of knowledge creation

Class-related exclusions

For climate change practitioners in Nicaragua, reaching the inhabitants of rural communities with information on climate change is a real and urgent concern. The fact that a communication specialist was hired in 2014 by an international organization in order to design a communication strategy on climate change for the rural communities of the 'Dry Corridor' illustrates this preoccupation. Significant efforts are deployed in order to contribute to localizing global climate change so that inhabitants of rural communities become capable of seeing climate change as a local problem with local solutions. However, the making of such an immaterial and distant issue as climate change (Slocum, 2004) a concrete everyday environmental preoccupation requires generating information and knowledge on climate change that speaks to people in the specific Nicaraguan context. In Nicaragua, this process of translation is marked by a class bias that renders smallholder farmers culprits and makes invisible the larger drivers of environmental and climate injustices.

During my conversations with inhabitants of El Nancite, people I talked with would relate climate change to deforestation, which they considered they themselves had caused. For example, Doña Francisca, a 22-year-old married woman, mother of two, explains:

> [Climate change] refers to the sudden changes in temperatures that we and our ancestors have caused because (...) [our] deterioration of land, tree felling, burning. All this affects us, all this brings us a change.
> (Interview Doña Francisca, El Nancite, August 12, 2014)

The idea of smallholder farmers' guilt is something that is widely present in my interviews about environmental changes. It mirrors a message presented in one of the radio programs, elaborated by a climate change adaptation project in order to translate and spread information on climate change. The program, made for the rural audience of the 'Dry Corridor', speaks directly to smallholder farmers through the figure of Aniceto Prieto, a funny, simple-minded but very popular Nicaraguan cabaret figure. The program is a conversation between Aniceto and his wife Lupita, in which Lupita explains that, among several technical and organizational adaptive solutions they are urged to adopt, smallholder farmers need to stop deforesting.

ANICETO: – Lupita, I'm worried.

LUPITA: – Why Aniceto?

ANICETO: – Goodness, in the first harvest we lost the maize production because of the lack of water, and in the second planting season, we lost the beans due to the heavy rains.

LUPITA: – It is climate change, Aniceto! That's why we have to adapt!

ANICETO: – And how do we have to adapt, Lupita?

LUPITA: – It's simple! By harvesting rainwater. By saving water. We don't have to burn or cut the trees. Rather we have to reforest. We have to protect rivers, we have to do diversification on the farm. In addition, we have to learn about adaptation and organize to adapt together with all men and women.

ANICETO: – Of course we do, my sweetheart!

The intention of building the message on the visible loss of the forest cover in the communities might be good. However, it blames smallholders and detracts attention from larger-scale phenomena including the structural reasons behind the drivers of deforestation. For example, the cattle-rancher who owns the largest herd and the biggest extension of land in El Nancite is the mayor of a municipality in the vicinity. Meanwhile, most inhabitants of El Nancite own very small plots or have no land and practice subsistence-level agriculture on a territory that has been degraded by the unsustainable exploitation of forest resources (Monachon and Gonda, 2011). These smallholders are the workforce for the large cattle-ranchers of the region and also the ones who clear land for them. For example, when in need of increased amounts of pastures for their cows, it happens that local cattle-ranchers give access to small-scale farmers or landless peasants to clear the remaining forested areas of their lands. These small-scale farmers and landless peasants gain the right to cultivate by deforesting for the cattle-rancher for two or three consecutive production cycles before they give it back as a planted pasture to its owner. This practice is widely known and mentioned in my interviews both by the farmers and the institutions. However, none of them think that it is possible to take any action in relation to this unsustainable practice, or at least not in the frame of climate change projects, which they present as apolitical and not responsible for remedying social and environmental inequalities. In the name of thoroughly imagined and naturalized women as well as a feminized Nature, climate change politics in Nicaragua erase political differences, among them the ones related to gender and class.

Ethnicity and age-based exclusions

Gender, discussed in the previous section, and class, discussed above, are not the only factors underlying exclusions in climate change politics. In particular, the politics of knowledge 'integration' on climate change is also marked by ethnicity-based oppressions as the process of knowledge 'integration' undervalues indigenous knowledge. An NGO worker explained to me the concept of 'harmonization between scientific and indigenous knowledge in climate change adaptation' that an NGO implemented between 2009 and 2011 in the 'Dry Corridor', in the following words:

> both knowledge that comes from outside [the community] and their knowledge from their own [indigenous] cosmovision as well as their wisdom can be harmonized with the ultimate goal of … well… a sustainable end.
> (Interview, Somoto, May 13, 2014)

When I asked members of the current Indigenous Government of Telpaneca (Collective interview, Telpaneca, April 28, 2014) how they recalled this 'harmonization', they told me that the efforts were limited to the inclusion of indigenous ceremonies at the beginning of project activities, together with an increased interest in recuperating ancestral knowledge on the use of medicinal plants (often with the participation of the elderly). The process was limited to knowledge that did not have the potential to compete with knowledge considered to be scientific. The approach also represents the essentialization of indigenous people and their knowledge. Indeed, the scientific approach to climate change neither includes religious aspects, nor does it prioritize discussions on medicinal plants. For all other types of knowledge, such as to evaluate possible productive strategies, the project put forward scientific and technological solutions instead.

Conversely, the 'indigenous resources' of a community for climate change adaptation, as established by the project, would include the socio-environmental advantages of the community rather than their knowledge systems. The project classified these 'indigenous resources' in soil, water, forest, human, food and agricultural resources. All the 'resources' were a list of what the community 'owned' such as rivers, forests, and the types of trained people in the community. The document did not list the significance and symbolic values of these. For the human resources, it recounted the number of shoemakers or health workers, while no mention was made of the elders, indigenous leaders, or the traditional healers. This shows how indigenous resources were valued only when knowledge was presented in a way that makes sense for external non-indigenous people.

This undervaluation of ancestral knowledge on climate change adaptation is also visible in the approach of the already discussed climate change adaptation project. The project organized a group of climate change promoters selected among the young people of the communities. During a training session, the project technician spelled out the reasons why young people were chosen: "You were chosen because you have a higher level of knowledge than … let's say … the producers" (Project facilitator, Telpaneca, May 29, 2014). The higher knowledge the project facilitator mentioned relates to the fact that the young people are more literate in the communities than older people. This view reflects how the effort of the integration of knowledge systems is asymmetrical: young people from the communities are trained to 'understand' climate change from the perspective of the climate change adaptation interventions, and then more highly valued by the system, while the local perspective based on ancestral knowledge held mainly by the elderly is seldom listened to. Meanwhile, only knowledge with important "folkloric" character, such as religious ceremonies and traditional healing practices, are made visible and acquire the attribute of indigenous knowledge in climate change.

While the gender, class, ethnicity and age biases generated by climate change adaptation projects might be qualified as "unintended" side effects, as James Ferguson who studied similar issues in development projects states, they still remain "at one and the same time instruments of what 'turns out' to be an exercise of power" (1994: 256). Just as development discourse in the work of Ferguson, the climate change discourse and the knowledge-generating practices of the projects in Nicaragua obscure the political dimensions of environmental degradation and climate change in general, as well as deforestation in particular, thus facilitating the expansion of the dominant non-indigenous, masculinist, and scientific knowledge and practice.

Conclusion

In this chapter I have shown that the discursive construction of nature as 'our own Mother' in post-neoliberal Nicaragua has contributed to giving women a primary place in climate change discourses and projects while the mainstream masculinist and science-oriented discourse is underlying the way climate change adaptation interventions are conceived. Whereas this paradigmatic change in comparison to previous neoliberal approaches has opened the floor to better include multi-dimensional (gendered) inequalities in climate change adaptation discourse and practice, the possibility has not been used for feminist purposes, which would require a system transformation. Post-neoliberal climate change politics essentialize and instrumentalize women by implementing policies that highlight women's 'natural connectedness to nature' and by reinforcing existing 'traditional' gender roles. This case study shows how gendered climate change politics, by giving a leadership role to women in climate change adaptation, can render invisible material and symbolic gender inequalities. While they intend to address both women's strategic needs (illustrated by the intentions of opening up participation spaces for women) and their practical necessities (shown by the intention of giving them access to means of production such as cows), climate change adaptation interventions in Nicaragua do not consider the necessity of system transformation, among other reasons because the current state still fits the interests of the most powerful minorities (the cattle-ranchers). These interventions encourage the return to a 'normal' state, understood as the one that eliminates the *additional* stressor of climate change, but do not question the multidimensional and relational stressors that (re)create (climate) vulnerabilities and that encourage the existence of current (sometimes unsustainable) practices. Climate change adaptation politics should be seen as a space for transformation where the persistence and reproduction of unequal power relations, value systems and institutions can be challenged (Pelling et al., 2014). In particular, climate change and feminist scholars, policymakers, and environmental and feminists activists should shift their focus from individual agricultural and climate change adaptation practices, and analyse and address better the (gendered) transformations needed in the social, political, economic and cultural contexts in which these practices emerge.

The Nicaraguan case has also highlighted a class bias in the politics of knowledge creation on climate change, through the pervasiveness of the discourse that constructs smallholder farmers as culprits. It has also shown how the process of knowledge integration between scientific and indigenous knowledge can undervalue the latter. I argue that feminist scholars and practitioners need to engage more systematically with gendered climate change politics, in particular by mobilizing the intersectional perspective that simultaneously addresses the multifaceted oppressions climate change politics may reproduce, even though they include 'gender concerns'.

Notes

1 In this chapter, I do not enter in conceptual discussions about post-neoliberalism. I understand post-neoliberalism as related to a rupture from a market-driven form of governance towards a new perspective on social, political and economic transformations (Simon-Kumar, 2011; Brand and Sekler, 2009). The Nicaraguan Government qualifies its regime as post-neoliberal. My intention is not to discuss what neoliberalism is, but to highlight some of the gendered place-based effects of the Nicaraguan post-neoliberal discourse on climate change.
2 One exception is the FAO report written by Yianna Lambrou and Sibyl Nelson. The report highlights how gender biases in humanitarian interventions can undermine rural women and men's adaptive capacities (Lambrou and Piana, 2006; discussed in Arora-Jonsson, 2011).
3 The names of the communities have been changed.
4 All names have been changed.
5 The analysis of the data presented in Table 11.1 is taken from an article (Gonda, 2016)
6 According to the 2014 community census.

References

Ahmed, Sara, and Elizabeth Fajber. (2009) "Engendering Adaptation to Climate Variability in Gujarat, India." *Gender & Development* 17 (1): 33–50.

Alston, Margaret. (2011) "Gender and Climate Change in Australia." *Journal of Sociology*, 47(1): 53–70.

Arora-Jonsson, Seema. (2011) "Virtue and Vulnerability: Discourses on Women, Gender and Climate Change." *Global Environmental Change* 21 (2): 744–51.

Bradshaw, Sarah. (2002) "Exploring the Gender Dimensions of Reconstruction Processes Post-Hurricane Mitch." *Journal of International Development* 14 (6): 871–79.

Bradshaw, Sarah, Ana Criquillion, Vilma Castillo A., and Goya Wilson. (2008) "Talking Rights or What Is Right? Understandings and Strategies around Sexual, Reproductive and Abortion Rights in Nicaragua." In *Gender, Rights and Development. A Global Sourcebook*, edited by Maitrayee Mukhopadhyay and Shamim Meer, 57–68. Amsterdam: Royal Tropical Institute (KIT) Publishers.

Brand, Ulrich, and Nikola Sekler. (2009) "Postneoliberalism: Catch-All Word or Valuable Analytical and Political Concept? – Aims of a Beginning Debate." *Development Dialogue* 51 (January 2009): 5–14.

Buechler, Stephanie. (2009) "Gender, Water, and Climate Change in Sonora, Mexico: Implications for Policies and Programmes on Agricultural Income-Generation." *Gender & Development* 17 (1): 51–66.

Campos Cubas, Victor M., Maura Madriz Paladino, Mónica López Baltodano, Iris Valle Miranda, and William Montiel Fernández. (2012) "Mapeo de Riesgos, Procesos, Políticas Públicas Y Actores Asociados Al Cambio Climático En Nicaragua [Mapping of Climate Change Risks, Processes, Public Policies and Associated Actors in Nicaragua]." Managua, Nicaragua: Centro Humboldt.

Cannon, Terry. (2002) "Gender and Climate Hazards in Bangladesh." *Gender & Development* 10 (2): 45–50.

Cupples, Julie. (2004) "Rural Development in El Hatillo, Nicaragua: Gender, Neoliberalism and Environmental Risk." *Singapore Journal of Tropical Geography* 25 (3): 343–57.

Denton, Fatma. (2002) "Climate Change Vulnerability, Impacts, and Adaptation: Why Does Gender Matter?" *Gender & Development* 10 (2): 10–20.

Elmhirst, Rebecca. (2011) "Introducing New Feminist Political Ecologies." *Geoforum* 42 (2): 129–32.

Ferguson, James. (1994) *The Anti-Politics Machine: "Development", Depoliticization, and Bureaucratic Power in Lesotho.* Minneapolis: University of Minnesota Press.

Flores, Selmira, and Sylvia Torres. (2012) "Ganaderas en la producción de leche: una realidad oculta por el imaginario social en dos zonas de Nicaragua." *Encuentro* 92: 7–28.

Ford, Jo, and Nicky Gregson. (1986) "Patriarchy: Towards a Reconceptualization." *Antipode* 18 (2): 186–211.

Ge, Jinghua, Bernadette P. Resurrección, and Rebecca Elmhirst. (2011) "Return Migration and the Reiteration of Gender Norms in Water Management Politics: Insights from a Chinese Village." *Geoforum* 42 (2): 133–42.

Gonda, Noémi. (2016) "Climate Change, 'Technology' and Gender: 'Adapting Women' to Climate Change with Cooking Stoves and Water Reservoirs." *Gender, Technology and Development* 20 (2): 149–68.

Harmeling, Sven, and David Eickstein. (2012) "Global Climate Risk Index 2013. Who Suffers Most from Extreme Weather Events? Weather-Related Loss Events in 2011 and 1992 to 2011." Briefing Paper. Bonn, Germany: Germanwatch.

Houtard, François. (2011) *From 'Common Goods' to the 'Common Good of Humanity.'* Brussels: Rosa Luxembourg Foundation.

IPCC. (2007) *Fourth Assessment Report of the Intergovernmental Panel on Climate Change.* Cambridge, New York: Cambridge University Press.

Kampwirth, Karen. (2008) "Abortion, Antifeminism, and the Return of Daniel Ortega: In Nicaragua, Leftist Politics?" *Latin American Perspectives* 35 (6): 122–36.

Lacombe, Delphine. (2014) "Les Données Trompeuses Du 'Global Gender Gap Report' [The Misleading Data of the 'Global Gender Gap Report']." Mediapart. fr. Available online at https://blogs.mediapart.fr/delphine-lacombe/blog/041114/les-donnees-trompeuses-du-global-gender-gap-report (accessed February 18, 2015).

Lambrou, Yianna, and Grazia Piana. (2006) *Gender: The Missing Component of the Response to Climate Change.* Rome: Food and Agriculture Organization of the United Nations (FAO).

Larracoechea Bohigas, Edurne. (2011) "¿Ciudadanía Cero? El 'Hambre Cero' y el Empoderamiento de las Mujeres. Los casos de Matiguás, Muy Muy y Río Blanco [Zero Citizenship? 'Hunger Zero' and Women's Empowerment. The Cases of Matiguás, Muy Muy and Río Blanco]." Matagalpa, Nicaragua: Grupo Venancia.

Lazar, Michelle M. (2007) "Feminist Critical Discourse Analysis: Articulating a Feminist Discourse Praxis." *Critical Discourse Studies* 4 (2): 141–64.

MacGregor, Sherylin. (2010) "'Gender and Climate Change': From Impacts to Discourses." *Journal of the Indian Ocean Region* 6 (2): 223–38.

Martínez, Saúl, and Melvin Rodríguez. (2015) "Árboles Que Dan Vida." *La Prensa*, September 19. Available online at www.laprensa.com.ni/2015/09/19/departamentales/1904418-arboles-que-dan-vida (accessed September 21, 2015).

Mollett, Sharlene, and Caroline Faria. (2013) "Messing with Gender in Feminist Political Ecology." *Risky Natures, Natures of Risk* 45: 116–25.

Monachon, David, and Noémi Gonda. (2011) "Liberalization of Ownership versus Indigenous Territories in the North of Nicaragua: The Case of the Chorotegas." CISEPA, AVSF, CIRAD, ILC.

Nelson, Valerie, Kate Meadows, Terry Cannon, John Morton, and Adrienne Martin. (2002) "Uncertain Predictions, Invisible Impacts, and the Need to Mainstream Gender in Climate Change Adaptations." *Gender & Development* 10 (2): 51–59.

Nicaraguan Government. (2010) "Estrategia Nacional Ambiental y del Cambio Climático. Plan de Acción 2010–2015 [National Environmental and Climate Change Strategy. 2010–2015 Action Plan]." Managua, Nicaragua: Nicaraguan Government.

Nicaraguan Government. (2012) "Plan Nacional de Desarrollo Humano 2012–2016 [National Human Development Plan 2012–2016]." Managua, Nicaragua: Nicaraguan Government.

Nightingale, Andrea J. (2011) "Bounding Difference: Intersectionality and the Material Production of Gender, Caste, Class and Environment in Nepal." *Geoforum* 42 (2): 153–62.

Pelling, Mark, Karen O'Brien, and David Matyas. (2014) "Adaptation and Transformation." *Climatic Change*, December 1–15.

Simon-Kumar, Rachel. (2011) "The Analytics of 'Gendering' the Post-Neoliberal State." *Social Politics: International Studies in Gender, State & Society*, August.

Available online at http://sp.oxfordjournals.org/content/early/2011/08/11/sp.jxr018. abstract. (accessed February 21, 2015).

Slocum, Rachel. (2004) "Polar Bears and Energy-Efficient Lightbulbs: Strategies to Bring Climate Change Home." *Environment and Planning D: Society and Space*, no. 22: 413–38.

Soares, Denise. (2006) "Género, Leña Y Sostenibilidad: El Caso de Una Comunidad de Los Altos de Chiapas [Gender, Wood and Sustainability: The Case of a Community in the Highlands of Chiapas)." *Revista Economía, Sociedad Y Territorio* 6 (21): 151–75.

Tschakert, Petra. (2012) "From Impacts to Embodied Experiences: Tracing Political Ecology in Climate Change Research." *Geografisk Tidsskrift-Danish Journal of Geography* 112 (2): 144–58.

Wilder Pérez R. (2012) "Nicaragua Ofrece Un Cambio Ambiental." *La Prensa*, September 27. Available online at www.laprensa.com.ni/2012/09/27/nacionales/117830-nicaragua-ofrece-un-cambio-ambiental. (accessed March 18, 2015).

World Economic Forum. (2014) "Global Gender Gap Report 2014." World Economic Forum.

12 Safeguarding gender in REDD+

Reflecting on Mexico's institutional (in)capacities

Beth A. Bee

Introduction

Over the last decade, Reducing Emissions from Deforestation, and Forest Degradation in developing countries (REDD+) has become one of the most controversial policy developments for mitigating climate change on the global stage. REDD+ and similar programs are often critiqued as a form of neoliberal environmental governance for its emphasis on cost-effectiveness, the commodification of nature, technocratic expertise, and individual behavior change as viable solutions to deforestation (Bakker, 2010; McAfee, 2012). Such concerns led to the development of broadly worded "safeguards" within the United Nations Framework Convention on Climate Change (UNFCCC), which are intended to establish procedures and approaches that promote transparency, inclusivity and conservation. In particular, countries are now required to develop a system for collecting and reporting on how they address and respect gender considerations in developing and implementing their national strategies and action plans (UNFCCC, 2011). Most recently, the 2015 Paris Accords failed to provide additional clarification, neither to how gender should be addressed and respected nor how it should be measured and reported. Consequently, the ways in which countries might demonstrate their compliance with the safeguards and other agreements is open for interpretation.

Mexico is the first Latin American country to develop a national REDD+ strategy and is considered to be among the more advanced and ambitious countries in their promotion and support of REDD+ (Hall, 2012). Furthermore, they are hailed as a pioneer in their incorporation of gender into the process (Aguilar, 2015). Indeed, the attention to gender issues in Mexico's Intended Nationally Determined Contribution (INDC) for the 17th United Nations Framework Convention on Climate Change-Conference of Parties (UNFCCC-COP) meeting in Paris, which was the first INDC to be submitted, was highlighted by international gender-specialists as a potential model for other countries to follow. The document explicitly states that it was elaborated with a gender equality and human rights approach in mind (Mexico, 2015). However, this statement is at odds with the conclusion of the national feminist meeting, held in Guadalajara in the same year, which concluded that many of the national efforts to mainstream gender into policy neglect the very real problem of the state's role in the human rights violations, including femicides (personal communication with Verónica Vázquez García, meeting attendee).

This chapter is an attempt to better understand how Mexico currently frames gender and women's participation within existing early REDD+ programs, as part of their compliance with international safeguards, and the results of such framings on the ground. I argue, specifically, that these framings represent a paradoxical relationship between discourse and the ways gender equity takes place in practice. To do this, I begin with a brief overview of Mexico's experience with mainstreaming gender into environmental policy. I then explore the current framing of gender in the recently published National REDD+ Strategy which is a planning and public policy document that outlines Mexico's contribution to reducing emissions. I then showcase the results of a small study from an early action REDD+ program that illustrates the contradictory ways that promoting gender equity in early REDD+ actions happens on the ground. This chapter aims to pose important questions about the mainstreaming of gender in REDD+, that are important to consider for policymakers and practitioners hopeful that REDD+ can bring about simultaneous sustainable and social development. Furthermore, it contributes to a small but growing body of research on how gender considerations are being taken up by various pilot REDD+ projects (Das, 2011; Di Gregorio et al., 2013; Khadka et al., 2014; Larson et al., 2015; Peach Brown, 2011; Westholm and Arora-Jonsson, 2015).

Gender, development and REDD+

Issues of gender equity have been central to development contexts since the 1980s. In particular, gender and development scholars have been especially critical of the gendered effects of neoliberal economic restructuring and capitalist policies in the Global South. Chief among these debates is the issue of mainstreaming gender (see, e.g. Porter and Sweetman, 2012 for a recent review). In the context of climate change policies, gender sensitive policies that draw from these lessons attempt to unravel gender-differentiated impacts of climate change and design appropriate prevention and response strategies. However, such attempts run the risk of uncritically adding women and gender-sensitive language to projects that do not challenge the underlying structural causes of gender inequities (Walby, 2005; Woodford-Berger, 2004). Moving forward, the challenge remains for programs and policies to explicitly analyse gender inequities and their causal structures.

Particularly relevant for newly emerging policies and programs in Mexico to combat climate change, is the issue of meeting international requirements for gender sensitivity through the targeting of women, either for specific programs or for specific positions within the program, as I will demonstrate.

A critique of programs that target women has been leveled against micro-finance and similar neoliberal development strategies, as a strategy for program efficiency and a means to realize development goals, which may not necessarily be in the interests of women (Razavi and Miller, 1995). Several scholars argue that women become targets because they are a) considered to comprise the poorest of the poor, b) are assumed to have a greater interest in benefiting their families and communities than men and c) have higher repayment rates than men so they are considered safe investments from the perspective of financial institutions (Isserles, 2003; Joseph, 2002; Kabeer, 1994; Rankin, 2001). Furthermore, microfinance programs that target women are often based upon precarious assumptions about female solidarity and women as a homogenous group, which only serves to instrumentalize women within programs (Bee, 2011).

Looking beyond microfinance to environment and development projects, Resureccion and Elmhirst (2008) similarly suggest that targeting women fails to address structural constraints affecting women and the poor and treats women as a homogenous group, thereby completely bypassing the issues that create and maintain gender inequities. At the same time, the act of justifying why women should be targeted relies upon gender myths that simplify and essentialize women, men and the complex ways that gender relations are produced, enacted and negotiated. Furthermore, Melissa Leach (2007) argues that the success of such projects occurs at the expense of women's labor demands, obscures the interest of women not targeted for the project, and ignores issues of power and poverty. As Rebecca Elmhirst (1998) argues, defining women as "target groups" is often based on rigid conceptions of gender that do not necessarily reflect what are at times fluid and shifting categories of identity. She suggests that poststructuralist feminist theories of gender performativity are useful for understanding how gender is produced and enacted through livelihood practices. However, it remains a challenge to incorporate such theoretical conceptualizations of gender into practical strategies, either to conduct gender-sensitive multidisciplinary research, as was the case for Elmhirst (1998), or in development programs and policies.

Additionally, a common feature of emerging climate change policies and programs is an emphasis on individual responsibility and behavior change, which is often argued to be a common characteristic of neoliberal development strategies. REDD+, for example, is considered by many to be a neoliberal development strategy that promotes local-scale environmental entrepreneurialism and pro-poor development through hybrid combinations of states, donors and civil society. Concerns about the extent to which REDD+ can improve livelihoods raise questions about the ability of these programs to reconcile the tenuous relationship between economic efficiency and social equity in their design and practice (e.g. Pokorny et al., 2013). Furthermore, "pro-poor" development has been the source of critique as it emphasizes local solutions and strengthening entrepreneurialism among low-income individuals, further legitimizing neoliberal restructuring and development aid discourse (St. Clair, 2006). Drawing on feminist philosophies of science, Bee et al., (2015) argue that programs born of such neoliberal climate governance, while framed in terms of climate protection, actually work to extend capitalist free-market economies onto individual bodies and de-emphasize collective forms of action. Therefore, it is imperative for feminist scholars to inquire into the ways in which governance structures and programs, like REDD+, construct particular kinds of subjects and subjectivities, such as gender and women, and the everyday spaces and situated ways of being and knowing in which action and responsibility are negotiated and enacted under highly uneven power relations (Bee et al., 2015). As I hope to demonstrate, Mexico's current vision of gender equity within REDD+ reflects a paradoxical relationship between the discourse and the practice of equity.

Mainstreaming gender in environmental policy and programs in Mexico

Mexico officially began efforts to mainstream gender into its programs shortly after the Beijing Conference in 1995, and in 1999, published a declaration of Policies for Gender Equity (Vázquez García, 2014). From 2002–2006, the Secretary of the Environment and Natural Resources (SEMARNAT), developed their first Program of Gender Equity, Environment and Sustainability, which included four strategic objectives: 1) mainstreaming and institutionalization of gender perspectives; 2) internal and inter-institutional coordination; 3) co-responsibility and social participation; 4) international cooperation (SEMARNAT, 2003). In 2007, SEMARNAT launched a new program to continue mainstreaming gender, although the differences between the first and second programs are small (for a detailed elaboration of these differences, see Vázquez García, 2014). However, although language exists at the national level for promoting gender equity in environmental programs, putting this language to practice has had mixed results.

Since the incorporation of neoliberal reforms in Mexico in the 1980s, the responsibility for carrying out social development projects has largely fallen under the purview of non-governmental organizations. Consequentially, the work of promoting gender equity has primarily been carried out by such NGOs. Among environmental NGOs, three of these have been the subject of institutional ethnographies that examine how the organizations incorporate gender equity into their work. These studies demonstrate that a favorable discourse of gender equity exists within these NGOs, but that there are significant limitations in practice (Ortiz, 2004; Rocha et al., 2006; Venegas, 2004). For example, these studies highlight confusion between women's work and gender perspectives, program evaluations that don't capture attitudinal perspectives like empowerment, a lack of financial resources that result in overburdening the workload of the position of "gender specialist", as well as a gendered division of labor within the organization itself. In another study of an NGO that promotes the production and marketing of organic coffee, fruit and forest products for indigenous women (Muñoz et al., 2010; Muñoz and Vázquez, 2012), the authors found that capacity-building focused too much on production and too little on women's rights and empowerment. For example, one-third of the women were illiterate and were often underpaid and excluded from leadership positions yet such issues remained unnoticed and unaddressed by the organization. Additionally, the organization failed to address the concern among many women that the organization did nothing to recognize or address their work at home as emphasis was placed on work outside the home, thus simply adding to women's already extensive responsibilities (Muñoz et al., 2010; Muñoz and Vázquez, 2012).

As the majority of efforts to promote women's participation in environmental projects in Mexico have been oriented towards economically productive projects (e.g. organic coffee, commercialized non-timber forest products, etc.), women's participation in forest governance in Mexico remains limited and community forest management in Mexico remains a highly masculinized enterprise. A fundamental part of the problem, argues Vázquez García (2015), is the high representation of men as landholders and representatives of their families, with corresponding rights (as landholders) to vote in community meetings. For example, in the majority (57%) of communities in temperate forests, women represent less than 20 percent of agrarian subjects with property rights (Vázquez García, 2015). An underlying cause of the exclusion of women in community forestry, she argues, is the gender discrimination that exists in the access to and rights over land (Vázquez García, 2015). In many communities throughout Mexico, only people with official title to land are allowed to participate in community-level decision-making and the vast majority of these individuals are men. Individuals with official access to and rights over land are considered to have the right to "voice and vote" in their communities. Nationally, women who have these rights make-up only 13.5 percent of all "agrarian subjects" as defined by the state (INEGI, 2007). Agrarian subjects, according to the federal government, include three categories of subjects in rural communities: 1) titled landholders with rights ("voice and vote"); 2) landholders who do not have title nor rights to their land; and 3) landless community members. In addition, women who are titled landholders with rights are 58 years old on average, which creates a class of women who are also much older and in some instances, no longer able to farm (INEGI, 2007). In addition, the exclusion of women from the production and sale of timber and the uneven distribution of benefits to those who are directly involved in these activities negatively affect women (Vázquez García, 2015).

For example, a review of forestry programs in existence from 2003–2007 found that out of 296 beneficiaries that participated, 94 percent were men and 6 percent women, however no women in humid-tropical forests participated (Tchikoué, 2008). Roughly 50 percent of men and women involved in the study were between 40 and 60 years of age. When asked to qualify the things that the programs have had the greatest impact upon, forestry bureaucrats ranked the inclusion of women and indigenous groups as 60 and 66 respectively on the impact scale of 1 to 100 (100 having the most impact). The study also found that for every 10 men involved in forest management programs, only one woman is involved (Tchikoué, 2008). In ecotourism, however, women's participation was significantly higher: 37 percent. But this work is often based in a care economy that is already feminized (e.g. cleaning and caring for areas, cooking, etc.), while the forestry sector is highly masculinized. At the same time, the study found that for women who are directly involved in forest management projects, they are paid much less than their male counterparts for the same job (Tchikoué, 2008). Similarly, in a recent study of Payment for Environmental Services (PES), a conservation program offered through the National Forestry Commission or CONAFOR, Galdámez Figueroa et al., (2015) found that women and community members without land are the two most excluded groups in conservation. Community members who receive payments from the program include only those that have "voice and vote" in the community. Yet, as mentioned above, this landed class tends to be primarily men, and represents a smaller percentage of community members as a whole.

Gender and REDD+ in Mexico

Now, as Mexico finds itself in a position to comply with international accords regarding climate change and specifically, the safeguard requirements for REDD+, Mexico's National REDD+ Strategy (ENAREDD+ by its Spanish acronym) might be the first indication that gender mainstreaming is being taken seriously in the forest sector (Vázquez García, 2015a). The strategy explicitly states an interest in promoting gender equality and guaranteeing the participation of women and in particular, male migration as a response to the lack of incomes in rural area, which makes the role of women in rural environments ever more preponderant (ENAREDD+, 2014). As financing is a key component of REDD+, the document specifically mentions that financing should be equitable between men and women and encourage sustainable forest management (ENAREDD+, 2014). Moreover, the current emphasis on women's participation is greatest in the context of developing and implementing financial systems that will incentivize conservation efforts. For example, it suggests gender equality will guarantee the socio-economic benefits of such a financial system (ENAREDD+, 2014). More specifically, it emphasizes the necessity to design a system that supports gender equity by promoting access to credit and loans for women, although it does not provide any details as to what the loans will be for or how they will be made (ENAREDD+, 2014).

As property rights have been a central concern among local and international activists wary of REDD+ policies, the ENAREDD+ document also mentions land tenure and property rights, specifically those pertaining to carbon. However, although it frequently mentions, "land owners", it very rarely acknowledges *forest users*, the majority of whom do not have fully realized property rights and are comprised predominantly of women and young community residents. Because of this, an earlier Gender Action Plan for REDD+ (PaGeREDD+, by its Spanish acronym), intended to guide practitioners on incorporating gender into REDD+ and drafted with the assistance of international gender experts, identifies property rights as a fundamental issue for REDD+ and for addressing gender equity (Aguilar y Castañeda, 2013). They propose acknowledging the rights of women as forest users regardless of whether they have property rights or not, and to design legal tools and economic incentives that increase women's property rights (Aguilar and Castañeda, 2013). However, such recommendations require constitutional reforms that are not impossible to achieve but require a broad interest and mobilization on the part of rural women and other landless groups. This would not be the first, nor will it be the last fruitless proposal because of a lack of working *with* instead of *on behalf of* rural women. Other efforts to increase women's property rights have had little success. As other feminist scholars suggest, urban women from elite classes who are often at the helm of such proposals do not necessarily represent the interests of rural women and so proposals developed *on behalf* of diverse groups of women, in lieu of *with* these women, stand little chance of success (Banana et al., 2012, emphasis mine).

The process for drafting the document began in 2008 in consultation with a technical committee, comprised of government and non-governmental organizations, academic institutions and civil society representatives. The incorporation of gendered language and gendered perspectives into the document was due to consultation with international gender specialists at the International Union of the Conservation of Nature (IUCN). Although the IUCN has a strong history of providing guidance and consulting on national gender policies worldwide, their presence does not necessarily represent national feminist interests. At the same time, the ENAREDD+ document also mentions an interest in including the National Institute for Women (INMUJERES, by its Spanish acronym), although they are not yet involved and it is not clear when they will become involved.

Although INMUJERES' institutional mandate is to promote gender equity in all areas of society (Vázquez García, 2015a), it is a national-level government agency that represents the interests of the state, as opposed to a civil society organization representing the interests of the people. Furthermore it has been the centre of critiques by Mexican feminists over the past decade for its lack of transparency and tangible support for women's human rights. Most recently, it was revealed that the director of INMUJERES fired 73 officials who did not align with her ideas or ways of working, has used public money to pay friends, and has blocked an initiative to save women's lives (Tourliere, 2016). Additionally, last September, the United Nations Special Rapporteur for the Committee for the Elimination of Discrimination against Women (CEDAW), Zou Xiaoqiao, determined that Mexico had not complied with five recommendations from the committee since 2012 (Tourliere, 2016).

Perhaps more importantly, to date, there are *no* other civil society groups with an explicit interest in gender equity involved in the ENAREDD+ process. While there are several civil society groups who have been involved, and others who have resigned their posts due to disagreements over policy and protections for forest users, none of these groups have an explicit gender focus in their work. Furthermore, according to some feminist scholars who have attempted to participate in workshops for mainstreaming gender into REDD+, there have been few opportunities to ask critical questions, for example, about whether or not REDD+ is even appropriate for communities. Such questions have been quickly silenced as such questions clearly contradict the priorities of the state (personal communication). For many, this is a serious shortcoming in the ENAREDD+ process as well as a contradiction to the expressed interest in promoting gender equity.

This represents a significant concern for feminist scholars who have been critical of the lack of human rights protections by the state and the lack of local civil society actors in shaping national policy. Furthermore, the details of the national plan were not presented formally to women from forest communities until 2014, after the draft for public consultation was released, in three regional forums organized by the National Commission on Forestry (CONAFOR, by its Spanish acronym) and the Secretary for the Environment and Natural Resources (SEMARNAT, by its Spanish acronym) (CONAFOR, 2015a). These forums were intended to provide the opportunity to understand how women are involved in diverse forest activities, such as collecting herbs for treating illness, conserving seeds, mangrove conservation, and payment for environmental services. However, women in forest communities are not necessarily interested or informed about the details of REDD+, despite the appearance of sophisticated institutional frameworks for gender equity (Vázquez García, 2015a).

In addition to the concerns over INMUJERES and the process for including women in the development of the ENAREDD+ document, SEMARNAT's budget was recently cut by one-third and 40 personnel who worked in the area of gender, sustainability, environmental education and indigenous issues were laid off (López Santiago, 2016). Activists denounce what they see to be a "double institutional standard" that the federal government manages, whereby the President's administration boasts advances of material with respect to gender and environment in Mexico during international summits, while at home, the administration works to dismantle the same structures built over the past 17 years, intended to promote gender equity (Tourliere, 2016). At the same time, activists deplore the "inert" attitude adopted by the director of INMUJER who neither spoke out nor acted to prevent the dismantling of such mechanisms that fortify the perspective of gender in environmental policy in the country (Tourliere, 2016). In particular, members of the Gender and Environment network sent an open letter to INMUJERES and to the President of the Republic to remind them that as a signer of the

1995 Beijing Platform, Mexico has an obligation to promote matters related to women and the environment (Tourliere, 2016).

As Vázquez García (2015a) argues, it is urgent to capture and systematize these and other issues and it is absolutely essential to do so before designing policies for reducing deforestation and degradation that try to have a focus on gender and other social differences. So, although the document is peppered with language about a desire to incorporate projects from a "gender perspective", it does not provide specifics as to what a gender perspective entails beyond specific mentions of the desire to increase women's participation and access to credit. Perhaps more problematically, the document also references constitutional rights as evidence of robust legal protections for human rights, land rights, and the like. However, scholars are critical of such references in a country, which has yet to resolve – narco violence, femicide, juvencidio (youth-icide, or the systematic killing of young people) – what many see as state-sponsored impunity that normalizes the rampant homicide of women and children, including the disappearance of 43 students in 2014 (Wright, 2013). As the national strategy sets out to strengthen the participation of women and "specific groups of attention" in CONAFOR projects, it is necessary to better understand how the institution and its actors already promote such participation and the contradictions that exist in its execution.

Promoting gender equity in early REDD+ programs

To understand how the gender considerations laid-out by ENAREDD+ will most likely be enacted on the ground, it's important to consider the ways in which gender equity is currently promoted within the existing early action program. CONAFOR, a sub-agency of SEMARNAT, first began incentivizing women's participation in the Payment for Ecosystem Service (PES) programs in 2010. The following year, CONAFOR reconfigured existing forestry and conservation programs (including PES) in 11 municipalities in the state of Jalisco as part of a broader umbrella program called "Programa Especial de Cuencas Costeras" (Special Programs for Coastal Watersheds). Financed by the World Bank, *Cuencas Costeras* was developed as a REDD+ Early Action strategy that Mexico implemented to comply with United Nations requirements. The program was intended as a pilot project to be replicated elsewhere and a first step towards consolidating programs within CONAFOR that meet the broad goals of REDD+. Since 2011, the program has expanded from 11 eligible municipalities to 35 in 2015. Early action programs now encompass five states in Mexico and CONAFOR hopes to replicate these programs in an additional three states by the end of 2016. Although these programs are not formally pilot projects, given that they do not currently incorporate carbon accounting, their role in establishing governance schemes, policy instruments and institutional arrangements will serve an important function in the national implementation of REDD+.

Currently, the program does not contain language about gender equity, but rather, includes mechanisms to incentivize the participation of women and young people for specific positions. The incentive comes from additional points awarded to applications that elect a "legal representative" who is a woman and/or between the ages of 18 and 25 (CONAFOR, 2015a). Applications can also garner additional points if the community soliciting funds have indigenous residents (CONAFOR, 2015a).

According to the CONAFOR rules of Operation (2015b), legal representatives need to provide evidence of their nomination by the community as well as evidence that the community is in agreement to solicit funds from CONAFOR. However, the only guidelines and/or requirements provided for the role of the legal representative are to attend a meeting on a given date to submit the application to CONAFOR and to be responsible for the signing of documents (CONAFOR, 2015b). As a result, the degree to which legal representatives *actually* participate in every aspect of the program vary greatly.

As I have documented elsewhere, electing female legal representatives is a common strategy among predominantly male forest technicians, who represent and assist communities, and communities themselves in order to gain additional points for their application (Bee, 2016). However, some technicians are unaware that the points system exists and therefore do not work with any communities who elected a female legal representative. The process for applying for CONAFOR programs is highly competitive as funds are limited, and the number of communities interested in funding grows each year. In addition, applications are often declined for missing or incorrect information by CONAFOR staff during the review process (personal communication). As a result, communities and technicians seek any advantage available to them to help the chances of having a successful application. As funding is awarded on a points system, any extra points a community can garner gives them an advantage over other applications.

In a previous study of legal representatives in three communities in Jalisco, I found that the strategy for electing women to be legal representatives simply for gaining extra points has paradoxical outcomes (Bee, 2016). In at least one instance, a community selected a young woman for the simple reason of gaining points (both for being a woman, and for someone between the ages of 18–35). The forest technician was very frank in his explanation that her only role is to sign paperwork and attend the CONAFOR meeting, when required, albeit always accompanied by her very protective father. Important to note is the fact that she was young, unmarried and part of the landless class of community members (although her father was a landholder with "voice and vote").

However, antithetical examples also exist where women assume positions of authority and leadership that are recognized and, indeed, supported by other members of their community (Bee, 2016). Yet, these opportunities seem to be only available to older women, who are part of the class of landholders with "voice and vote", and who are also already active members in their communities (Bee, 2016). Furthermore, and consistent with other studies (Arora-Jonsson, 2012; Mohanty, 2004; Sijapati-Basnett, 2008), the degree to which she is able to participate depends a great deal on the influence of supportive male community members and forest technicians who encourage a more active role for the legal representative (Bee, 2016). So, her active participation is made possible through the support of male community members and forest officials, as well as her age and class. This is significant for understanding the exclusionary politics of gendered power relations not only in local communities, but also in CONAFOR policy and the potential future for safeguarding gender in REDD+.

Discussion and considerations

How women are conceptualized as agents and participants in REDD+ policy and programs, how this conceptualization is enacted on the ground, and the contradictions and contestations that result are key issues in analysing the possibilities and pitfalls for promoting gender equity in REDD+. In the context of *Cuencas Costeras*, CONAFOR constructs women, indigenous members, and youth as instruments through which communities gain access to funding and programs. Women's participation can be enacted through their role as a "legal representative", which in very basic terms, translates into someone who signs documents. Both the instrumental construction of these gendered, racialized and generationally disparate subjects, and their enactment of this position are produced through the instrumentalization of gender and other categories of difference in REDD+ as something to be measured and verified. As Bee and Sijapati Basnett (2016) suggest, the lack of guidance for states to address gender equity in REDD+ as a requirement to comply with international safeguards raises concerns about the potential for gender to be treated as a bureaucratic obligation as it is currently, that merely serves to legitimize the safeguards.

The contradictory outcomes of promoting gender equality signify the conceptual and practical challenges Mexico faces for addressing gender within REDD+. For example, the interest in promoting gender equity by increasing access to credit and loans for women, as stated in the ENAREDD+ document, can be seen as a reduction of women's contribution to the anti-poverty work of REDD+ through their productive labor, while their reproductive labor and potential contributions to decision-making within the program remain ignored, as has been shown in other early REDD+ programs (Westholm and Arora-Jonsson, 2015). This framing ignores women's reproductive responsibilities and risks burdening their labor demands and perpetuating the already uneven division of labor between men and women. Furthermore, targeting of women for credit-related projects is often based on essentialist assumptions about women's ability to empower their community, or notions of female solidarity that construct women as economically rational and homogenous groups, which does little to address the underlying cause of gender inequities (Rankin, 2001; Bee, 2011).

The paradox between the discourse and the practice of gender equity in Mexico is something feminist activists have recently drawn attention to as funding cuts and continued impunity in cases of violence against women and children continue to plague efforts to promote gender equity in the country. Additionally, the lack of participation in the development of the ENAREDD+ framework on the part of local, gender-centred civil society groups can be viewed as a significant shortcoming, especially considering the expressed interest in promoting gender equity throughout the process. Although IUCN gender experts acted as consultants, these so-called "experts" do not necessarily represent local interests. And despite the fact that there is an expressed interest in involving INMUJERES, feminists have been critical of the ability of elite women's groups to effectively represent the interests of local women (Cornwall, 2003). Moreover, INMUJERES has been under fire for several years due to their inattention to women's human rights and what are alleged to be corrupt and unethical practices.

While the Mexican state continues to receive praise in international circles for its appropriation of gender equity discourse within the ENAREDD+ process, the institutional capacity to implement gender equity in REDD+ remains precarious and unstable. Yet there is an opportunity for women's organizations to become involved in shaping the process and the outcomes on the ground. What is clearly needed is careful consideration of the complexity of gender and design and adapt each initiative to specific needs and contexts. Addressing these challenges is critical if Mexico hopes to build a REDD+ strategy that can bring about environmental sustainability in socially equitable ways.

References

Aguilar, L. 2015. *Cómo le hicimos y no morimos en el intento: género en las convenciones ambientales de la ONU.* Presentation at the First Inernational Meeting of the Gender, Society and Environment Network, Merida, Mexico.

Aguilar, L. and Castañeda, I. 2013. *Plan de acción de género para redd+ México PAGeredd+*. IUCN: Switzerland.

Arora-Jonsson, S. 2012. *Gender, Development and Environmental Governance*. London: Routledge.

Bakker, K. 2010. The limits of 'neoliberal natures': Debating green neoliberalism. *Progress in Human Geography*, 34(6): 715–735.

Banana, A., Bukenya, M., Arinaitwe, E. and Birabwa, B. 2012. *Gender, tenure and community forests in Uganda*. (Working Paper No. 87). Bogor, Indonesia: CIFOR.

Bee, B., 2011. Gender, solidarity and the paradox of microfinance: reflections from Bolivia. *Gender, Place and Culture*, 18(01): 23–43.

Bee, B.A. 2016. La construcción de género en REDD+: Un estudio de caso en la Sierra Occidental de Jalisco, Mexico. En V. Vásquez García, M. Velázquez Gutiérrez, D. M. Sosa Capistrán, & A. De Luca (coords), *Transformaciones ambientales y igualdad de género en América Latina: Temas emergentes, estrategias y acciones*, 121–145. Cuernavaca, Mexico.

Bee, B.A. and Sijapati Basnett, B. 2016. Engendering social and environmental safeguards in REDD+: Lessons from feminist and development research. *Third World Quarterly,* 1–18.

Bee, B.A., Rice, J. and Trauger, A., 2015. A feminist approach to climate change governance: everyday and intimate politics. *Geography Compass,* 9(6): 339–350.

CONAFOR. 2015a. *Linamiento de operación para el Programa Especial de Áreas de Acción Temprana REDD+.* Consejo Nacional Forestal, Guadalajara: Mexico.

CONAFOR. 2015b. *Reglas de operación del Progama Nacional Forestal.* Consejo Nacional Forestal, Guadalajara: Mexico.

Cornwall, A. 2003. Whose voices? whose choices? reflections on gender and participatory development. *World Development,* 31(8): 1325–1342.

Das, N. 2011. Can gender-sensitive forestry programmes increase women's income? Lessons from a forest fringe community in an Indian province. *Rural Society,* 20(2): 160–173.

Di Gregorio, M., Brockhaus, M., Cronin, T., Muharrom, E., Santoso, L., Mardiah, S. and Büdenbender, M. (2013). Equity and REDD+ in the media: a comparative analysis of policy discourses. *Ecology and Society: a journal of integrative science for resilience and sustainability,* 18(2).

Elmhirst, R., 1998. Reconciling feminist theory and gendered resource management in Indonesia. *Area,* 30(3): 225–235.

ENAREDD+. 2014. Estrategia Nacional para REDD+ (ENAREDD+). CONAFOR, México.

Galdámez Figueroa, DY; Vázquez Garcia, V; Perezgrovas Garza, R; Fierro González, AM. 2015. Convervar, ¿cómo y para quién? Pago por servicios ambientales en Chiapas, Méxcio. In Press in *Maderas y Bosques.*

Hall, A. 2012. *Forests and climate change: The social dimensions of* REDD *in Latin America.* Cheltenham: Edward Elgar.

INEGI, 2007. Censo Ejidital 2007 Ejidatarios y Comuneros segun Sexo y Disposición de Parcela individual Aguascalientes. Institución Nacional de Estadistica, Geografía e Informática, Mexico.

Isserles, R.G. 2003. Microcredit: The rhetoric of 'empowerment', the reality of development as usual. *Women's Studies Quarterly* 313(4): 38–57.

Joseph, M. 2002. *Against the Romance of Community.* Minneapolis, MN: Minnesota Press

Kabeer, N. 1994. *Reversed realities: gender hierarchies in development thought.* London; New York: Verso.

Khadka, M., Karki, S., Karky, B. S., Kotru, R. and Darjee, K. B. 2014. Gender equality challenges to the REDD+ initiative in Nepal. *Mountain Research and Development,* 34(3): 197–207.

Larson, A.M., Dokken, A.E., Duchelle, A., Atmadjai, S., Resosudarmo, I. A.P, Cronkelton, P., Cromberg, P., Sunderlin, W., Awono, A. and Selaya, G. 2015. The Role of women in early REDD+ implementation: Lessons for future engagement. *International Forestry Research,* 17(1): 1–23.

Leach, M. 2007. Earth mother myths and other ecofeminist fables: How a strategic notion rose and fell. *Development and change,* 38(1): 67–85.

López Santiago, D. 2016. Graves retrocesos en género, medio ambiente y derechos humanos. Proceso. Available online at www.contralinea.com.mx/archivo-revista/index.php/2016/02/21/graves-retrocesos-en-genero-medio-ambiente-y-derechos-humanos/ (accessed January 12, 2017).

Mexico. 2015. Intended Nationally Determined Contribution. Mexico City, Mexico. Available online at www4.unfccc.int/submissions/INDC/Published%20 Documents/Mexico/1/MEXICO%20INDC%2003.30.2015.pdf (accessed January 12, 2017).

McAfee, K., 2012. The contradictory logic of global ecosystem services markets. *Development and Change, 43*(1): 105–131

Mohanty, R. 2004. Institutional dynamics and participatory spaces: The making and unmaking of participation in local forest management in India. *IDS Bulletin, 35*(2): 26–32.

Muñoz, C. and Vázquez, V. 2012. El Estado neoliberal y las mujeres indígenas. Un estudio de caso de la Sierra Negra de Puebla. *Espiral, 19*(53): 91–121.

Muñoz, C., Vázquez, V., Zapata, E., Quispe, A. y Vizcarra, I. 2010. Pobreza real y desarrollo de capacidades en mujeres indigenas de la Sierra Negra de Puebla. *La Ventana, 4*(31): 64–99.

Ortiz, A. S. 2004. Planeación ambiental, participación social e inequidades de género en San Antonio Nuevo Paraíso, Oaxaca. Tesis de maestría. Universidad Autónoma Metropolitana, México.

Peach Brown, H. C. 2011. Gender, climate change and REDD+ in the Congo basin forests of central Africa. *International Forestry Review, 13*(2): 163–176.

Pokorny, B., Scholz, I. and de Jong, W. (2013). REDD+ for the poor or the poor for REDD+? About the limitations of environmental policies in the Amazon and the potential of achieving environmental goals through pro-poor policies. *Ecology and Society, 18*(2): 3.

Porter, F. and Sweetman, C. (2005). Introduction [special issue on gender mainstreaming). *Gender and Development, 13*(2): 2–10.

Rankin, K. 2001. Governing development: Neo-liberalism, microcredit, and rational economic woman. *Economy and Society* 30: 18–37.

Razavi, S. and Miller, C., 1995. *From WID to GAD: Conceptual shifts in the women and development discourse* (vol. 1). Geneva: United Nations Research Institute for Social Development.

Resurrección, B. and Elmhirst, R. (eds). (2008). *Gender and natural resource management: Livelihoods, mobility and interventions.* London: Earthscan

Rocha, M. M., Zapata, E., Vázquez, V. y Martínez, B. 2006. La transversalidad del enfoque de género en organizaciones no gubernamentales: EDUCE, un caso mexicano. *Agro Nuevo, 2*(3): 45–66.

SEMARNAT. 2003. *Informe de actividades y resultados,* Unidad Coordinadora de Participación Social y Transparencia, Dirección General de Participación Social y Equidad, Dirección de Equidad de Género, Semarnat: México.

Sijapati-Basnett, B. 2008. *Gender, institutions and development in natural resource management: A study of community forestry in Nepal* (Unpublished doctoral dissertation). London: London School of Economics and Political Science.

St. Clair, A. L. 2006. Global poverty: development ethics meets global justice. *Globalizations, 3*(2): 139–158.

Tchikoué, H. 2008. *Evaluación externa del Programa para el Desarrollo Forestal (PRODEFOR).* UniversidadAtútoma de Chapingo: División de Ciencias Forestales.

Tourliere, M. 2016. La corrupción encabeza el INMUJERES. *Proceso.* Available online at www.proceso.com.mx/431530/la-corrupcion-encabeza-el-inmujeres (accessed January 12, 2017).

UNFCCC. 2011. *The Cancun agreements: Outcome of the work of the ad hoc working group on long-term cooperation under the convention, decision 1/CP.16, FCC/CP/ 2010/7 add.1.* Bonn, Germany: United Nations Framework Convention on Climate Change.

Vázquez García, V. 2015a. Género y cooperación internacional en México: Análisis critic y lecciones aprendidas. En revisión, *Cooperación internacional y política ambiental en México*, Gustavo Sadot and Simone Lucatello (eds), México: Instituto Luis Mora.

Vázquez García, V. 2015. Manejo forestal comunitario: Gobernanza y género en Hidalgo, México. *Revista Mexicana de Sociologia* 77(4): 611–635.

Vázquez García, V. 2014. Gender Mainstreaming en la política ambiental mexicana. Balance y perspectivas. *Sociedades Rurales, Producción y Medio Ambiente*, *14*(28): 17–45.

Venegas, F. 2004. Género, sustentabilidad y salud reproductiva. El proceso de empoderamiento de mujeres en la Selva El Ocote, Chiapas. Tesis de maestria. Universidad Autonoma Metropolitana, México.

Walby, S. 2005. Gender mainstreaming: Productive tensions in theory and practice. *Social Politics: International Studies in Gender, State & Society*, *12*(3): 321–343.

Westholm, L. and Arora-Jonsson, S. 2015. Defining solutions, finding problems: Deforestation, gender, and REDD+ in Burkina Faso. *Conservation and Society*, 13(2): 189–199.

Woodford-Berger, O. 2004. Gender mainstreaming: What is it (about) and should we continue doing it? *IDS bulletin*, 35(4): 65–72.

Wright, M. W. 2013. Feminicidio, narcoviolence, and gentrification in Ciudad Juárez: The feminist fight. *Environment and Planning D: Society and Space*, 31(5): 830–845.

13 'Women and men are equal so no need to develop different projects'

Assuming gender equality in development and climate-related projects

Virginie Le Masson

Introduction: attention to gender in the context of disaster risks and climate change

The occurrence of natural hazards and the impacts of climate change in mountain areas generate phenomena (e.g. storms, floods, landslides) that local communities must continually cope with and adapt to. Sometimes, such risks lead to disasters when people who are affected do not have the means to protect themselves and recover from loss and damages. In other words, the vulnerability of communities (including their exposure to risk and the extent to which they have to cope with everyday challenges such as securing incomes and access to basic services) determines the way people are affected and respond to natural hazards and environmental changes.

Both Disaster Risk Reduction (DRR) and Climate Change Adaptation (CCA) aim at supporting communities to address these challenges by simultaneously reducing their vulnerabilities and enhancing their capacities (Schipper and Pelling, 2006; O'Brien *et al.*, 2008; Kelman and Gaillard, 2010; Mitchell *et al.*, 2010). The emphasis on vulnerability and capacity recognises that not everybody can avoid natural hazards and deal with climate change impacts in the same way (Lewis, 1999; Cannon, 2000). Those marginalised in their daily life tend to lack adequate economic income; they might live in remote or hazards-prone areas; they might belong to groups that are discriminated against and their voices might be disregarded and not acted upon (Gaillard *et al.*, 2009; Wisner *et al.*, 2012). They tend to be disproportionately affected by disasters and climate change because their marginalisation prevents them from accessing adequate resources that would help them cope with hazards and adapt to climate change.

Differences between men and women constitute a typical example where the marginalisation of one group compared to the other is frequently underestimated (Fordham, 2012). Despite the common assumption that women have equal rights with men, gender studies have repeatedly stressed that men and women face different everyday constraints, which often privilege the position of men and exclude women (Moser and Levy, 1986; Razavi and Miller, 1995; Harper *et al.*, 2014). The latter find themselves with lower status, less access to decision-making and overall more limited livelihood opportunities, which render them more vulnerable (Momsen, 2010). Gender thus constitutes a crucial variable to compare people's marginalisation and therefore their ability to cope with hazards and their opportunities to adapt to long-term climate changes.

Based on field research conducted in the Himalayan province of Ladakh in India, this chapter documents the process and outcomes of using a gender perspective when undertaking vulnerability and capacity assessments to address development and environmental, including climate, issues. The first section provides an overview of the geographical context of the case study. The second section describes the gender-sensitive approach to the methodology used for this research. In the third section, the analysis contributes to provide empirical evidence of (i) the way gender equality is perceived in local Ladakhi communities; (ii) gender-differentiated vulnerabilities facing environmental changes; (iii) the way gender is conceived in approaches inscribed in Non-Governmental Organisations (NGO) practices; and (iv) the discrepancies existing between practitioners' perspectives on gender and beneficiaries' gendered realities and needs. The conclusion discusses the relevance of using a gender perspective when implementing DRR and CCA to acknowledge the views of local communities and respond to their needs.

Case study: climate change within the wider development context in Ladakh

Mountain communities not only live in geographical areas extremely sensitive to global environmental change, they have also historically suffered from economic and political marginalisation which exacerbates their vulnerability to hazards and undermines their capacities to cope with environmental shocks and trends (Kohler *et al.*, 2010; Gaillard and Kelman, 2012). The context of Ladakh illustrates rapid changes occurring in mountain regions of the Global South. Located in the Indian state of Jammu and Kashmir and bordering with Pakistan and China, Ladakh (Leh, 34.10°N, 77.35°E) is a politically strategic region that was closed to foreigners until 1974. For the last four decades, this remote high-altitude desert has been increasingly targeted by the Indian government's development projects while becoming a world-renowned tourist destination. With this context in mind, Ladakh faces a combination of societal changes that impact on the sustainable development of local communities. Population dynamics, economic growth and mass tourism have negative impacts on the environment by generating more pressure on natural resources (Geneletti and Dawa, 2009). Agricultural systems are losing their diversity and no longer sustain higher demands for water, land, food and energy (Nüsser *et al.*, 2012). Increasing rural to urban migration affects familial relationships and the socio-economic characteristics of mountain villages and nomadic communities (Goodall, 2004). People increasingly depend on the modern wage economy, controlled by the global market, over which they have no control (Norberg-Hodge, 2000). Finally, increasing interactions with western cultures through tourism are transforming local mountain communities in both positive and negative ways (Michaud, 1996). Climate change comes as an additional challenge that is likely to increase the risk of disasters and exacerbate existing development issues linked to natural resource access. Climate projections for Ladakh suggest a global rise in temperatures and a decrease in snowfall, which could diminish the replenishment of glacial areas and alter water runoffs (Angmo and Heiniger, 2009). As the majority of the population in Ladakh relies on irrigation-based agriculture and animal husbandry, they are highly sensitive and vulnerable to water fluctuations. Increasing episodes of water shortages could lead to a decline in agricultural land, crop failures and loss of livestock, which could seriously undermine food security. Additionally, shifts in precipitation patterns attributed to climate change, could potentially exacerbate hydrological-related hazards in the form of intense rainfall, flash floods and mudflows. Ladakh experiences recurrent deadly flood events triggered by heavy and sudden precipitation as happened for instance in 2006 and 2010. The combination of these socio-economic and environmental changes have different impacts on men and women given their distinct roles and experiences of their environment (Leduc and Ahmad, 2009; Hewitt and Mehta, 2012). Whereas research in the Hindu Kush-Himalayan region seldom addresses gender in relation to the environment (the special issue of Mountain Research and Development from 2014 is an exception, see Verma *et al.*, 2014), this chapter contributes to enhance the comprehension of mountain development by integrating a gender dimension to responses to disaster risks and climate change.

Methodology: a gender sensitive approach to conduct fieldwork

This chapter draws on the findings of a three-year doctoral thesis that explored the gender dimension of DRR and CCA, focusing on the case of Ladakh, India (see Le Masson, 2013). The methodology has used a gender-sensitive approach in order to assess the social impacts of development projects and their relevance to address men's and women's needs and experiences of their environment. Such an approach recognises gender and its intersection with other social identities (age, ethnicity, class, religion, etc.), and gender relations, as significant variables in understanding environmental issues, and aims to promote greater gender equality both in the research process and in analysing the impacts of development projects (Leduc, 2009). The fieldwork, conducted in Ladakh between 2010 and 2011, sought to collect gender-disaggregated data, give both men and women an equal chance to participate in the research, make gender differences visible and give attention to voices that otherwise might be ignored. The chapter draws on two main sources of information:

First, the research relied on a gender-sensitive assessment of mountain communities' vulnerabilities and capacities to hazards and environmental changes. Vulnerability and capacity assessments analyse the risks, vulnerabilities, and capacities of people and institutions in a given location and in the context of climate change and/or multi-hazards (Morchain *et al.*, 2015). When the process is participatory and community-based, it combines community knowledge and scientific data to yield greater understanding about local impacts of climate change and other phenomena (CARE, 2010; ICIMOD, 2011). The objective is to highlight men's and women's different experiences of their environment and distinct needs, with the view to inform the design of development and climate-related projects to enhance communities' capacities to deal with stresses and shocks. For this research, an assessment was based on 89 semi-structured interviews (44 women and 45 men) conducted with inhabitants of both rural and urban communities in Ladakh. Questions aimed at identifying people's roles and responsibilities within the household and the community, their access and control over resources and decision-making processes, and their views of environmental changes.

The research team comprised two researchers who lived for a few months in Leh between 2010 and 2011 and two local interpreters, based in Leh. Having a gender-balanced team helped to conduct simultaneous interviews disaggregated by gender. Interviews were typically organised with a male researcher interviewing men while a female researcher would usually interview women. This proved a useful approach to both respect the position of the family head or the head of an organisation (usually a man) and access other members of households or organisations such as female aid workers. This approach also guaranteed that participants' answers would not be greatly influenced or constrained by the presence of their kin (see Le Masson, 2014 for a broader analysis of using a gender-sensitive methodology). Data collected was disaggregated by gender, age and geographical location and triangulated thanks to mixed focus group discussions organised with both male and female participants.

Second, the research examined the way NGO practitioners conceived gender when designing and implementing their projects. Twenty-three semi-structured interviews were conducted with representatives of thirteen NGOs (nine of which were local). Wherever possible, one male and one female representative of the organisation were interviewed, in order to compare people's perspectives and triangulate information.

The following analysis draws on the gender-sensitive assessment of natural hazards and climate change occurring in Ladakh, to discuss the relevance of development approaches pursued by NGO practitioners in regard to people's vulnerabilities, capacities and views of both gender equality and environmental changes.

The assumption of gender equality

The general discourse of villagers, both men and women, highlighted differences in daily gender roles but considered there to be an equality of status between both sexes. The majority of respondents pointed out the difference between the discrimination against women that prevailed in the past and the equal status now characterising men and women, particularly as a result of the improved access to education for girls. They also considered that Ladakhi women tend to enjoy an equal if not higher position than men within the household or in the village, especially when compared with the rest of India. Interviewees' opinions about gender equality may be that they genuinely perceived gender relations to be equal, but it is also possible that they were keen to represent Ladakh as relatively modern or morally superior to other parts of India (to Western researchers). The use of a gender-balanced team of researchers also meant that participants were perhaps more inclined to stress that women are equal to men. Overall, villagers' initial response echoed previous studies that have documented Ladakhi women's relatively high status, the equal share of fieldwork activities by both sexes and the complementarities of gender roles (e.g. Crook, 1980; Norberg-Hodge, 2000).

However, the triangulation of data, thanks to the use of alternative interview questions and mixed-focus group discussions, questioned the alleged gender equality in Ladakh. For example, while the majority of respondents would consider that men and women are equal, both in principle and in reality, this picture was nuanced when asking them 'who has a harder life between men and women?' The majority of female participants expressed having a harder life because their amount of work is heavier than men's and remains constant throughout the year. Data also suggest that women in Ladakh are considered as being less 'advanced' (according to both male and female interviewees' expressions) than men because their literacy rate is lower (43 per cent of female interviewees never attended school compared to 26 per cent of men) and because the majority of them are primarily associated with domestic and agricultural activities which were not valued as important by half of their male counterparts. Findings from the field also suggest that women are much more marginalised politically than men within their communities. The absence of women from the position of elected political representatives within the Ladakh Autonomous Hill Development Council (LAHDC) means that half of the population might not have their needs and opinions acted upon by policies. This echoes the view of Ladol (2013: 7) according to which the lack of recognition of forms of oppression among Ladakhi communities has sustained patriarchal structures.

Despite overall preconceptions in Ladakh about men as the main breadwinners, Ladakhi women, most of whom are subsistence farmers, have a crucial role in providing daily sustenance. During focus group discussions in which both male and female participants were brought together, women highlighted that their role and activities give them legitimate power in their family and community even if some of their male counterparts will not admit it. They further stressed that some of their stereotyped gender roles, such as cooking or taking care of their family members, constitute an intrinsic part of their life that cannot be questioned or considered as a burden. According to their views, those responsibilities need to be fulfilled for the proper functioning of the household and the community and they should be valued as fundamentally important activities performed by women. Such views, raised during the mixed group discussions, initiated a debate between men and women in order to discuss what men's and women's main daily challenges were.

A first lesson to underline here is the opportunity brought by a gender-sensitive approach to triangulate people's responses and challenge initial results of interviews to better understand gendered roles and positions. In the absence of questions pertaining to equality (both perceived and experienced) between men and women, several aspects of villager's vulnerabilities and capacities might have been overlooked, such as women's disproportionate workload and lower literacy rate.

Gender differentiated vulnerabilities

Men's and women's different status and roles affect the availability and stability of their access to resources and therefore their experience of dealing with natural hazards and climate change (Enarson and Morrow, 1998; Masika, 2002). This section focuses on three main examples of differences between men's and women's views and means to cope with environmental shocks and stress.

First, issues related to water were particularly underscored when giving attention to women's voices.

As Ladakhi people rely on irrigation, adequate allocation of water is paramount during the first months of the growing season (April and May) when temperatures are not high enough to induce the melting of glaciers and when irrigation relies exclusively on snow and ice-melted water. When villagers were asked what their primary daily challenges were, women interviewees were more concerned with water access than men. In the village of Phuktse, every interviewed woman stressed a deep concern to access sufficient water during the sowing and growing season while only a minority of their male counterparts raised this issue. Women identified the bad weather and the cold temperatures that prevent the glacier melting on time and providing a sufficient amount of water to every villager. As the ones responsible for collecting water, irrigating the fields and looking after vegetable gardens and livestock, women are likely to be the first to experience pressures linked with the availability of natural resources, particularly water. The lack of water constitutes a major issue as it prevents them from securing their yields, thus impacting negatively on their food stocks, incomes from cash crops, fodder for animals and, therefore, on the physical resources of the household as a whole. It exacerbates families' vulnerability by undermining their food security and limiting their stock capital for the winter or in case of a crisis. This situation becomes particularly difficult for widows who have to rely on incomes earned by their children or other family members, as a coping mechanism.

Second, and drawing on the previous example, the impacts of climate change are not perceived similarly by men and women. In the village of Phuktse, most women actually considered the predicted global warming of temperatures as a positive phenomenon which would help them receive more melt-water and earlier in the year, and which would make their daily activities easier. Most women seemed less informed about the concept of climate change than men but were still worried about the potential impacts on natural resources. Conversely, male respondents were overall better informed of climate projections and understood well what the impacts could mean for their daily lives without necessarily raising water access as their primary daily challenge (see Figure 13.1). Overall however, the majority of people pointed out the role of issues other than climate change such as pollution, urbanisation and rising individualism among neighbourhoods to explain negative changes in the environment.

Third, Ladakhi women's greater economic marginalisation constitutes an additional example of people's different access to resources. Socio-economic changes in Ladakh, including imports of subsidised food and manufactured goods, have increased the need for cash (Norberg-Hodge, 2000). Many

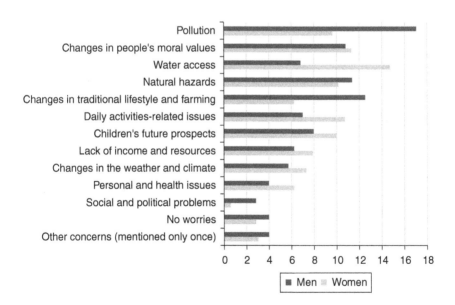

Figure 13.1 Ranking of men's and women's main sources of worry (regrouped by themes) and expressed in percentage of the total number of male and female answers.

households try to diversify their economic resources, particularly those living near the capital Leh, and close to tourist attractions. However, women indicated that they have fewer opportunities than men to engage in income-generating activities or earn as much income as men because of their limited formal education, greater household responsibilities and lack of spare time. In contrast, male interviewees were involved in a wider range of money-making activities and only a minority stated participating in domestic tasks. With the transition from subsistence agriculture to a market-based economy, men seek employment in the army or elsewhere in the government and take advantage of tourism by becoming travel agents, guides, cooks or taxi drivers while agriculture, devalued and considered as non-lucrative, was mostly associated with women or the elderly. Two decades ago, Hay (1997) had already sought to understand and document the impact of 'modernisation' on the social construction of gender roles and relations. Her research stressed how the shift from a traditional subsistence economy has increased gender-based inequities by reinforcing and revising local religious and cultural patriarchal traditions and by introducing new ones. 'Modernisation was shown to increase women's workload, diminish their mobility, curtail their decision-making and access to resources, and subject them to restrictive and foreign gender norms and ideals' (Hay, 1997: 192). Hay further suggested that wide-scale reductions in men's traditional childrearing roles caused by their employment in the army, tourism sector and the paid labour force in general, could further shape gender relations in Ladakh.

Women's reproductive responsibilities take up a significant amount of their time, which limits their opportunities to engage in paid activities, and therefore their ability to earn and save money. This makes them more finan-cially dependent on other household members who earn sufficient and stable income and who are usually men. Most female interviewees indicated that they were not responsible for managing their household's budget and many women often resort to borrowing money to make ends meet. Unavailable and insufficient economic resources exacerbate people's vulnerability to environ-mental shocks and longer-term climate change and diminish their capacities to recover in the aftermath of a disaster (Wisner *et al.*, 2004). In Ladakh, men who earn salaries and receive pensions, provide their household with financial resources to buy food from the market, increase and diversify their diet and improve their health status. Such households also have a better chance to recover from damages as their income is less likely to be disrupted and still enables them to pay for basic services without necessarily resorting to borrow-ing money in times of stress.

Overall, any interventions supposed to help people reduce their vulnerabili-ties to disaster risks and climate change must therefore acknowledge people's social identity as it shapes their different perspectives and experiences of their environment as well as their opportunities to access and control resources that they can mobilise in times of crisis.

Contradictions between discourses and practices among NGO practitioners

The previous section provides evidence of the importance in considering gender when assessing people's vulnerabilities and capacities in order to acknowledge differences among individuals and to address their distinct realities and needs. Yet, this research uncovered contradictory approaches of organisations in terms of how they recognise the gender dimension of people's vulnerabilities and capacities.

First, interviews suggest that in the majority of NGOs being researched, senior representatives' discourses differ from that of their staff. This is especially true when comparing a man at the top of the hierarchy with a female staff member (see Table 13.1). According to the majority of heads of NGOs, there are no gender disparities in Ladakh, echoing the initial discourse of villagers, and therefore they saw no real need to address gender issues. This view was also shared by women leaders of international organisation #2 and local organisation #10. Conversely, almost every coordinator or project officer stressed the importance of targeting women in their interventions because they are more marginalised than men. This could be explained by the fact that staff members were mostly women, although quotes from Table 13.1 show that the gender of the respondent might not be the reason. For instance, while the female senior representative of local organisation #10 insisted that gender equality exists in Ladakh, her male counterpart thought otherwise. The lack of representation of women in decision-making processes was also highlighted by the woman head of local organisation #7 but their programmes clearly target the empowerment of women. This is the sole example where a senior representative acknowledged the necessity to address gender biases and the underrepresentation of women's voices in Ladakh.

Second, there seems to be a gap between interviewees' discourse underestimating the importance of gender and organisations' actual practices. Despite the assumption of gender equality in Ladakh by senior representatives, literally every organisation's documentation includes a component on women's empowerment. Booklets and websites emphasise the need to encourage gender equality, to advance rural women's health and education, and to improve decision-making power. Interviewees easily advertised their projects, such as awareness campaigns and the creation of Self Help Groups (SHG), to enable women to develop income generation activities (e.g. making and selling handicrafts or developing homestays[1]) to earn cash and improve their livelihoods.

One climate-related project is the construction of Passive Solar Houses (PSH) in order to provide people with a warmer and less polluted habitat, reduce their fuel consumption and limit the pressure on local resources (preventing desertification and reducing carbon emissions). There is a strong gender component to the project. Women, children and the elderly are those who spend most of their time indoors and therefore are the main beneficiaries of

Table 13.1 Examples of answers provided by NGO representatives in Ladakh regarding the integration of gender in their approaches

NGO	Quotes from senior representatives	Quotes from other staff members
#8	*In Ladakh women are equal to men they have the same rights in decision-making and do an equal share of the work.* (male)	*More work should be done towards helping women. It is necessary to focus on women because they have fewer opportunities.* (female)
#9	*There is no real difference between men and women. We simply promote education and learning to both men and women.* (male)	*Everybody thinks there is gender equality in Ladakh but there is actually a gender discrimination against women. When you look at the council, you find many issues linked to male domination at the political level and in terms of job access.* (female)
#11	*Women are more empowered in many villages. There is no disparity. [...] It's not my opinion, it's a fact, you can't deny it. Look, since, there is no gender bias we don't think it* [a gender policy] *is important to give importance to that issue.* (male)	*Men have more opportunities for jobs. There is gender equality in daily life in the city, but in villages women don't have so much power and decision-making. Even if the father or the husband is away, women wait for them before deciding anything, like for spending money, they need permission.* (female)
#2	*Gender is not a big issue in Ladakh compared to other parts of India thanks to Buddhism.* (female)	
#10	*Women's status in Ladakh is much better. When we go to villages we don't think that we should focus on women. [...] Men and women are equal.* (female)	*Women still have problems in Ladakh. We have this democratic system but there are no women in practice. It should be half-half, because half of the population is women. [...] We definitely need more activities that empower women.* (male)
#7	*Most projects are not gender balanced. There is a big gender gap at the governmental level. There are 2 seats for women at the council. If women stand up at the election, they are most welcomed, but women are shy and they are not enough educated so they don't dare enter in politics.* (female)	

increased indoor temperatures and reduced air pollution. Moreover, as the main fuel collectors, women who live in PSH spend less time collecting wood or cow dung. The project thus relies on this time gained to encourage the creation of SHG whereby women have more time and a more comfortable space to engage in 'productive' activities. The creation of SHG was positively evaluated by the majority of both male and female interviewees among local communities. Most women who were members of one of these groups stressed the economic and social benefits of this project and most respondents, both men and women, also considered it beneficial to improve women's livelihoods.

Discrepancies between NGOs agenda and local realities

As the vulnerability and capacity assessment emphasised women's greater economic marginalisation compared with men's, development practices and climate-related projects in Ladakh that provide women with more opportunities to earn incomes might contribute to reduce their vulnerability to hazards and enhance their capacities to adapt to environmental changes. However, such projects do not necessarily draw on women's views of what constitute their most pressing challenges, such as access to water. Nor do they guarantee women the control of economic resources. For instance, the coordinator of local organisation #11 highlighted that "*Women don't earn much income but they have a lot of activities to do, so we try to improve their economic status because we have seen that whatever incomes it goes to men's pockets*". This raises the question of whether it is possible to promote the emancipation of women based on income-generating activities without challenging power structures within the household.

Moreover, to encourage women to develop income-generating activities seems to overlook their heavier workloads and limited spare time highlighted previously. Many female interviewees living in more remote villages stressed that they could not reconcile joining a group and fulfilling their daily duties. This suggests that the creation of SHG sometimes does not necessarily appear commensurate with women's daily challenges and needs. Furthermore, promoting the empowerment of women by encouraging them to generate income replicates the idea that the improvement of one's social status reflects one's ability to engage in productive work. The conceptualisation of women as illiterate, poor, subordinate and unproductive has long justified development efforts to help women become economically productive (Apffel-Marglin and Simon, 1994). This fails however, to address local women's own views and desires, including the necessity for men and people in general to value their reproductive roles even though they do not generate cash.

Conclusion and policy implications

Numerous studies have stressed that gender remains a separate discipline or field of policy that is not systematically mainstreamed as a cross-cutting issue within DRR or CCA (e.g. Masika, 2002; Christensen *et al.*, 2009; ICIMOD, 2011). Findings from this research echo the literature and provide some explanations and ways forward.

First, the use of a gender-sensitive methodology to conduct this research and particularly the vulnerability and capacity assessment, has highlighted that differences exist between men's and women's views, vulnerabilities and capacities to environmental risks despite the popular beliefs that natural hazards and climate change impact everyone regardless of their social identity. The assessment uncovered gender-differentiated vulnerabilities in Ladakh such as women's lesser access to stable financial resources and lower participation in decision-making processes. Attention to women's voices also enables highlighting issues related to water scarcity, which underscores men's and women's different experiences of environmental changes. This supports the relevance of a gender-sensitive approach to better capture people's different perspectives about what constitute the more pressing needs of marginalised people to efficiently address environmental issues.

Second, findings highlight that gender equality was not incorporated by NGO practitioners in the design of development strategies because they consider gender equality to be a reality in Ladakh and therefore thought it was irrelevant to develop gender sensitive approaches. A few NGO practitioners, especially men, seemed reluctant to link their approach with any recognition that gender differences exist in Ladakh. This might be explained by their worry that this could be misinterpreted as intolerance, sexism or misogyny. On the one hand, this does not prevent NGOs from implementing projects that address the particular problems of individuals according to their gender (e.g. women's lower economic resources). The construction of PSH, for instance, benefits women by improving their health and reducing their workload. On the other hand, assuming that men and women are equal in their status and needs might prevent NGOs from evaluating the impacts of their projects from a gender perspective. This might also prevent them from conducting gender-sensitive needs assessment with the risk of designing projects that might not be appropriate to women's specific vulnerabilities and capacities when addressing environmental shocks and trends. NGOs should systematically use a gender-sensitive approach to help identify and implement projects that are relevant, socially accepted and therefore sustainable to ensure the well-being of mountain communities.

Third, although the necessity to mainstream gender within development is already being advocated by international institutions (e.g. UNDP, 2010; ICIMOD, 2011), this study highlights a gap between policies, discourses and practices. The approach followed in this research suggests that the process of first separating men and women using gender mixed researchers, is useful to highlight gender differentiated vulnerabilities and experiences of natural hazards and climate change as well as discrepancies between the discourse and practices of NGO practitioners. However, it does not address the issue of gender equality being assumed both by local communities and NGO workers as a reality and therefore the mainstream consideration in the development sector in Ladakh, that gender-sensitive approaches are not needed. The assumption of gender equality prevents NGOs from conducting gender-sensitive needs assessment and evaluating the impacts of their projects with a gender perspective. This runs the subsequent risk of designing projects that might not be appropriate to people's specific needs, vulnerabilities and capacities. More generally, gender-blindness is a way to avoid any conflicts associated with the reallocation of access and control over resources (Vainio-Matilla, 2011).

Instead, creating a space where both women and men are confronted with different viewpoints such as in the case of convening mixed focus group discussions as a second round in interviewing and data collection processes, helps to raise awareness and challenge gender assumptions. This strategy should complement the process of discussing with gender groups separately in order to uncover differences and potential divisions and imbalances. Adopting a gender perspective when conducting research and when designing and implementing development projects thus contributes to addressing men's and women's different constraints, needs and interests in order to achieve sustainable development (Leduc and Ahmad, 2009). The process of confronting gender groups' perspectives could be one crucial step for NGO workers to foster the recognition that gender does matter when dealing and responding to climate change issues and to challenge the dominant gender-neutral approach to development and disaster risk reduction.

Note

1 Homestays constitute a major livelihood for inhabitants of Leh and villages along trekking paths. Families allocate rooms within their homes, or build an extension to their house, to accommodate tourists and trekkers.

References

Angmo, T. and Heiniger, L.E. (2009) *Impacts of climate change on local livelihoods in the cold deserts of the western Indian Himalayan region of Ladakh, Lahaul and Spiti.* Leh: Geres.

Apffel-Marglin, F. and Simon. S. (1994) 'Feminist orientalism and development', in W. Harcourt (ed.), *Feminist Perspectives on Sustainable Development.* London: Zed Books, 26–45.

Cannon, T. (2000) 'Vulnerability analysis and disasters', in D.J. Parker (ed.), *Floods.* London: Routledge.

CARE (2010) What is adaptation to climate change? CARE International Climate Change Working Brief.

Christensen, H., Breengaard, M. and Oldrup, H. (2009) 'Gendering climate change', *Kvinder Kon Forskning*, 3–4: 3–9.

Crook, J. (1980) 'Social change in Indian Tibet', *Social Science Information*, 19 (1): 139–166.

Enarson, E. and Morrow, B. (eds), (1998) *The Gendered Terrain of Disaster: Through Women's Eyes.* Westport, CT: Praeger Publishers.

Fordham, M. (2012) 'Gender, sexuality and disaster', in B. Wisner, J.-C. Gaillard and I. Kelman (eds), *The Routledge Handbook of Hazards and Disaster Risk Reduction.* New York: Routledge, pp. 711–722.

Gaillard, J.-C. and Kelman, I. (2012) 'Foreword: Mountain, marginality and disaster', *Journal of Alpine Research*, 100(1): http://rga.revues.org/1649

Gaillard, J.-C., Maceda, E.A., Stasiak, E., Le Berre, I. and Espaldon, M.A.O. (2009) 'Sustainable livelihoods and people's vulnerability in the face of coastal hazards', *Journal of Coastal Conservation* 13 (2–3): 119–129.

Geneletti, D. and Dawa, D. (2009) 'Environmental impact assessment of mountain tourism in developing regions: A study in Ladakh, Indian Himalaya', *Environmental Impact Assessment Review* 29: 229–242.

Goodall, S.K. (2004) 'Rural-to-urban migration and urbanization in Leh, Ladakh', *Mountain Research and Development*, 24(3): 220–227.

Harper, C., Nowacka, K., Alder, H. and Ferrant, G. (2014) *Measuring Women's Empowerment and Social Transformation in the Post-2015 Agenda*. London: Overseas Development Institute.

Hay, K.E. (1997) Gender, modernization, and change in Ladakh, India. Master of Arts Thesis, Norman Paterson School of International Anairs, Carleton University Ottawa, Ontario.

ICIMOD (2011) Gender Experiences and Responses to Climate Change in the Himalayas. Kathmandu, Nepal: International Centre for Integrated Mountain Development.

Kelman, I. and Gaillard, J.-C. (2010) 'Climate change adaptation and disaster risk reduction: Issues and challenges community', *Environment and Disaster Risk Management*, 4: 23–46.

Kohler, T., Giger, M., Hurni, H., Ott, K., Wiesman, U., Wymann Von Dach, S. and Maselli, D. (2010) 'Mountains and climate change: A global concern', *Mountain Research and Development* 30(1): 53–55.

Ladol, C. (2013) *Contemporary Ladakh. A Subtle Yet Strong Patriarchy*. Issue Brief # 237. New Delhi: Institute of Peace and Conflict studies.

Leduc, B. (2009) *Guidelines for Gender Sensitive Research*. Kathmandu: ICIMOD.

Leduc, B. and Ahmad, F. (2009) *Guidelines for Gender Sensitive Programming*. Kathmandu: ICIMOD.

Le Masson, V. Exploring Disaster risk reduction and climate change adaptation with a gender perspective: Insights from Ladakh. Thesis from Brunel University. Available online at http://bura.brunel.ac.uk/handle/2438/7504 (accessed 15 January 2017).

Le Masson, V. (2014) 'Seeing both sides: Ethical dilemmas of conducting gender-sensitive fieldwork', in J. Lunn (ed.), *Fieldwork in the Global South. Ethical Challenges and Dilemmas*. London: Routledge.

Lewis, J. (1999) *Development in Disaster-prone Places: Studies of Vulnerability*. London: Intermediate Technology Publications.

Masika, R. (2002) *Gender, Development, and Climate Change*. Oxford: Oxfam.

Michaud, J. (1996) 'A historical account of modern social change in Ladakh (Indian Kashmir) with special attention paid to tourism', *International Journal of Comparative Sociology*, 37: 286–301.

Mitchell, T., van Aalst, M. and Villanueva, P.S. (2010) 'Assessing progress on integrating disaster risk reduction and climate change adaptation in development processes', *Strengthening Climate Resilience Discussion Paper 2*. Brighton: Institute of Development Studies.

Momsen, J. (2010) *Gender and Development* (2nd edn). London: Routledge.

Morchain, D., Prati, G., Kelsey, F. and Raon, L. (2015) What if gender became an essential, standard element of Vulnerability Assessments? *Gender and Development*. http://policy-practice.oxfam.org.uk/publications/what-if-gender-became-an-essential-standard-element-of-vulnerability-assessments-582264

Moser, C.O. and Levy, C. (1986) A theory and methodology of gender planning: meeting women's practical and strategic needs. Development Planning Unit, Bartlett School of Architecture and Planning, University College London.

Norberg-Hodge, H. (2000) *Ancient futures. Learning from Ladakh* (Revised edn). London: Rider.

Nüsser, M., Schmidt, S. and Dame, J. (2012) 'Irrigation and development in the Upper Indus Basin', *Mountain Research and Development*, 32(1): 51–61.

O'Brien, K, Sygna, L., Leichenko, R., Adger, W. N., Barnett, J., Mitchell, T., Schipper, L., Thanner, T., Vogel, C. and Mortreux, C. (2008) Disaster risk reduction, climate change and human security – a study for the Foreign Ministry of Norway. GECHS (Global Environmental change and human security) Project. Available online at www.gechs.org/downloads/GECHS_Report_3-08.pdf (accessed 23 March 2012).

Razavi, S. and Miller, C. (1995) *From WID to GAD: Conceptual Shifts in the Women and Development Discourse* (vol. 1). Geneva: United Nations Research Institute for Social Development.

Schipper, L. and Pelling, M. (2006) 'Disaster risk, climate change and international development: Scope for, and challenges to, integration', *Disasters* 30: 19–38.

UNDP (United Nations Development Program) (2010) Gender, Climate Change and community-based adaptation. New York: UNDP.

Vainio-Mattila, A. (2001) Navigating Gender. A framework and a tool for participatory development. *Ministry for Foreign Affairs,* Department for International Development Cooperation, Helsinki, Finland.

Verma, R., Molden, D., Hurni, H., Zimmermann, A. and Wymann von Dach, S. (2014) Special issue: Gender and sustainable development in mountains – transformative innovations, tenacious resistances. *Mountain Research and Development* 34(3): 185–187.

Wisner, B., Blaikie, P., Cannon, T. and Davis, I. (2004) *At Risk: Natural Hazards, People's Vulnerability, and Disasters* (2nd edn). London: Routledge.

14 Co-housing

A double shift in roles?

Lidewij Tummers

Introduction

Comfort-standards in European housing have been going up: we demand more space, hot water, room temperature comfort and electricity for appliances for security and information in the twenty-first century than any century before. The amount of square meters, kWh and tons of material used for dwellings have increased steadily in the second half of the twentieth century (see for example Brounen *et al.*, 2012). New houses are built according to energy efficiency standards, but the existing stock, especially the mass-produced post-war flats, form an 'energy leak'. Their heating systems are responsible for a large share of CO_2 emissions (European Environmental Agency, 2015). To cater for post-war deficiencies in the reconstruction period housing was industrialised and designed for economic and household efficiency, with shared domestic spaces, for example, washing machines in the basement. Neighbourhoods offered housing types for all generations of households, from young families to seniors, as well as services close to home, for example, shops, repairs, schools and medical care. Nowadays, the industrial flat-buildings in post-war estates are seen as one homogenous type and receive much criticism as being too small, too cheap, too standardised, noisy and offering too little comfort. The small-scale everyday services have disappeared, making way for large supermarkets, educational and health services.

Considerations about the energy consumption in housing only started after the UN Committee on Environment and Development's report introduced the term of 'sustainable development' (Brundtland *et al.*, 1987). At present, EU policies are built on the presumption that excessive energy consumption of houses is no longer viable in light of climate change. Since 1995 housing built in European cities needs to respond to a benchmark for the energy performance of housing: the Energy Performance Coefficient (EPC). This does not only require the increase of insulation, but also the development of new technologies for energy supplies, sanitation and waste disposal equipment and the infrastructure that services it (for heating, domestic activities such as cleaning and maintenance, ventilation, waste-treatment and so on). The building industry has designed technical innovations for such utilities, often supported by governmental programmes. Retrofitting is one important strategy to reduce energy demand, but the housing stock continues to expand due to demographic change and higher standards of living. New models need to be developed, and while the architectural solutions are available, current housing provision structures are slow in implementing them. This condition, as well as inaccessible housing markets in urban Europe, lead groups of households to initiate housing projects aiming for 'low-impact living'. Co-housing is an umbrella term for a residents-led housing practice that aspires to respond to contemporary standards of living with low-impact solutions (see Figure 14.1). These include sharing rooms, artefacts and devices, exchange services such as transport; as well as creating low-impact buildings by using environmental, recyclable and energy-efficient materials and technology (Pickerill, 2015). Co-housing projects consider that equal social relations, including gender equality (Vestbro and Horelli, 2012), are part of sustainability. They therefore include a range of alternative diverse lifestyles (Wohnbund, 2015).

Co-housing initiatives represent an alternative model whose implementation nevertheless needs to follow mainstream trajectories, such as applying for building licences. The final project often needs to make compromises, not necessarily fulfilling the ideal design. What 'gets lost' or needs to be 'compromised' on the way may be frustrating to the participants, who follow a different

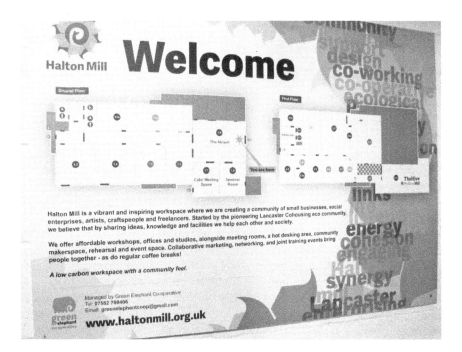

Figure 14.1 'Low-impact' example from Lancaster, UK co-housing

narrative (prioritising, for example, social interaction and energy-saving) to that of established planning and housing institutes (who might prioritise cost-efficiency and standardisation). These different narratives are interesting to explore for researchers, as they reveal how mainstream or institutional practices such as the housing market, energy-calculation or town planning, contain assumptions such as gender relations in the household. Housing plans, for example, could be different depending on whether they are designed in projects with the nuclear/breadwinner family as target group or if they are planned by initiatives supporting gender equality, such as Frauen-Werk-Stadt in Vienna (Damyanovic *et al.*, 2013). Likewise, devices for an intelligent use of renewable energy (for example, to use the washing machine while the solar panel works) could be designed as a built-in mechanism rather than assuming that someone (a mother, a pensioner) is at home all day to push the button. Alternatively, and suitable for homes of commuting career-people, domestic systems could be operated from an app. This chapter therefore looks at the planning characteristics, rather than the ideals or the social interaction, of the co-housing model, to understand how the gender-equality and low-impact ideals can be put into practice.

Aim, research questions and method

This chapter explores the following questions to examine how far the co-housing initiatives succeed in materialising feminist and environmental ideas:

- Does co-housing offer a different model from other low-carbon housing proposals?
- How are concepts of gender equality and low-impact living influencing co-housing design?
- When do these differences challenge and cause friction with the general (housing, energy) offer?

This chapter focuses on the mitigation aspect of climate change, namely the necessary transition from relying on fossil fuels to developing renewable energies, and how that energy transition affects housing and vice versa. There have been ample technical studies in this respect, but in this case the focus lies on looking at design features through a gender lens.

The chapter is based on a combination of two strands of research: on the one hand, literature and empirical research into co-housing as part of a PhD project that included an in depth study of nine co-housing projects in the Netherlands, exploring the building archives, interviewing of residents and some of the professionals that worked with them (Tummers 2015a, 2015b), and on the other hand participating in international working groups and seminars as well as numerous consultancies on gender planning. Both themes involve almost 10 years of looking at criteria for urban design, analysing housing plans, policy documents and toolboxes, as well as speaking with planners and officials about best practices, a selection of which figures below.

Both strands also provide a reflection on 20 years of practice in environmental engineering in the construction sector. Writing from the perspective of an engineer (rather than from a social science background) focuses on the 'hardware': the buildings, their structure, physics and layout. These are partly the consequences of the social architecture of co-housing and have an impact on energy consumption of households (see for example Kido and Nakajima, 2011; Chatterton, 2014; Stevenson *et al.*, 2013; Marckmann, 2012). This chapter will rather raise instrumental questions, for example around effectiveness and operational systems in practice.

The chapter is structured as follows: the following section frames the questions for co-housing, gender-aware planning and climate change, exploring how connecting these fields might advance thinking about climate change strategies in a European context. The next section then looks at co-housing projects in general: what they are, why they emerge, and, following this, the key characteristics are specified. The focus then turns to the extent to which co-housing initiatives conceive energy transition and gender equality respectively. To understand what makes co-housing different from mainstream clustered housing projects, representative empirical material is presented, reflecting on the overlaps between the strategies for climate change and gender equality as found in co-housing practices. This enables a reflection on whether co-housing can be seen as a feminist, ecological practice of housing. Finally the conclusions resume the lessons learned to derive new, gender sensitive criteria for spatial planning in light of climate change policies.

Links between climate change, gender and co-housing

Identifying the planning difficulties that arise when realising a co-housing project may be helpful to smooth the complex planning and building process for new co-housing initiatives. In addition, lessons learned can extend beyond the project level to the higher scales of city or region, such as housing policies and strategic development plans. For example: how do self-managed water recycling systems function? For what reasons is one sustainable building material preferred over another? If co-housing experiences become part of such policy and planning documents, the social and sustainable models that co-housing is based on also enter mainstream domains. Identifying assumptions of gender-roles or stereotypes may then help to advance 'gender-sensitive' planning. The central issue for gender mainstreaming is the combination of care and waged work.[1] In spatial planning this is often interpreted as the need to create better conditions for women as a 'disadvantaged' group. This can lead to partial solutions, such as safe streets: in itself a useful goal, although most gender-based violence occurs in the home. In complex planning practice, gender planning is regularly confronted with such gaps between gender mainstreaming objectives and the need to make strategic choices. Looking back on the first gender planning toolkit, Moser qualifies the "simplification and perceived 'technification' of gender planning (w)as a conscious decision in the highly hostile climate in which it was developed both to reach practitioners, and to provide operational tools they could implement." (Moser, 2014: 20). A more holistic approach is developed in the compact city or 'city of proximity'. This concept, which will be further explained below, argues that a (re-)mix of functions such as housing, shops, education and recreation has a positive impact on reconciliation as well as safety, and the environment (Sanchez de Madariaga and Roberts, 2012). Among others, Alber (2015:33) has pointed out that gender planning and climate change come together in the concept of the city of proximity.

Moser's dilemmas for gender-sensitive or feminist planning strategies are similar to those which Resurrección quotes regarding climate change: "Feminists, in short, have had to embrace simplification of identities and interests in order to insert gender agendas into institutions that otherwise have different priorities (Cornwall, Harrison, and Whitehead, 2007)" (Resurrección, 2013: 37). MacGregor elaborates further and signals four shortcomings of the mainstream narrative around gender and climate change which:

1. victimises poor women in the Global South;
2. tries to fit into the quantitative discourse;
3. cannot avoid confirming the bi-polar gender divide, and finally
4. does not take a normative position

(MacGregor, 2010: 226).

MacGregor concludes that the root causes of climate change are not challenged, and gender-sensitive approaches remain locked in a circle of addressing women as victims, with empowerment strategies that do not challenge the powers that be, nor the human exploitation of the planet.

Looking at co-housing provides a new perspective for these four deadlocks:

1) Climate change and gender inequality are concepts often primarily associated with specific groups of women in the Global South. Co-housing as an emerging European phenomenon relates to a mixed, albeit predominantly middle-class population (Bresson and Denèfle, 2015; Krokfors, 2012). Both men and women take action, stepping out of the passive role of consumers, within a context of western culture and standard of living. Thus, co-housing not only concerns environmental activism but also the re-consideration of consumerism, and the division of labour in waged work and domestic tasks.

2) So far the co-housing movement is based on qualitative assessment, and contrary to many climate change policies it hardly attempts to assess its impact in quantitative terms. Whereas this gives room for so-called 'soft' or 'fluid' approaches, there are some drawbacks such as the lack of insight in effectiveness in reducing CO_2 emissions or energy-demand, and

difficulties in comparing project results. Claims that solutions are inclusive and low-impact cannot be fully verified. This imbalance is not typical for co-housing, but it illustrates the need for a combination of quantitative and qualitative research.

3) Co-housing discourse promotes gender equality, often emphasising the interests or the contributions of women, thereby involuntarily maintaining the dual gender divide and the male standard (see for example Sangregorio, 1995; Vestbro and Horelli, 2012). Most projects aim to create a mixed group regarding income, age, ethnicity, household composition, and treat all members equally. However, often there is a gap between intention and practice, due to external factors such as housing allocation rules, and time-budgets.

4) Co-housing networks and projects clearly take positions and act upon them. Part of that position is: think global (an awareness of the 'root causes') act local (an awareness of the scale of change they can achieve on a daily basis), expressed in concepts such as 'Fair Trade'.

With these considerations, the following sections see co-housing as an instructive practice, that could produce lessons to advance thinking about the importance of a gender perspective in housing design or in climate mitigation.

The re-emergence of co-housing in Europe

'Co-housing' is a term for groups of households that together plan and build a housing project. This form of housing provision is not new, but since 2000 is re-emerging in most EU countries (Wohnbund, 2015). Even when including a large variety of project-types under the umbrella of collaborative housing, the absolute number of co-housing projects is small: estimated between 0.5–17 per cent of housing stock in different EU-countries (Tummers, 2015a). Nevertheless, since the turn of the century their number is increasing rapidly, raising the interest of researchers and local administrators (Droste, 2015; Locatelli *et al.*, 2011). In co-housing projects (future) residents are the initiators, and remain in charge of the project-management, although not always as homeowner and often in collaboration with institutional partners such as housing corporations, planning departments and funding agencies. Another feature that makes co-housing different is sharing common activities and spaces, with variable degrees of intensity (Krokfors, 2012; Tummers, 2015a; Fromm, 2000). Some are based on pragmatic considerations, others on idealistic principles, but all identify with environmental concerns, at least in their initial goals and external profile (such as website or criteria for new members). The articulation of the projects' objectives is specific for each generation of initiatives. Some projects have been initiated more than thirty years ago, and in the 1980s 'climate change' was not part of general discourse. The new term 'sustainability' for environment conservation strategies began to spread only after the earlier mentioned UN report (Brundtland, 1987).

In the meantime, the context has changed: the technical understanding and equipment of home-engineering has become more sophisticated, leading for example to housing models such as 'energy-neutral' or '*Passivhaus*'. Energy-targets have been agreed by EU member states and been translated into national policies (Maldonado, 2012). Calculation methods such as the now mandatory Energy Performance Coefficient (EPC), a model to estimate the building-related energy-demand during the lifespan of the building, did not exist. This means that comparison between projects for energy use is not straightforward, and always needs to be seen against the standards of the period in which projects have been built. Also, over the course of time different 'green' criteria apply, which makes it complex for different generations of projects to establish if their claim for being sustainable is founded.

A review of post-2000 co-housing literature revealed that there are many expectations attached to co-housing (Tummers, 2015a). The approach of the initiatives is to create healthy and empowering environments, and many case studies highlight these qualities, emphasising especially the social benefits, for example sharing care for young children or the elderly (see, e.g. Jarvis, 2011; Labit, 2015). Their ambitions correspond to contemporary urban policies such as social cohesion, care for the elderly and neighbourhood regeneration (Fromm, 2012). Gradually the potential of co-housing is being recognised by local authorities. There are examples of resident-led urban development, such as the former French Quarters in southern Germany, through the mobilisation of *Baugruppen*; expansion of new-town Almere in the Netherlands based on small-scale lots; or integration of co-housing initiatives in French 'eco-quartiers' (Bresson and Tummers, 2014). These developments demonstrate that self-managed housing does not mean that there is no role for planning departments. Droste even argues that in the absence of a municipal framework, co-housing initiatives risk becoming defensive, introverted and elitist instead of being socially inclusive and technically low-impact (Droste, 2015). At the same time, the relevance of co-housing does not lie in accommodating large quantities of households, but in opening up institutional planning, transforming the production of housing towards bottom-up or resident-involved processes and presenting a more ecological and human model (Tummers, 2015b).

Residents' groups as clients are relatively new in housing provision and the planning process. Therefore, responsibilities and decision-making of partners involved in design and engineering as well as operation and administration of housing need to be redefined. For example: Housing Associations as clients have more investment capacity, and are bound to housing distribution regulations, whereas inhabitants are more involved with the neighbourhood and often accumulate more knowledge about the everyday environment. Since the co-housing model re-introduces intermediate space as a buffer or meeting zone, the regulations and building methods that are based on strict division between 'private property' and 'public space' need to be re-interpreted

(Williams, 2005). Mixed tenure or collective property requires new forms of finance, liability and interaction (Fenster, 1999). Mutualisation and joined resources open possibilities not only for common gardens and playgrounds but also for installing local circuits for water recycling or so-called 'cascading': waste energy (for example surplus heat out of industrial processes) is re-used in nearby locations (for example to heat glasshouses for local production) (Tillie *et al.*, 2014). Seen as niche innovators (Seyfang and Haxeltine, 2012), co-housing could point the way to potential innovations of future housing provision to meet objectives to mitigate CO_2 emissions. Some of these innovations will be further explained below.

Typical co-housing planning characteristics

Although every co-housing project is unique, there exist some common characteristics. To allow for comparison, the projects referred to in this chapter are located in medium-sized towns (in the Netherlands that means between ±125,000–350,000 inhabitants), and different regions. European metropoles have a specific history of co-housing: as part of a strong squatters' movement (Hamburg, Amsterdam), historically representing a large share of subsidised housing (Berlin, Vienna) or as a response to high demand driving up prices on the housing market (Budapest, London). Rural (remote) self-sustaining 'eco-villages' are not included, because their characteristics in terms of connection to the energy-grid, and energy used for transport, or food-production, are very different.

Most of the projects concern new constructions, but re-use of buildings also occurs. This leads to a tentative profile for co-housing (see also Figure 14.2): In Europe, a typical co-housing project consists of approximately 20–40 individual, fully equipped units. Except for co-housing for seniors, there is a mix of household size and composition resulting in mixed typology and sometimes mixed tenure. Occasionally smaller subgroups are embedded in the larger project. The majority of projects are low-rise, with terraced two- to three-floor houses and/or a three- to four-floor apartment block, built around a garden or courtyard for common use. A modular design system can bring cost efficiency creating a collective basic structure and optional individual extras (this also allows for mixed income). Besides homes, some offices or workshops for creative industry or care may be included. All projects include shared spaces, such as gardens, playgrounds, laundries, guest-rooms, car parks, bicycle storage, community kitchens, meeting rooms and/or study-music-play rooms, in the build volume or as a separate 'common house'.

Building materials vary from one country to another, but residents seek to use eco-materials, as far as possible within budget constraints (Bresson and Denèfle, 2015; Chatterton, 2014; Locatelli *et al.*, 2011). The field-study on nine projects in the Netherlands showed that institutional partners are not always prone to experiment. Also, pre-investment in common utilities is not

Figure 14.2　Bongerd (Zwolle, The Netherlands): mixed typology; mixed use; mixed tenure but harmonised architecture

often supported by banks and projects fall back on individual gas-heaters as feasible, low-risk options (interviews in the Netherlands, and, for the UK, see also Chatterton, 2014). Building regulations and requirements may also limit the possibilities, especially for applying new materials before they have secured mandatory health and safety certificates. Recently, several projects in Germany, France, the UK and the Netherlands have started to experiment with straw-bale construction[2] but most others use more standard methods to achieve high insulation values. Almost all use solar energy, and further renewable options may include a common heatpump, heat-power-exchange or re-use of waste-warmth.

The re-use of rainwater in various forms, and on-site water purification if the plot-size allows, are among preferred and successful options observed during the field-study. One Dutch project, built around a 'common' in 1995, formed an association of homeowners around the public land. The residents' association manages the common garden as a whole and receives a small yearly sum from the municipality for maintaining the part that is public property (see Figure 14.3). In exchange the garden is open to the neighbourhood. This does occasionally cause tensions, for example when children outside the project use the ecological playground the residents have built, they may not be aware that the vegetation is also used for water purification, and vulnerable to pollution. The residents' association of this project maintains an internal drainage network that captures rainwater and re-distributes it for toilet flushing and washing machines. Every household around the common is connected both to the general drinking water grid but uses about 50 per cent through the internal rainwater against the same fee. The financial benefits, after maintenance costs, are used to maintain the common house.[3]

Figure 14.3 'The green common', built in 1995 (purification and playground in garden)

Co-housing and energy-transition

The fossil fuels-based economy is centralised in one-sided networks, whereby supply departs from large plants, large supply companies and often state-controlled delivery hubs.[4] Such networks have a high rate of loss in transport and accumulated power through extensive distribution networks. Renewable energy production is characterised by decentralised sources and requires networks that can absorb fluctuations in supply and demand at the same time (so-called 'smart grids'). The EU funds extensive research into new appliances and software that can regulate such 'smart grids' and helps the consumer with efficient equipment, smart houses and self-reading meters.[5] Until recently, such programmes were exclusively technical; meanwhile there is more understanding that new technology is not effective when it does not involve end-users. This shows the relevance of co-housing initiatives as 'living labs'.

Since the 1980s, housing collectives have been responding in different ways to the environmental challenges. Co-housing initiators, operating from the user perspective, take an integrated view, not only towards considering the built volume, but also mobility, food and goods consumption and other components of the environmental 'footprint' of everyday life. The emphasis as well as the intensity and the standards the project aspires to may vary, however there is a general concern, if only to reduce the energy bill (Baborska *et al.*, 2014; Marckman *et al.*, 2012; Locatelli *et al.*, 2011). Following the environmental ambitions of co-housing initiators, many co-housing projects have become micro-laboratories for energy transition in housing (see for example, Buis, 2000). The energy *'prosumers'* join forces to not only renegotiate with suppliers (to obtain greener energy for example) but also to experiment with new technologies such as solar collectors in the 1980s and airtight 'passivhaus'[6] in the early 2000s. Besides, a number of projects organise educational activities and share their experiences beyond the project membership (for example on the annual European Cohousing Open Door Day in May). Energy labels are designed for the individual unit, and depart from standardised patterns of use, for example a heating pattern following a 9–5 job, and lower temperatures for bedrooms even if these are often places for homework. This makes it difficult to assess if co-housing is more or less energy-demanding. Although there is to date very little systematic research into the engineering aspects of co-housing, there is ample evidence of applying 'new' technologies, such as solar collectors when these were first introduced, more than average insulation materials, collective heating systems, and rainwater recycling systems. Building on current experiences, there is real potential for energy saving as well as implementing renewable energy sources. But there are also reports of failing technology, or replacement of experimental by conventional systems where the underlying motives are not fully clear or known to the residents.

The residents-steered initiatives fundamentally change the classic relation between producer, energy supplier and consumer. They exchange services and goods that are otherwise supplied by retail, industry or institutions, for example food, clothing, rides, care-assistance, education or (solar) energy. Most of such activities are 'neighbourly' exchanges, and still rely on the main energy grids in order to function. When the neighbourhood exchange grows, it meets a threshold from where it will be considered at par with established suppliers. It then enters an institutional realm of safety regulations, reliability, taxation and feed-in tariffs. In the present situation, this realm offers mostly solutions for the individual consumer and large-scale industry, which are not always suitable for self-managing collectives (see, e.g. Tillie *et al.*, 2014; Rahimian, 2015).

Considering co-housing initiatives as '*prosumers*' is promising for a number of reasons: first, from an engineering point of view, co-housing as a cluster forms the intermediate level between grid and individual unit. In a network of decentralised and unstable energy production, such medium-sized hubs are potentially a place to mediate between peaks in demand and peaks in supply. Second, technically, the storage of energy is an important issue for renewable sources, because they tend not to be constant in supply and not synchronised with use. Peaks of supply (for example, on sunny days) need to be buffering for peaks in demand (when the lamps go on at night). Cities like Rotterdam, Oslo and Milton Keynes promote electrical transport for this reason, to consume the solar power surplus while reducing noise and emissions by conventional transport. Most co-housing projects have shared cars, laundries and so on which could achieve similar results.

Third, like homeowners, co-housing initiatives are in charge of their living environment: maintaining gardens, buildings, and utilities, including the energy system, and often seek to do this with low environmental impact. Decisions of a community can have more impact than individual decisions, and they may concern energy allocation. Examples are: peak times management by shared e-cars, priorities of daytime washing to take advantage of photovoltaic power (PV), and cooling food supplies in common storage. Finally, the dynamic of sharing creates opportunities for learning or developing together, but this strongly depends on the efforts of the co-housing members (Baborska *et al.*, 2014). For example, technicians argue that 'residents can't handle the ventilation system' to explain the results that fail to meet expectations (Gram-Hanssen, 2014). Co-housing residents have proven the contrary: they are willing and capable of organising the technical management mutually. The applied technologies and engineering do not in themselves need to be innovative. Rather it is the specific combination, produced by the location, the patterns of use and sharing, and the steering residents group, that makes the project effective. Low-tech mechanisms that can be handled easily by residents can be more effective than electronics whose functioning very few people are capable of controlling or supervising.

Co-housing, then, makes visible how closely social and technical aspects of housing are intertwined. But where is the gender-perspective?

Co-housing and gender

Based on the Swedish history of co-housing, Horelli writes:

> Literature on the history of cohousing from the gender perspective (see Vestbro and Horelli, 2012) provides evidence that cohousing increases equality between women and men by making the domestic chores visible and thus sharable by both sexes.
>
> (Horelli, 2013: 49)

Apparently in Scandinavian *cohousing*, gender equality is reached through the re-division of domestic labour (though it is usually through cooking – seldom through cleaning). There is, however, no systematic survey available to support this. Other aspects of gender-equality, such as body integrity, a voice in decision-making and equal pay, are even less discussed in academic literature (Tummers, 2015a). As clearly, co-housing cannot achieve a re-division of labour, roles and inequality by new planning criteria alone, gender equality needs to be explicitly on the collectives' agenda. Does it question the assignment of gender-specific skills, for example, in dividing maintenance tasks? What is the gap between rhetoric and practice? How do the projects benefit from breaking with gender stereotypes? Or are questions such as 'who does the cleaning' considered marginal in light of the major issues, such as climate change? Illustrative is Metcalf's observation that 'Within most intentional communities, however, we find traditional gender roles being followed by women and men' (Metcalf, 2004: 100). More recently, Jarvis (2012) showed that this applies to *Christiania* (Denmark) and Pickerill (2015) draws a similar conclusion based on empirical research into a number of eco-housing initiatives in different countries.

The collectivisation of domestic labour has been associated with gender-equality since the first planning discussions. Already in the 1900s, Charlotte Perkins Gilman and later European architects and reformers such as Alva Myrdal proposed the collective kitchen (Hayden, 1980; Vestbro and Horelli, 2012). The idea of 'eating out' as an alternative for home-cooking has been taken further by modernist architects such as Le Corbusier (Jarvis, 2011; Denèfle *et al.*, 2006). Dolores Hayden includes a co-housing model in the image of a 'non-sexist city' (Hayden, 1980), while Roberts discusses the British 'public kitchens' in the post-war period (Matrix, 1984). In the 1990s a Scandinavian working group 'New Everyday Life' proposed co-housing as a model with the potential to alleviate women from domestic tasks (Horelli and Vepsä, 1994; Sangregorio, 1995). Since then, several projects have been realised following the model of the 'kollektivhus' (Sweden) 'Cohousing' or 'Centraal Wonen' (The Netherlands) model,[7] in which the sharing of cooking and common meals structures the community (Vestbro and Horelli, 2012). Men and women in these projects participate equally in these visible, formal-ised tasks, according to empirical evidence (Vestbro, 2010). However, shared meals in co-housing outside Scandinavia are less common practice. Moreover, domestic work comprises much more activities, such as cleaning, childcare, and washing. How far co-housing makes these visible and thus leads to sharing of reproductive responsibilities is an impact that has not been demonstrated yet, although Horelli and Wallin claim that

> "The New Everyday Life-approach, which sought to embed the self-work model of cohousing in the neighbourhood context, still seems to be valid. It is currently being applied in a number of gender-aware neighbourhood projects in Germany, Spain, Austria, Italy and Finland"
>
> (Horelli and Wallin, 2013).

The collectivity and scale of co-housing mediates between the private, intimate sphere, and the public domain. The culture of interaction and attitudes associated with activities in each of the domains is different, and gendered to a high degree (Rosaldo, 1980). To bridge those spheres, co-housing environments potentially offer a new learning and negotiation ground. Nevertheless, co-housing offers insight in the spatial implications of the (partially) collective household model. Besides (often professionally equipped) common kitchens, this includes laundries, playgrounds, food cooperatives, maintenance tools and garden sheds, transport and 'taxi-services'. And domestic tasks do not stop at the front door. At present Gender Mainstreaming[8] emphasises the importance of 'reconciling work and home' and creating equal access to the labour market through equal division of domestic tasks. Gender Mainstreaming promotes the visibility of 'care' and its unpaid contribution to the economy. For this reason, gender-aware planning promotes the 'city of proximity concept' mixing jobs, residential and amenities at close distance to reduce travel times and facilitating the combination of job and home (Gilroy and Booth, 1999). Its interests overlap with environmental concerns, in reducing CO_2 emissions from transport and creating accessible public space (see, e.g. Lehmann, 2016). In other words, at a different scale, gender-aware approaches to urban planning share two central concepts with co-housing: everyday life as a constituent process, and the intermediary level of collective or shared spaces between private and public, as important conditions for sustainable spatial development. As Jarvis puts it, there is much to learn from co-housing about the interaction between social and spatial structures:

> By drawing attention to the multiple temporalities that shared amenities and collective decision making open up, (...) I reject the suggestion, often made from architectural observations alone, that proximity and social contact are sufficient to cultivate conviviality and cooperation between residents.
>
> (Jarvis, 2011: 573)

On the other hand, gender-aware approaches for planning make it clear that not only the household chores, but all aspects of housing are 'gendered': for example the choice of location (that involves mobility questions); mixed use; ownership and tenure; decision-making and participation dynamics; priority criteria in design, and maintenance, and so on (Kennett and Chan, 2011). Associating the care economy solely with everyday and small-scale practices at project-level does not yet reveal how co-housing responses for gender equality may be relevant for the housing provision institutes. The implicit gender assumptions of spatial planning remain unchallenged.

Conclusions: co-housing, lessons learned

Allowing children to grow up in a protected low-carbon footprint area with friendly neighbours and all urban facilities nearby seems to be the ideal model for European young households today. Co-housing initiatives realise this ideal by collective action, rather than individual consumption. Co-housing is an invitation to move out of the passive house and the passive role: it 'empowers' rather than 'victimises'. This makes co-housing different from other low-carbon housing proposals, but how far is the co-housing concept able to address climate change and equal rights structurally? Foremost, co-housing projects need to be contextualised to understand the gaps between reality and practice. One project does not change the building industry, but there are important lessons in the misfits with planning requirements that can be brought to a structural level.

Even if each inhabitant in every co-housing project throughout Europe fully achieved the low footprint they aimed for, quantitatively this only means a small indent on climate threats. Instead, the relevance of co-housing initiatives lies in its attempt to put into practice what, for most policies, remains on paper. The trial and error process can be highly instructive to connect national and global strategy concepts to the everyday needs, aspirations and realities of urban households. Co-housing demonstrates that the application of new technologies and renewable fuels cannot be seen as separate from the domestic practices if its impact is to be optimised. This social aspect, assuring the understanding and enabling self-management of the installations, is often ignored in the design engineering. Technology-based approaches such as '*passivhaus*' include sophisticated technologies, and require a specific knowledge as well as active handling by its inhabitants. Co-housing residents are motivated to engage with technology, and to participate in experiments with innovative energy-systems. Institutional housing partners are often more dependent on legal and technical structures and less prone to experiment.

The building and engineering industry is still very male dominated, and it appears that the redivision of domestic tasks has been more successful than the redistribution of technical maintenance. Even when more women participate, the type of knowledge that circulates and the ways it is shared do not necessarily correspond to the priorities of residents, or allow for different ways of knowledge and communicating. But different attitudes towards technology and the collective learning process in co-housing do not automatically break with gender roles. Further studies are needed on issues of visibility, leadership and decision-making in collectively self-managed housing, and on stereotyped gender roles in particular. Such data would provide insight into the limitations of spatial strategies and their impact on social relations. Providing shared spaces and utilities creates possibilities that are not available in the current single-unit-based planning.

Housing and planning professionals must be aware of how the built environment enables or hinders the choice of lifestyle, yet ultimately how these options are used depends on the inhabitants. For co-housing to represent a 'double shift in roles' and to challenge the 'root causes of climate change', it will be necessary not only to overcome technocratic approaches but also to strengthen the normative stand on gender relations. It is vital that new theory is developed on the gender dynamic in housing practices, as gender equality may be the key to resolving the challenges that climate change presents in a just manner.

Notes

1 http://standard.gendercop.com/about-the-standard/what-is-gender-mainstreaming (accessed 2 May 2016).
2 See for example Netherlands: www.iewan.nl/ecologisch/; UK: www.lilac.coop/; France: https://leszecobatisseurs.wordpress.com/ Germany: Sieben Linden www.siebenlinden.de/index.php?id=52&L=2
3 www.middenhuis.nl/vwz.html
4 http://ec.europa.eu/energy/en/topics/infrastructure
5 see for example: http://ec.europa.eu/energy/en/topics/markets-and-consumers/smart-grids-and-meters/smart-grids-task-force
6 for technical standards see for example www.passivhaus.org.uk/
7 www.kollektifhus.nu; www.cohousing.org.uk; www.lvcw.nl (accessed 2 May 2016).
8 "Gender mainstreaming is the integration of the gender perspective into every stage of policy processes – design, implementation, monitoring and evaluation – with a view to promoting equality between women and men", http://ec.europa.eu/social/main.jsp?catId=421&langId=en (accessed 12 March 2015), for more information see also http://eige.europa.eu/gender-mainstreaming/what-is-gender-mainstreaming (accessed 26 July 2016).

References

Alber, G. (2015) *Gender and Urban Climate Policy Gender-Sensitive Policies Make a Difference*. Bonn: GIZ; UN Habitat; GenderCC,
Baborska-Narozny, M., Stevenson, F. and Chatterton, P. (2014) A social learning tool – barriers and opportunities for collective occupant learning in low-carbon housing. *Sustainability in Energy and Buildings*: 10.
Bresson, S. and Denèfle, S. (2015) Diversity of self-managed cohousing initiatives in France. *Journal of Urban Research and Practice*, 8(1): 5–16.

Bresson, S. and Tummers, L. (2014) L'habitat participatif autogéré en Europe : vers des politiques alternatives de production de logements ? *Metropoles Politiques alternatives de développement urbain*(15). Available online at http://metropoles. revues.org/4622 (accessed 23 January 2015).

Brounen, D., Kok, N. and Quigley, J.M. (2012) Residential energy use and conservation: Economics and demo graphics. *European Economic Review*, 56(5): 931–945.

Brundtland, G. *et al.*, (1987) *Our Common Future. Report of the World Commission on Environment and Development*. United Nations. Available online at http://conspect. nl/pdf/Our_Common_Future-Brundtland_Report_1987.pdf (accessed 2 July 2014).

Buis, H. (ed.), (2000) *Duurzaam bouwen een kwestie van willen en weten. (report demonstration projects sustainable building*. Utrecht: Nationaal DuBo Centrum.

Chatterton, P. (2013) Towards an agenda for post-carbon cities: Lessons from Lilac, the UK's first ecological, affordable cohousing community. *International Journal of Urban and Regional Research*, 37(5): 1654–1674.

Damyanovic, D., Reinwald, F. and Weikmann, A. (2013) *Handbuch 'Gender Mainstreaming in der Stadtplanung und Stadtentwicklung'*. Werkstattbericht 130, Viena: Magistrat der Stadt Wien, Stadtentwicklung und Stadtplanung. Available in english from: www. wien.gv.at/wienatshop/Gast_STEV/Katalog.aspx?__jumpie#magwienscroll (1 August 2016).

Denèfle, S., Bresson, S., Dussuet, A. and Roux, N. (2006) *Habiter Le Corbusier. Pratiques sociales et théorie architecturale*. Rennes (Fr): Presse Universitaire de Rennes.

Droste, C. (2015) German Co-Housing: An Opportunity for Municipalities to Foster Socially Inclusive Urban Development? *Journal of Urban Research and Practice*, 8(1) (March): 79–92.

European Environmental Agency (EEA) (2015) *The European Environment – State and Outlook 2015*. Copenhagen: EEA.

Fenster, M. (1999) Community by covenant, process, and design: Cohousing and the contemporary common interest community. *Journal of Land Use & Environmental Law*, 15(1): 3–54.

Fromm, D. (2000) Introduction to the cohousing issue. *Journal of Architectural and Planning Research*, 17(2): 91–93.

Fromm, D. (2012) Seeding community: Collaborative housing as a strategy for social and neighbourhood repair. *Built Environment*, 38(3): 364–94.

Gilroy, R. and Booth, C. (1999) Building infrastructure for everyday lives. *European Planning Studies*, 7(3): 307–324.

Gram-Hanssen, K. (2014) New needs for better understanding of household's energy consumption – behaviour, lifestyle or practices? *Architectural Engineering and Design Management*, 10(1–2): 91–107.

Hayden, D. (1980) What would a non-sexist city be like? Speculations on housing, urban design, and human work. *Signs: Journal of Women in Culture and Society*, 5(3), Supplement (spring 1980): S170–187.

Horelli, L. (2013) The role of shared space for the building and maintenance of community from the gender perspective - a longitudinal case study in a neighbourhood of Helsinki. *Proceedings of the 11th conference of the International Communal Studies Association*, Social Sciences Directory, 2(5): 47–63.

Horelli, L. and Vepsä, K. (1994) In search of supportive structures for everyday life, in Irwin Altman and Arza Churchman (eds), *Women and the Environment*. New York: Plenum Press, pp. 201–226.

Horelli, L. and Wallin, S. (2013) Towards an architecture of opportunities, in L. Horelli (ed.), *New Approaches to Urban Planning–Insights from Participatory Communities. Aalto University Publication Series.* Helsinki: Aalto University, pp. 153–161.

Jarvis, H. (2011) Saving Space, Sharing Time: Integrated Infrastructures of Daily Life in Cohousing. *Environment and Planning,* 43(3): 560–577.

Jarvis, H. (2012) Against the 'tyranny' of single-family dwelling: Insights from Christiania at 40. *Gender, Place & Culture: A Journal of Feminist Geography,* 20(8): 939–959.

Kennett, P. and Kam Wah Chan, K.W. (eds) (2011) *Women and Housing: An International Analysis.* London; New York: Routledge.

Kido, H. and Nakajima, Y. (2011) Predicted Energy Conservation by Use of Common Areas in Cohousing, 6. Helsinki, Finland: Kogakuin University, 20111008. Available online at www.irbnet.de/daten/icond (accessed 16 January 2017).

Krokfors, K. (2012) Co-housing in the Making. *Built Environment,* 38 (3): 309–314.

Labit, A. (2015) Self-managed co-housing in the context of an ageing population in Europe. *Urban Research & Practice,* 8(1): 32–45.

Lehmann, S. (2016) Transforming the City towards low-carbon Resilience, the second lecture in the 3rd season of the Habitat UNI Global Urban Lectures: http://unhabitat.org/urban-knowledge/urban-lectures/

Locatelli, D., Desrues, F. and Biry, J.-M. (eds) (2011) *Guide Pratique de l'Auto-Promotion.* Strasbourg: Association Eco-Quartier Strasbourg et CAUE Bas-Rhin.

MacGregor, S. (2010) Gender and climate change: From impact to discourses. *Journal of the Indan Ocean Region,* 6(2): 223–238.

Maldonado. E. (ed.), (2012) *Implementing the Energy Performance of Buildings Directive (EPBD) – Featuring Country Reports 2012.* Paneuropean: CA-EPBD, pp. 279–286. Available online at www.epbd-ca.eu/archives/669 (accessed 24 March 2016).

Marckmann, B., Gram-Hanssen, K. and Christensen, T.H. (2012) Sustainable living and co-housing: Evidence from a case study of eco-villages. *Built Environment,* 38(3): 413–429.

Matrix (1984) *Making Space: Women and the Man-Made Environment.* London and Sydney: Pluto Press.

Metcalf, W. (2004) *The Findhorn Book of Community Living.* Forres, UK: Findhorn Press.

Moser, C. (2014) *Gender planning and development: Revisiting, deconstructing and reflecting.* DPU60 Working Paper Series: Reflections, London: The Bartlett Development Planning Unit. Available online at www.bartlett.ucl.ac.uk/dpu (accessed 7 December 2015).

Pickerill, J. (2015) Bodies, building and bricks: Women architects and builders in eight eco-communities in Argentina, Britain, Spain, Thailand and USA. *Gender, Place & Culture* 22(7): 901–919.

Rahimian, M.L.D.I. and Cardoso Llach, D. (2015) The case for a collaborative energy sharing network for small scale community microgrids, in *Proceedings Architecture and Resilience on the Human Scale* Sheffield 10–12 September, pp. 173–180.

Resurrección, B.P. (2013) Persistent women and environment linkages in climate change and sustainable development agendas. *Women's Studies International Forum,* 40(September): 33–43.

Rosaldo, M.Z. (1980) The use and abuse of anthropology: Reflections on feminism and cross-cultural understanding. *Signs,* 5(3): 389–417.

Sanchez de Madariaga, I., and Roberts, M. (eds) (2013) *Fair Shared Cities. The Impact of Gender Planning in Europe.* Farnham: Ashgate.

Sangregorio, (1995) Collaborative housing: the home of the future? in L. Ottes, E. Poventud, M. van Schendelen and G. Segond von Banchet (eds), *Gender and the Built Environment*. Assen, The Netherlands: Van Gorcum, pp. 101–114.

Seyfang, G. and Haxeltine, A. (2012) Growing grassroots innovations: exploring the role of community-based initiatives in governing sustainable energy transitions. *Environment and Planning C: Government and Policy*, 30: 381–400.

Stevenson, F., Baker, H. and Fewson, K. (2013) Cohousing case studies in the UK: Is sharing facilities really resourceful? *PLEA* 7, Munchen.

Tillie, N., van den Dobbelsteen, A. and Carney, S. (2014) Planning for the Transition to a Low-Carbon City: A New Approach, in S. Lehman (ed.), *Low Carbon Cities: Transforming Urban Systems*. London: Earthscan, pp. 173–190.

Tummers, L. (2015a) The re-emergence of self-managed co-housing in Europe: A critical review of co-housing research. *Urban Studies*, 23(May): 1–18.

Tummers, L. (ed.) (2015b) Taking apart co-housing. Towards a long-term perspective for collaborative self-managed housing in Europe (Introduction to the Special Issue). *Urban Research and Practice*, 8(1) (March).

Vestbro, D. (ed.) (2010) *Living Together: Co-housing Ideas and Realities Around the World*. Stockholm: Royal Institute of Technology division of urban studies in collaboration with Kollektivhus NU.

Vestbro, D. and Horelli, L. (2012) Design for gender equality: The history of co-housing ideas and realities. *Built Environment*, 38(3): 315–335.

Williams, J. (2005) Designing neighbourhoods for social interaction: The case of cohousing. *Journal of Urban Design*, 10(1): 195–227.

Wohnbund, E.V. (ed.), (2015) *Europa Gemeinsam Wohnen | Europe Co-Operative Housing*. Berlin: Jovis.

15 Integrating gender and planning towards climate change response

Theorising from the Swedish case

Christian Dymén and Richard Langlais

Introduction

Planning guidelines for responding to climate change are only beginning to enter mainstream planning practice in Sweden. How any such guidelines can best affect municipal planning decision-making processes and their implementation is a complex question that has yet to be satisfactorily studied. The addition of a gender perspective to such deliberations greatly amplifies the extent to which much remains to be achieved. This chapter considers some of what has been learned so far, and uses the Swedish case to exemplify some of the progress, and the challenges, of integrating a gender perspective into climate change responses at the level of municipal planning.

While climate change mitigation has been on the agenda of spatial planning practitioners in Sweden for over two decades, adaptation has only become influential in recent years (Langlais, 2009b; Sweden, 2007, 2008, 2010). Considering climate change from those different angles, even without a gender perspective, is therefore only beginning to be explored. The revised planning and building law from 2011 clearly states that municipalities must consider climate change in their planning work, both as an issue of mitigation and as a motive for adaptation, but clear guidelines for that, let alone for including a gender perspective, await development. The co-existence of a strong municipal planning monopoly and weak regional authorities calls for new approaches to climate change responses, ones where municipalities do not feel alone in their mitigation and adaptation work (Dymén and Langlais, 2012). In reflecting that situation, Andersson *et al.*, (2015), at the Swedish Meteorological and Hydrological Institute, state the importance of strengthening municipal responses to climate change, as well as ascribing to the regional level the permanent task of developing far-reaching climate change adaptation strategies and more involvement by the national level. In addition to that official activity, calls are proliferating for citizen participation, whatever their gender, in policy-making (Bang, 2005, as cited in Montin 2009). Björnberg and Hansson (2011), Larsen and Gunnarsson-Östling (2009), Montin (2009) and Mannberg and Wihlborg (2008) explicitly suggest that participatory decision-making may have the potential to produce better results.

Montin (2009) specifically argues that participation of non-governmental actors is imperative in addressing insecurities in climate change policy. Such insecurities include:

1) cognitive insecurity, which implies that knowledge regarding the causes and effects of climate change is lacking;
2) strategic insecurity, which implies that stakeholders with different interests are involved and that no single individual or organisation can be held responsible for societal problems;
3) institutional insecurity, which implies that decisions about coping with climate change are made at different levels, in different sectors, and in different arenas, and, just as in strategic insecurity, no single individual or organisation can be held responsible for societal problems; and,
4) value insecurity, which implies that it is difficult to develop a common understanding of what risks are acceptable, unacceptable, and dangerous.

In parallel, a body of knowledge has emerged that focuses on the relationship between gender and concern for environmental and climate change. One of its main arguments is that women generally are more concerned and proactive with respect to environmental issues. In desiring to learn more about whether it is meaningful to speak of gender-based perspectives in planned municipal responses to climate change, and how that insight can be useful for planners, we were inspired by a particular series of studies. That literature, which is presented below, discusses empirical studies of different kinds of behaviour, concern and values with regard to the environment and its exploitation.

With respect to behaviour, there are suggestions that everyday life patterns in, for example, transportation and energy-use differ between women and men (see, e.g. European Institute for Gender Equality, 2012; Sandow, 2008; Johnsson-Latham, 2007; Transek, 2006a, 2006b; Polk, 2003 1998; Krantz, 2000; Carlsson-Kanyama et al., 1999). In addition, research in, for example, environmental sociology and social psychology, suggests that gender differences do exist regarding concerns, values and perceptions related to the environment (Norgaard and York, 2005; Zelezny et al., 2000; Bord and O'Connor, 1997; Davidsson and Freudenberg, 1996). In general, women are seen as being more concerned about the environment, and as adopting behaviour that is more environmentally friendly than that of men. Norgaard and York (2005) and Villagrasa (2002) argue that women's involvement was imperative to the development of Local Agenda 21 programs, and in UN negotiations regarding the Kyoto Protocol. Yet other studies, somewhat contradictorily, argue that gender differences primarily occur in private, not public, life, where the latter include public expressions of environmental concern and attendance at public meetings (see, e.g. Hunter et al., 2004; and Tindall et al., 2003). While reviewing the literature, we became concerned that focusing on differences between women and men would solely contribute to a quantification of the ways in which women and men use the environment. We find it more useful to argue, similarly to Henwood *et al.*, (2008), that future research must move beyond focusing on gender differences, to studying the effects of the interplay between various aspects of gender and power.

This emerging situation calls for a better understanding of the relationship and possible synergies between planning theory, as extended to environmental planning and participation, and gender theory. This chapter draws on a theoretical discussion of the intersection between gender and planning theory to demonstrate how an understanding of feminine and masculine values and perspectives (cf. Kurian, 2000) contributes to and enhances the participatory planning approach, which is essential in achieving well-informed climate change response (Montin, 2009). In concluding that demonstration, it is argued that the inclusion of a gender perspective in environmental planning contributes an added power dimension; it systematically raises awareness of how masculine values and perspectives tend to gain power over feminine values and perspectives. This is a loss, since it is precisely the latter that might otherwise prove equally fundamental in developing well-informed climate change response.

Our argumentation, in this chapter, explores three main dimensions of municipal planning as climate change response, namely (1) spatial planning and participation, (2) environmental planning and participation, and (3) the characterisation of feminine and masculine values and perspectives in planning for climate change. The exploration is followed by an analysis and integration of those three dimensions in order to contribute not only to gender and planning theory, but to its more informed practice, as well as provide a platform for further study.

On spatial planning and participation

The section above touched on several aspects of how participatory spatial planning and decision-making can make a contribution to climate change responses. The present section seeks to expand that understanding. This is an essential background for being able to enrich the notion of planning by including a gender perspective. The notion of public participation, for instance, is by definition gender neutral – the "public" is without gender – as are "citizens". Below, we describe our interpretation of two main spatial planning approaches that we consider to be important, generally, for understanding municipal environmental planning and, specifically, its extension to climate change responses and, even further, its relation to a gender perspective. The two approaches are the rational planning approach and the communicative planning approach (which subsumes a participatory approach).

Since the 1900s, cities have been the product of rational planning practices, which situate the planner as an expert who knows what is best for the city. Until the 1980s, public authorities, especially local authorities – municipalities – were assigned powerful roles in developing, managing, and planning cities (Healey *et al.*, 2002).

More recently, regarding planning as a rational process has been criticised for being founded on the false assumption that science can produce the best possible plan. That critical discourse therefore presents alternative indications that planning can be a communicative process that acknowledges stakeholders, citizens, and other aspects of civil society, as experts. The planner is then instead regarded as a mediator (Strömgren, 2007).

The communicative approach follows a trend among local authorities in which planners increasingly delegate power to civil society. One of the pioneers of communicative planning and political decision-making is John Forester. He follows Friedmann (1969) by arguing that to achieve progress in preparing and implementing a plan, the planner should have the capacity to learn from others, to communicate in an empathic way, and to negotiate and make compromises in a contested environment.

Healey (1997) notes that the relationship between knowledge and action has been contested, in the sense that knowledge not only has an objective existence in the external world, but is also shaped by social and interactive processes (see, e.g. Latour, 1987, as cited in Healey, 1997). This intellectual trend has been increasing since the 1970s, and is further developed by Healey, who has developed approaches for what she terms collaborative planning. Her starting point is the question of how to address collective concerns when different communities and groups co-exist in shared spaces, and often have different priorities and different perceptions of things (1997: 310).

Strömgren (2007), on the other hand, argues that citizen participation and consultation are rarely part of contemporary planning processes; the planner dominates and controls the entire process of planning the city. Based on everyday empirical observation, we would nonetheless modify Strömgren's argument that citizen participation and consultation are not aspects of a rational planning process. Even judicially speaking, according to Swedish planning and building law, regular consultations must be made within spatial planning activities. Moreover, engaging in consultation de facto exists within the rational planning paradigm, since many spatial planning processes actually use consultations, even if occasionally only as tokenism, to varying degrees (see, e.g. Arnstein, 1969). More generally, the adoption of a communicative, or, in Healey's sense, collaborative, approach requires the delegation of some power to civil society. Perhaps, indeed, it may be argued that being communicative is, itself, the result of a choice that is rational, from a functional perspective.

In the section, "Integrating a gender perspective into environmental planning", below, this discussion is returned to, for the purpose of analysis and further argumentation. Before that, however, a more developed understanding of environmental planning and the place of a gender perspective in planning for climate/ environmental change needs to be considered.

On environmental planning and participation

"Environmental planning" is a term that is used in this chapter for analysing and discussing those forms of Swedish spatial planning that focus on environmental issues, including climate change. Rydin and Pennington (2000) suggest that environmental planning, when public participation is involved, can be understood as a combination of three different approaches, namely, environmental management, environmental governance, and collaborative environmental planning. Deciding which, or which mix, of these approaches to adopt depends on individual incentives either to participate, or to adopt non-cooperative behaviour, such as free riding and rent-seeking. "According to public choice, the exceptions to this logic of non-public-participation tend to occur in small-group situations, where the potential participants know each other and there is the prospect of strategic bargaining in an iterative social context" (Rydin and Pennington, 2000: 257). One reflection here is to ask, somewhat provocatively, whether a non-public-participation approach might actually allow more space for gender perspectives.

Collaborative environmental planning is more distinctive than the other two approaches, which are closer to each other in style. In collaborative environmental planning, participation is an end, not a means to an end. The aim is to allow for new modes of democratic governance. Participatory exercises are applied, so as to avoid conflicts that would otherwise jeopardise policy delivery. Participation for the purpose of gathering information can also be considered as legitimate. Conversely, in environmental management, public participation is regarded as a means to an end, and the state has the role of controller. Environmental governance occupies an intermediate position between those other two forms of environmental planning. Policy delivery is the goal, but in contrast to environmental management, the state acts as a facilitator rather than a controller (Rydin and Pennington, 2000).

Environmental planning and Sweden

Applying the above to the Swedish case, our empirical evidence (Dymén, Andersson and Langlais, 2013; Dymén and Langlais, 2012) indicates that the challenges that Swedish municipalities face in integrating climate change into spatial planning are closely related to the lack of a gender perspective and the insecurities – cognitive, strategic, institutional and value (Montin, 2009) – presented and summarised at the start of this chapter. The approaches to environmental planning developed by Rydin and Pennington (2000) provide a basis for our argument that environmental governance, which is driven by policy delivery and delineates the role of the state as a facilitator, is a suitable environmental planning approach for Swedish municipalities.

We adopt this position because the types of planning challenges reported by Swedish municipalities are usually addressed with the assistance of participatory decision-making (Dymén and Langlais, 2012). Recall that, as mentioned in the introduction, the demand for citizen participation in policymaking is increasing in Sweden, and that participation may resolve insecurities related to climate change responses (Montin, 2009). However, in that context, there is a growing call for the participation not only of citizens, but also of a multitude of other stakeholders in such processes.

Specifically, the challenges and insecurities faced by Swedish municipalities, and that, to some extent, correspond to those highlighted by Montin (2009), are related to the following aspects:

- mapping climate change risks at a comprehensive level is straightforward, but when legally-binding decisions are made on a detailed level, conflicts of interest become concrete and economic aspects tend to prevail (as expressed, for example, in value insecurity and strategic insecurity);
- conflicts of interest emerge among the social, economic, environmental and gender dimensions of spatial development (value insecurity, strategic insecurity);
- the effects of climate change are not always clear (cognitive insecurity);
- municipal planners are often isolated in their efforts to adapt to and mitigate the effects of climate change, which has led them to call for cooperation across municipalities and within regions, which in turn requires additional inter-regional cooperation and national assistance (institutional insecurity); and
- municipalities struggle with national guidelines that are not always commensurable (institutional insecurity) (Dymén and Langlais, 2012; Dymén, Andersson and Langlais, 2013).

In other situations, where non-cooperative behaviour, such as free riding and rent-seeking, is a challenge, the adoption of an environmental management approach may be necessary. In those circumstances, the state is compelled to adopt a more controlling role, but the main goal remains policy delivery. When the efforts of Swedish municipalities to address climate change through public participation, as part of an environmental governance approach, are weakened by powerful and complex conflicts of interest, they may desire increased guidance, structure, and assistance from the regional and national governments (cf. Dymén and Langlais, 2012).

In our research, we have found that, in spite of their efforts, municipalities can rarely achieve climate change goals in isolation. Increasing the responsibility that is assumed by regional and national governments is needed (Dymén and Langlais, 2012). Letting the regional and national levels introduce long-term goals and guiding principles may minimise the challenges and insecurities experienced by municipal planners (see the list of points above).

In Dymén and Langlais (2012), we describe Swedish cases where a goal conflict can be observed between mitigation measures, on the one hand, and adaptation issues, on the other. A concrete example of a climate change response is the increasing use of air conditioning as a way of coping with uncomfortably warm temperatures. It can be argued that this is an adaptation measure, but it can also be considered as increasing the need for mitigation. In other words, if the electricity used for air conditioning is generated from the burning of fossil fuels, with the resultant emission of greenhouse gases, the need for mitigation measures is increased. According to Howard (2009), synergies in effort can differ in extent, ranging along a scale from a point where adaptation and mitigation measures are synonymous, to where adaptation measures are mitigation neutral (2009: 28). The latter is illustrated by returning to the example of the installation of more air conditioning as an adaptation measure. If renewable energy is used to power it, it might be considered to be mitigation neutral. An example of a synonymous measure would be the implementation of LEED-certified (Leadership in Energy and Environmental Design) building standards, since it can be seen as both a mitigation and an adaptation measure (Howard, 2009). We have termed these synonymous measures, when they are synergetic, 'adaptigation' (Dymén and Langlais, 2012; Langlais, 2009a).

As argued above, the establishment of regional and/or national guidelines and standards can mitigate conflicts of interest and insecurities at the municipal level, in relation not only to adaptation and mitigation, but also to other dimensions of municipal spatial development. Yet another illustration is useful here, this time from Norway.

In 2008, the Norwegian government adopted new regulations for the construction of shopping malls. Such construction is now limited to those areas that have been designated in regional development plans. The purpose of the regulation is to ensure regional coordination of shopping malls, and to avoid urban sprawl that leads to increased automobile use (Dymén, Brocket and Damsgaard, 2009). The regulation also minimises unsustainable competition among municipalities; one municipality with aggressive sustainability goals might reject applications from shopping mall developers, while a neighbouring municipality might accept such offers. In other words, the municipalities may also in some cases be in the undesirable position of having to accept applications for construction that contradict their own motives.

As indicated in Norway, guidelines and principles developed at the national level are not a guarantee that effective climate change policies will be implemented. Municipalities might have more aggressive climate change agendas than the national government, leaving them with the impression that their efforts are being undermined.

The regional level can play an intermediate role in this respect, as interlocutors, since regional governments are generally in closer touch with the bottom-up and grass-roots sensitivities present in municipalities than the national level is. Another way that different levels of authorities may play varying roles in environmental planning is in the incorporation of a gender perspective into their planning for climate change.

On feminine and masculine values and perspectives in planning for climate change

In this section, as well as the following one, we discuss in depth how the integration of a gender perspective into environmental planning can assist planners in developing well-informed responses to climate change.

A definition of gender for use in an analytical framework is called for here. Such a definition is crucial to an understanding of how a gender perspective can relate to spatial planning and environmental planning. By a gender perspective, we mean an awareness of and an ability to be responsive to gendered aspects and gendered dimensions of climate change responses. By gendered aspects and dimensions, we refer to something either, or both, produced and influenced by women's and men's different experiences of everyday life, as they arise in power structures that generally subordinate women. By indicating certain values and perspectives according to sex, they become gendered (see Arora-Jonsson, 2013).

Adopting a gender perspective is the engagement of an analytical framework. Such a framework can be useful when we, as researchers, or practitioners, work towards knowing more about, creating, and implementing well-informed climate change responses. Adopting a gender perspective is, however, only one of several analytical perspectives that influence policymaking and power structures. As Kurian (2000) argues,

> attitudes and perceptions of people differ on issues dealing with environment, development, and cultural values. These differences, while mediated by the realities of class, race or community, are further delineated according to gender. Gender values and world-views that are privileged get institutionalized in ways that have implications for decision-making and policy analysis generally.
>
> (2000: 26)

On the basis of this knowledge, Kurian (2000) introduces an analytical framework consisting of masculine and feminine values and perspectives, or attributes (see Table 15.1 below).

Kurian's framework was originally designed for use in evaluating and analysing the World Bank's environmental impact assessment (EIA) policies in the 1990s. More specifically, this involved analysis of the World Bank's own EIA literature and documents, as well as interviews of a number of its officials and of groups of people affected by the bank's projects. The framework was derived from a thorough review of gender scholarship (especially as extended to policymaking), in the fields of psychoanalysis, sociology and anthropology, where gender is often understood as socially-constructed (Diesing, 1962; Duerst-Lahti and Kelly, 1995; Gilligan, 1982; Kathlene, 1989). The framework is also inspired by gender theory deriving from ecofeminism (Shiva, 1988).

Table 15.1 Masculine and feminine attributes according to Kurian (2000).

Masculine attributes	Feminine attributes
(1) Main focus is development defined as economic growth	(1) Main focus is sustaining a way of life
(2) Sees nature and humans as separate	(2) Sees humans as part of nature
(3) Ignores the significance of culture and cultural norms, or sees them as unimportant	(3) Stresses cultural norms and values
(4) Utilitarian	(4) Preservationist (including cultural preservationist)
(5) Technical and economic rationality	(5) Political and social rationality
(6) Stresses hard sciences, quantitative techniques	(6) Stresses the social sciences, unquantifiable values
(7) Only top-down sources of knowledge are considered	(7) Bottom-up sources of knowledge are given legitimacy
(8) Only scientific sources of knowledge are considered	(8) Non-scientific sources of knowledge are also considered
(9) Process controlled by experts	(9) Process stresses participation by citizens

An analysis of how being aware of and responsive to gendered aspects and gendered dimensions, whether they be socially constructed or biological, can contribute to well-informed climate change responses, as is discussed in the next section. Such analysis is assisted by an analytical framework that derives from a gender perspective (see, e.g. Kurian, 2000; Duerst-Lahti and Kelly, 1995; Shiva, 1988; and Gilligan, 1982).

Integrating a gender perspective into environmental planning

As emphasised in the introduction, there is evidence that the inclusion of a gender perspective in municipal planning is important for developing well-informed responses to climate change. In other research undertaken with Andersson, we stipulate that, for the purposes of planning climate change responses, as in many other areas of life, differences exist between women and men regarding concerns, attitudes and behaviours related to the environment. We do not reflect on the origins of these differences, but adopt a pragmatic approach and argue that the perspectives of both women and men should be included in the work and results of spatial planning. Gender differences were observed among selected planners in a number of municipal administrations and in their interactions with civil society. Some empirical evidence indicates that planners may have tendencies to perceive differences between female planners and male planners with respect to attitudes and perceptions regarding climate change (Dymén, Andersson and Langlais, 2013). However, in a relatively gender-equal society, such as Sweden, one would expect feminine and masculine attributes to be relatively dissociated from the professional planning practice of women and men. Magnusdottir and Kronsell (2015) find that the mere presence of a critical mass of women does not necessarily lead to gender-sensitive climate policymaking. This conclusion is based on analysis that shows that even though women and men are equally represented (or even in some cases women are in majority), in Swedish administrative and political units involved in climate policymaking, climate policy remains gender unaware.

Knowledge about this issue remains a topic for further research. What is important, however, is that aspects of planning, and the departments where it is practiced, that have been historically and often still are associated with feminine attributes, have been less powerful and influential than those associated with masculine attributes (Sandercock, 1998). That planning departments, both in education and in practice, have to some extent indicated some awareness of this situation reveals at least a modicum of progress.

Returning to Rydin and Pennington (2000) discussed above, we propose that a gender perspective is compatible with, and adds a useful dimension to, not only collaborative environmental planning, but also environmental management and environmental governance, although in different ways. In the case of environmental management, planning power resides with planners and decision makers, embodying the presence of the state.

Environmental management corresponds well with the rational planning approach. To achieve policy outcomes that are well-informed (from the planners' perspective), a rational postulate is that both women's and men's experiences should be taken into account. Municipalities can encourage this by ensuring that women and men are equally represented among planners. Notwithstanding that this is at least a step forward, The Swedish National Board of Housing, Building and Planning (2007), stipulates that merely achieving gender equity among planners does not guarantee that a more feminine approach will be pursued in the ensuing years. If the training that professional female planners receive is within a masculine tradition, feminine attributes may not be represented. Friberg and Larsson (2002) say much the same thing, by emphasising that, just because women planners are employed in a planning organisation, there is no certainty that they will actually be interested in gender equality issues. Alternatively, planners could gather gender-sensitive empirical evidence from citizens, NGOs, stakeholders, and so forth, as a pro-active, compensatory and conscientious measure. Adopting a gender perspective to achieve well-informed policy does not necessarily mean that power is delegated. The planner, as the expert, must either include women and men and, or, assure the presence of feminine and masculine attributes in working groups, in order to make sound decisions (cf. Dymén *et al.*, 2014; and Dymén, Andersson and Langlais, 2013).

In environmental governance, the issue of climate change would most likely be considered to be a collective concern, even if communities have different priorities and perspectives, particularly relative to other social and economic priorities of urban and regional spatial planning. As Healey (1997) argues, "collaborative planning efforts ... seek to reframe how people think about winning and losing. It looks for an approach which asks: can we all get on better if we change how we think to accommodate what other people think?" (1997: 312). This approach (environmental governance) to environmental planning is closely related to a communicative planning approach. Note that in this approach, regional and national governments are not prohibited from intervening by, for example, establishing boundaries, guidelines, and rules. On the contrary, as we found during interviews with planners in a number of Swedish municipalities, respondents argued that such intervention is necessary (Andersson *et al.*, 2015; Dymén and Langlais, 2012).

A gender perspective can strengthen the communicative approach to environmental governance. Planning preparations that include evaluation of spatial planning processes from a gender perspective, based on Kurian's (2000) work, have the potential to identify issues, perspectives, values and actions that have not been the norm in spatial planning practice – all what Kurian terms as feminine attributes – and that might otherwise be neglected (see, e.g. Dymén *et al.*, 2014). The purpose of such analyses is not to essentialise women or men, but rather to identify what aspects gain power in spatial planning processes, to the detriment of other aspects. Sandercock (1998), among others, has clarified this.

The history of spatial planning practice has been, and largely remains, the story of a modernist and rational planning project developed by white, middle-class men. Sandercock (1998) voices the concern that the practice of planning is dominated by the modernist and rational approach, and that academia has contributed to this situation. One of Sandercock's (1998) main points is that the history of planning is "... a narrative about the ideas and actions of white middle-class men, since women and people of colour were, at least until recently, systematically excluded from the profession" (1998: 7). Sandercock (1998) also emphasises that

> in revisiting planning history we discover an 'official story', which keeps being repeated – the story of the modernist planning project, the representation of planning as the voice of reason in modern society, the carrier of the enlightenment mission of material progress through scientific rationality.
>
> (1998: 1–2)

Since spatial planning has traditionally been and largely still is a male-dominated profession, projecting masculine values, perspectives and beliefs onto society (see, e.g. Listerborn, 2007, 2015; Greed, 1994, 2006; Sandercock, 1998), it is difficult to contend that the introduction of a communicative (participatory) approach and of a gender perspective into spatial planning practice and theory is anything but a challenge.

Nevertheless, when one considers the so-called feminine attributes that Kurian identified, they are well in line with what, for example, Forester (1987) would propose as a communicative, or argumentative, approach to planning. Based on that comparison, we do contend that, because a communicative and argumentative dimension may be useful in environmental governance at the municipal level, ensuring the presence of Kurian's (2000) more feminine attributes in planning will be helpful in working with a gender perspective, which in turn aids the achievement of well-informed climate change response. A more detailed consideration of how the feminine attributes are well aligned with, and add a dimension to, communicative or argumentative attributes, strengthens the latter assertions.

Feminine/communicative attributes in comparison to masculine/rational attributes

Forester (1987) argues that several problems must be addressed once we recognise that planning practice is communicative and argumentative. Power is a central issue in this respect. Whether power is considered or ignored will affect whether a process is more or less democratic and more or less technocratic (1987: 28). One of Forester's main conclusions is that when political organisation and debate are ignored, although the outcome in solving a certain issue might be technically correct, there still remains the question of whether the correct questions and problems have been considered, and potential alternative solutions ignored. Furthermore, by ignoring politics, an action may not be achievable, and public mistrust could result.

Table 15.2, below, is based on Forester's attributes of communicative and argumentative planning, and presents a juxtaposition with Kurian's framework of attributes, for comparison. By extracting some of the attributes presented by Kurian (2000),[1] one can clearly observe that the feminine attributes are aligned with a communicative and argumentative approach. This especially relates to attributes five through nine in Table 15.1, above. Combining the feminine attributes with the communicative and argumentative attributes reveals an important dimension, namely, that the communicative/feminine attributes are not the norm in spatial planning. As such, they require special attention and effort.

When a municipal organisation acknowledges the special attention and effort required for ensuring the presence and workings of communicative/feminine attributes, it becomes more likely to succeed in strengthening its climate change response.

A spatial planning process requires both environmental management (especially at the regional and national levels) and environmental governance (preferably at the municipal level). Environmental management adopts a more rational approach that corresponds to more masculine attributes, meaning that the process is controlled by experts and is primarily top-down. However, the adoption of a more rational approach does not imply that feminine attributes (such as focusing on a sustainable way of life, regarding humans as

Table 15.2 The two left-hand columns represent Forester's (1987) rational planning approach and Kurian's (2000) masculine attributes. The two right-hand columns represent Forester's communicative/argumentative approach and Kurian's feminine attributes.

Rational attributes	Masculine attributes	Communicative/ argumentative attributes	Feminine attributes
Technical expertise	Technical and economic rationality Process controlled by experts	Nonprofessional contribution	Political and social rationality Process stresses participation by citizens
Formal procedure	Only top-down sources of knowledge are considered	Informal consultation and involvement	Bottom-up sources of knowledge are given legitimacy
Strict reliance on databases	Only scientific sources of knowledge are considered	Careful use of trusted resources, contacts, and friends	Non-scientific sources of knowledge are also considered
Formally rational management procedures	Only top-down sources of knowledge are considered	Internal and external politics and the development of a working consensus	Bottom-up sources of knowledge are given legitimacy
Solving an engineering equation	Stresses hard sciences, quantitative techniques	Complements technical performance with political sophistication	Stresses the social sciences, unquantifiable values

Source: Our analysis, based on Forester (1987) and Kurian (2000)

part of nature, and stressing cultural norms and values) are not relevant; to the contrary. What is stressed here is that attributes that are typically related to the spatial planning process (see Table 15.2 above) are more consistent with the masculine/rational attributes. Environmental governance, on the other hand, adopts a more communicative approach, implementing the feminine attributes.

Thus, one potential solution, while being aware that what has historically been associated with masculinity tends to gain power at the expense of femininity, is to reflect on how the categorisation of spatial planning processes by using communicative/feminine and, or, rational/masculine attributes, can contribute to the development of well-informed policy without dichotomising differences between women and men. A well-designed process, we contend, combines both.

Discussion and future research

The attributes presented in Table 15.2 are, in this chapter, primarily discussed in relation to environmental planning and, more specifically, climate change response. However, their usefulness as an analytical framework needs to be tested more thoroughly in spatial planning and research more generally.

A first indicative attempt is made here, by drawing on knowledge gained from a study of experiences of fear of crime, and safety from it, in urban environments (cf. Dymén and Ceccato, 2012). From a masculine perspective, being afraid of crime is not rational, as in quantitative terms, being fearful is not related to actual crimes. Still, why can being fearful not be considered rational? Ultimately, fear of crime is a particular hindrance to women's ability to freely use public spaces.

One lesson learned in that context is that matters associated with rationality and masculinity are often regarded as precise and quantifiable. However, our hypothesis is that for planners to successfully plan for urban environments where especially women do not need to feel restricted because of fear of crime, the adoption of a perspective of political and social rationality is needed (cf. with Table 15.2 above). From a feminine perspective, political and social rationality are "rational". Social rationality is about understanding interpersonal relations and social relations and, ultimately, why individuals and groups make certain choices in a complex world; political rationality is about the rationality of decision-making, with order provided by discussion and decision (Diesing, 1962; Kurian, 2000). Planners would have to adopt a perspective of social rationality to be able to detect and understand fear of crime and how women (and also men) experience and cope with it. On the other hand, for urban planners to be able to plan for urban environments where people feel safe, a perspective of political rationality is needed. Through political rationality, the planner reflects upon how knowledge and participation by those who experience the urban environment should be considered. Inviting women, especially, to participate in urban planning and in the control of the design of public spaces can decrease the feeling of being excluded from, and fearful towards, public space (Dymén and Ceccato, 2012).

In conclusion, Table 15.2, above, is useful in understanding and analysing the relationship between planning and gender theory, on the one hand, and between masculine and feminine attributes, on the other. Initial prospects indicate that the framework discussed here is also useful for analysing and reflecting on other aspects of spatial planning that are not directly related to environmental planning. The study of fear of crime in urban public spaces is an example of such an attempt. However, there is a risk, when applying the framework in Table 15.2, that we, as researchers and planners, reiterate the notion that masculinity is rational whereas femininity is not. We therefore wonder whether new terminology should be adopted so as to avoid dichotomising femininity and masculinity. We contend that the decision to ensure the consideration, inclusion and application of communicative/feminine attributes in planning is also, ironically, a rational choice. A result of rethinking such terminology would be that the environmental management approach, as defined by Rydin and Pennington (2000), would no longer be termed according to usage that should already be rapidly becoming obsolete, that is, as rational planning; rather, the use of a new term would be pro-active and more appropriate. A term such as 'central and formal management of planning' may well serve in this regard. By avoiding the term 'rational planning' (as used by Forester, 1987) altogether, we challenge the dichotomy between masculinity, as being rational and scientific, on the one hand, and femininity, as being communicative and non-scientific, on the other. Recall that even the communicative planning approach (see Table 15.2) is also driven by specific forms of rationality, albeit different ones, namely political and social rationality. More generally, the latter are constituents of the grand project of rationality known as social science. By informing planners of these distinctions, the framework discussed above promotes planning practices that, to the extent they implement a gender perspective, are appropriate in responding to climate change.

Note

1 For a more detailed understanding of the origins of Kurian's attributes, see Dymén *et al.*, (2014) and Kurian (2000).

References

Andersson, L., Bohman, A., van Well, L., Jonsson, A., Persson, G. and Farelius, J. (2015) Underlag till kontrollstation 2015 för anpassning till ett förändrat klimat. (Foundation for "control station 2015" regarding climate change adaptation) (SMHI Klimatologi Nr 12, SMHI, SE-601 76) Sweden: Swedish Meteorological and Hydrological Institute.

Arnstein, S. (1969) A ladder of citizen participation. In J. Hillier and P. Healey (eds), *Critical Essays in Planning Theory: Foundations of the Planning Enterprise* (2008). Aldershot: Ashgate.

Arora-Jonsson, S. (2013) *Gender, Development and Environmental Governance. Theorizing Connections.* New York: Routledge.

Björnberg, K.E. and Hansson, S.O. (2011) Five areas of value judgement in local adaptation to climate change. *Local Government Studies*, 37(6): 671–687.

Bord, R.J. and O'Connor, R.E. (1997) The gender gap in environmental attitudes: The case of perceived vulnerability to risk. *Social Science Quarterly*, 78(4): 830–840.

Carlsson-Kanyama, A., Lindén, A.L. and Thelander, A. (1999) Gender differences in environmental impacts from patterns of transportation: A case study from Sweden. *Society and Natural Resources*, 12 (4): 355–369.

Davidsson, D. and Freudenberg, W. (1996) Gender and environmental risk concerns: A review and analysis of available research. *Environment and Behaviour*, 28 (2): 302–339.

Diesing, P. (1962) *Reason in Society.* Westport, CT: Greenwood Press.

Duerst-Lahti, G. and Kelly, R.M. (eds), (1995) *Gender Power, Leadership, and Governance.* Ann Arbor: University of Michigan Press.

Dymén, C. and Ceccato, V. (2012) An international perspective of the gender dimension in planning for urban safety. In V. Ceccato (ed.), *Urban Fabric of Crime and Fear.* Dordrecht: Springer Science, pp. 311–339.

Dymén, C. and Langlais, R. (2012) Adapting to climate change in Swedish planning practice. *Journal of Planning Education and Research*, 33(108): 108–119.

Dymén, C., Andersson, M. and Langlais, R. (2013) Gendered dimensions of climate change response in Swedish municipalities. *Local Environment: The International journal of Justice and Sustainability*, 18(9): 1066–1078.

Dymén, C., Brockett, S. and Damsgaard, O. (2009) Framtidens nordiska stad (Nordregio Working Paper 2009:1). Stockholm: Nordregio

Dymén, C., Langlais, R. and Cars, G. (2014) Engendering climate change: The Swedish experience of a global citizens consultation. *Journal of Environmental Policy & Planning*, 16(2): 161–181.

European Institute for Gender Equality (2012) *Review of the Implementation in the EU of area K of the Beijing Platform for Action: Women and the Environment*. Lithuania: Gender Equality and Climate Change.

Forester, J. (1987) *Planning in the Face of Power*. London: University of California Press.

Friberg, T. and Larsson, A. (2002) Steg framåt. Strategier och villkor för att förverkliga genusperspektivet i översiktlig planering (Rapporter och notiser 162). Lund: Institutionen för Kulturgeografi och Ekonomisk Geografi.

Friedmann, J. (1969) Notes on societal action. *Journal of the American Institute of Planners*, 35(5): 311–318

Gilligan, C. (1982) *In a Different Voice, Psychological Theory and Women's Development*. Cambridge, MA: Harvard University Press.

Greed, C. (1994) *Women and Planning: Creating Gendered Realities*. London: Routledge.

Greed, C. (2006) Making the divided city whole: Mainstreaming gender into planning in the United Kingdom. *Tijdschrift voor Economische en Sociale Geografie*, 97(3): 267–280.

Healey, P. (1997) *Collaborative Planning. Shaping Places in Fragmented Societies*. New York: Palgrave.

Healey, P., Cars, G., Madanipour, A. and De Magalhaes, C. (eds), (2002) *Urban Governance, Institutional Capacity and Social Milieux*. Aldershot: Ashgate.

Henwood, K.L., Parkhill, K.A. and Pidgeon, N.F. (2008) Science, technology and risk perception: from gender differences to the effects made by gender. *Equal Opportunities International*, 27 (8): 662–676.

Howard, J. (2009) Climate change mitigation and adaptation in developed nations: A critical perspective on the adaptation turn in urban climate planning. In S. Davoudi, J. Crawford and A. Mehmood (eds), *Planning for Climate Change: Strategies for Mitigation and Adaptation for Spatial Planners*. London: Earthscan, pp. 19–32.

Hunter, L., Hatch, A. and Johnson, A. (2004) Cross-national gender variations in environmental behaviours. *Social Science Quarterly*, 85 (3): 677–694.

Johnsson-Latham, G. (2007) A study on gender equality as a prerequisite for sustainable development (Report to the Environment Advisory Council, Sweden 2007: 2). Stockholm: Ministry of the Environment.

Kathlene, L. (1989) Uncovering the political impacts of gender: An exploratory study. *The Western Political Quarterly*, 42(2): 397–421.

Krantz, L.G. (2000) Rörlighetens mångfald och förändring. Befolkningens dagliga resande i Sverige 1978 och 1996 (Diversity and change of mobility. Citizen's daily travel patterns in Sweden between 1978 and 1996) (PhD dissertation). Gothenburg: University of Gothenburg.

Kurian, A.K. (2000) *Engendering the Environment? Gender in the World Bank's Environmental Policies*. Aldershot: Ashgate.

Langlais, R. (2009a) Adaptigation. *Journal of Nordregio*, 9(4): 2.

Langlais, R. (2009b) A climate of planning: Swedish municipal response to climate change. In S. Davoudi, J. Crawford and A. Mehmood (eds), *Planning for Climate Change: Strategies for Mitigation and Adaptation for Spatial Planners*. London: Earthscan, pp. 262–271.

Larsen, K. and Gunnarsson-Östling, U. (2009) Climate change scenarios and citizen-participation. Mitigation and adaptation perspectives in constructing sustainable futures. *Habitat International*, 33 (3): 260–266.

Listerborn, C. (2007) Who speaks? And who listens? The relationship between planners and women's participation in local planning in a multi-cultural urban environment. *GeoJournal*, 70(1): 61–74.

Listerborn, C. (2015) Medborgarinflytande – om makt, genus och stadsutveckling (Citizen participation – on power, gender and urban development) In T. Lindholm, S. Oliveira e Costa and S. Wiberg (eds), Medborgardialog – demokrati eller dekoration? Tolv röster om dialogens problem och potential i samhällsplanering (Citizen dialogues – democracy or decoration? Twelve voices about the dialogue's problem and potentials in spatial planning) (Arkus skrift no. 72). Stockholm: Arkus.

Magnusdottir, G.L. and Kronsell, A. (2015) The (in)visibility of gender in Scandinavian climate policy-making. *International Feminist Journal of Politics*, 17(2), 308–326.

Mannberg, M. and Wihlborg, E. (2008) Communicative planning – friend or foe? Obstacles and opportunities for implementing sustainable development locally. *Sustainable Development*, 16(1): 35–43.

Montin, S. (2009) Klimatpolitiken och kommunerna. In Y. Uggla and I. Elander (eds), *Global uppvärmning och lokal politik* (pp. 15–42). Stockholm: Santérus Academic Press Sweden.

Norgaard, K. and York, R. (2005) Gender equality and state environmentalism. *Gender & Society*, 19 (4): 506–522.

Polk, M. (1998) Swedish men and women's mobility patterns: issues of social equity and ecological sustainability. In *Proceedings from Women's Travel Issues*, 2nd national conference, 23–26 October 1996, Baltimore, MD (pp. 185–211). Washington, DC: Federal Highway Administration.

Polk, M. (2003) Are women potentially more accommodating than men to a sustainable transportation system in Sweden? *Transportation Research Part D*, (8): 75–95.

Rydin, Y. and Pennington, M. (2000) Public participation and local environmental planning: The collective action problem and the potential of social capital. *Local Environment*, 5(2): 153–169.

Sandercock, L. (1998) Framing insurgent historiographies for planning. In L. Sandercock (ed.), *Making the Invisible Visible: A Multicultural Planning History*. Berkeley, CA: University of California Press, pp. 2–33.

Sandow, E. (2008) Commuting behaviour in sparsely populated areas. Evidence from Northern Sweden. *Journal of Transport Geography*, 16(1): 14–27

Shiva, V. (1988) *Staying Alive: Women, Ecology, and Survival*. London: Zed Books.

Strömgren, A. (2007) Samordning, hyfs och reda – Stabilitet och förändring i svensk Planpolitik 1945–2005 (PhD dissertation). Uppsala: Uppsala Universitet.

Sweden (2007) Sweden facing climate change – threats and opportunities. Final report from the Swedish Commission on Change and Vulnerability (Swedish Government Official Reports SOU 2007:60). Stockholm: Ministry of the Environment.

Sweden (2008) Svensk klimatpolitik (Swedish climate change policy) (Swedish Government Official Reports SOU 2008:24). Stockholm: Ministry of the Environment.

Sweden (2008/2009) En sammanhållen klimat- och energipolitik – Klimat (Cohesive policy for climate change and energy) (Proposition 2008/2009:162). Stockholm: Ministry of the Environment.

Sweden (2010) Plan- och Bygglagen (Planning and Building Law) (SFS 2010:900, 2§). Swedish Parliament.

Swedish National Board of Housing, Building and Planning (2007) Jämna steg: checklista för jämställdhet i fysisk planering (2nd edn). Karlskrona: Boverket.

Tindall, D., Davies, S. and Mauboule, C. (2003) Activism and conservation behaviour in an environmental movement: The contradictory effects of gender. *Society and Natural Resources*, 16(10): 909–932.

Transek (2006a) Mäns och kvinnors resande – Vilka mönster ses i mäns och kvinnors resande och vad beror dessa på? (Men's and women's travels – what patterns can be observed and what are the reasons for them) (Report 2006:51). Stockholm: Transek.

Transek (2006b) Jämställdhet vid val av transportmedel (Gender equality when choosing transport modes) (Report 2006:13). Stockholm: Transek.

Villagrasa, D. (2002) Kyoto protocol negotiations: Reflections on the role of women. *Gender and Development*, 10(2): 40–44.

Zelezny, L.C., Chua, P.P. and Aldrich, C. (2000) Elaborating on gender differences in environmentalism. *Journal of Social Issues*, 56(3): 443–457.

16 A gender-sensitive analysis of spatial planning instruments related to the management of natural hazards in Austria

Britta Fuchs, Doris Damyanovic, Karin Weber and Florian Reinwald

This chapter discusses 'gender+' aspects in disaster risk reduction and in particular on decision-making processes and instruments concerned with spatial and socio-economic challenges. In order to account for the intersectionality of gender, the authors apply the term and definition of 'gender+' (Verloo, 2006). 'Gender+' includes the differences between individuals and within groups in terms of age, race, their phase of life, life situation, physical ability, social and cultural background (Verloo *et al.*, 2011).

These aspects are of growing importance in the prevention of and adaptation to climate change impacts and natural hazards. A case study of a Gender Impact Assessment of a major debris flow in Austria reveals how women and men (can) participate in decision-making processes in the prevention phase of natural hazards and where the obstacles and opportunities for a gender-sensitive participation process lie.

Climate Change Adaptation (CCA) and Disaster Risk Reduction (DRR): the Austrian context

According to the IPCC Assessment Report 2014 it is predicted that "extreme precipitation events will become more intense and frequent in many regions" (IPCC, 2014: 58). Austria is part of the Alps, which significantly influence the climate of its non-alpine regions.

> The climate in the Alpine region is expected to be considerably affected by 21st century global warming. This refers not only to rising temperatures, but also to changes in the seasonal cycle of precipitation, global radiation and humidity, to changes in temperature and precipitation extremes, and closely related impacts like floods, droughts, snow cover and natural hazards.
>
> (Gobiet *et al.*, 2014: 1149)

Although the complexity of the Alpine region poses considerable challenges to climate models, which translate to uncertainties in climate projections (Gobiet *et al.*, 2014: 1138 and 1139), recent research indicates that climate change can likely lead to temperature and precipitation extremes. These extremes are closely related to the occurrence of natural hazards. Floods, droughts, debris flows, permafrost degradation as well as snow cover are changing in terms of frequency, magnitude, seasonality and spatial patterns (Gobiet *et al.*, 2014: 1149; Blöschl *et al.*, 2011; Frei *et al.*, 2006; Schmidli *et al.*, 2007; Köplin, 2014: 2575, 2576; Rajczak *et al.*, 2013; Stoffel, 2010; APCC, 2014: 568).

Despite the uncertainties regarding the extent and intensity of these changes, which call for further research, anticipated changes – as well as the current situation – pose challenges to the management of natural hazards (Fuchs, 2010: 169) and DRR. This is particularly the case if the societal and demographic changes, as well as the changes in land use and spatial patterns, are to be taken into account.

The gender+ dimension of climate change and natural disasters

Climate change and natural hazards are not gender-neutral. Women and men are affected by climate change in different ways (e.g. Rathgeber, 2009; Weber, 2005; Mehta, 2007). But it is not gender alone: different groups of people – defined by their gender as well as by age, socio-demographic background and spatial circumstances – have different capacities in dealing with climate change (e.g. Balas *et al.*, 2011; Prettenthaler *et al.*, 2008; McCallum *et al.*, 2013; Terry, 2009; Ibarrarán *et al.*, 2009). Social identities affect – in a positive and negative way – people's capacities to deal with climate change and disasters associated with natural hazards at both individual and community levels (e.g. UNISDR, UNDP, IUCN, 2009; Le Masson, 2013; Weber, 2015). Gender+ recognises the above-mentioned aspects, which define and produce differences between individuals and groups (Verloo *et al.*, 2011). It includes knowledge about the complexity of gender as intersected with other structural inequalities (Verloo *et al.*, 2011) and other indicators like income, age, ethnic background, handicaps, life stage and situation etc. – in short, the differences between women and men (Chávez-Rodríguez, 2013) and within women and men in different situations.

In the field of climate change adaptation as well as in DRR, gender and gender-related roles, social norms and inequality between men and women influence the exposure to risk, the perception of risks and the ability to respond to risk and rescue, as well as various psychological and physical consequences and the potential for recovery and reconstruction after disasters (Baćanović, 2015: 7). Often women's disadvantages may increase with the change in or loss of natural resources associated with climate change (UNDP, 2009: 55). "It is inequities in the everyday, and not just in times of disaster, that create greater risk and reduce life chances for women and girls" (Bradshaw and Fordham, 2013: 36). As women and girls tend to have less access to or control over assets, including the resources necessary to cope with hazardous events, such as information, education, health and wealth, their vulnerability is in general relatively greater than men's (Bradshaw and Fordham, 2013: 3).

But, neither can women be attributed with a 'universal vulnerability' (Bauriedl, 2014: 9) irrespective of their age, life phase or situation, nor can they be dubbed 'the key to successful climate change mitigation and adaptation' without questioning existing roles, power relations and responsibilities. Bradshaw (2015: 54), for example, points out that the 'feminisation of responsibility' can reinforce rather than challenge traditional gender relations. "The construction of women affected by disasters as both an at-risk group and as a means to reduce risk suggests similar processes of feminisation [of responsibility]" (Bradshaw, 2015: 54). In general, unequal gendered power relations are rarely discussed or addressed in the climate change debates and research (Bauriedl, 2014: 9). But, as Bradshaw states, "if disasters are to learn from development, the key lesson is not that gender matters, but that how gender is addressed matters" (Bradshaw, 2015: 70).

Gender+ aspects of climate change and natural disaster in research and policies

In general, there remains a lack of gender-sensitive research on disasters and climate change although international policies and frameworks have stressed the importance of a gender perspective, and the production of gendered disasters knowledge has made large advances (Bradshaw, 2015: 61). So far, research on the gender dimensions of climate change and disaster has mainly been conducted on major catastrophes in the Global South (e.g. Bradshaw, 2004; Le Masson, 2013; Chávez-Rodríguez, 2013). Research on that topic has also been conducted in Canada (Enarson, 2008, Gender Mainstreaming in Emergency Management), the USA (David and Enarson, 2012 on hurricane Katrina) and Australia (see Shaw *et al.*, 2012 on the floodings in Queensland and Victoria in 2011). Also in the European context, research has produced more knowledge about the gender dimension of disaster risk management (e.g. Chávez-Rodríguez, 2013 in a comparative case analysis on flooding in Germany; Damyanovic *et al.*, 2014 and Weber, 2015 on a debris flow in Austria 2012, Baćanović, 2015 on the flooding in Serbia 2014).

These gender analyses reveal a more detailed picture of gender-specific vulnerability and capacities. Baćanović (2015), for instance, identified that single mothers, households without male members, as well as elderly households were particularly affected by the floods in Serbia in 2014. In their survey on a flooding in Canada, Enarson and Scanlon (1999) found a "strongly gendered division of labour in all phases of the flood" (similar Weber, 2015); much of women's work was taking place behind the scenes, providing childcare for others and providing meals and other provisions for volunteers, while men's activities were more visible and focused on responding to the hazard itself, e.g. sand-bagging or building up dykes. In their analysis of the floods in Australia in 2011, Shaw *et al.*, (2012) were able to draw a detailed picture of the economic impact of the disaster on women and men. For example, women undertook longer-term unpaid work, often in addition to extra income-generating activities and extra caring duties (Shaw *et al.*, 2012: 28).

Recent initiatives to mainstream gender into DRR have been successful at least on the policy level. The Manila Declaration for Global Action on Gender in Climate Change and Disaster Risk Reduction[1] states that 'women and men must equally participate in climate change and disaster risk reduction decision-making processes at community, national, regional and international levels'. Furthermore the Hyogo Framework[2] (HFA) states that a gender perspective should be integrated into all disaster risk policies, plans and decision-making processes, including those related to risk assessment. Women, if given equal opportunities, can perform multifunctional roles well – as participants, managers, decision makers and leaders in the field of disaster risk reduction. The 'Sendai Call to Action on Gender and Diversity in DRR'[3] emphasises the calls in the HFA and formulates recommendations and actions for states to implement the framework. For example, states should 'ensure the participation of a minimum of 30 per cent women and 30 per cent men from diverse sectors, training and expertise, and of diverse ages and family formations, in all national, prefectural, and municipal policymaking and decision-making bodies dealing with Disaster Risk Reduction' or 'ensure accessibility of information and services through all phases of disaster prevention, response, recovery and reconstruction to all members of society'.

However, "the gendering of the dominant disasters discourse has advanced slowly. Where the rapid advances have been made is not in the integration of women into DRR, but the inclusion of women in post-disaster reconstruction activities" (Bradshaw, 2015: 262). Yet the way to reduce disaster risk is through gender equality, and the only way to promote gender equality is through specific programmes and projects that look to address unequal gendered power relations during, after and – most critically – before a disaster (Bradshaw, 2015). It is before a disaster that fair access to decision-making processes related to DRR is important. For example, decision-making processes related to the spatial development of communities determine not only the risk exposure but are also means of knowledge transfer about the environmental and spatial attributes of an area. In an Austrian context these processes include the development of Hazard Maps, the Zoning Maps and the Local Development Plans. All three address and affect spatial and land-use development. Therefore, the number and proportion of women and men involved in decision-making processes in developing social safety nets and protection programmes can be used as a gender-specific indicator (UNISDR, UNDP, IUCN, 2009: 117).

The following case study of a disaster in a small village of Austria aims at revealing how women and men can participate in spatially relevant decisions and projects in the prevention phase. To understand better the Austrian context, the following section provides an introduction to the Austrian spatial planning instruments and hazard mapping.

Zoning Map, Local Development Plan, Hazard Map: dealing with spatial challenges and natural hazards at the local level in Austria

Because climate change and its impacts strongly affect land use and land-use development (Pütz *et al.*, 2011: 1) planning instruments like Zoning Maps or Local Development Plans, and hazard assessments like Hazard Zoning are of growing importance for the adaptation to climate change and DRR on a local scale. To understand their legal framework, role and impact on the spatial development in Austria we provide a short introduction to each (Damyanovic, 2007; Weber, 2015).

In Austria each federal province (in total nine 'Länder') has its own planning laws and policies for spatial development. At the local level, the municipalities are in charge of local spatial planning and hence responsible for the Zoning Map and Local Development Plans. The Zoning Plans for municipalities are the most important instrument of spatial planning. They determine the possible use of land of municipalities' territory and define zone codes for land that can be built upon, land dedicated to preserving natural resources and areas dedicated to transport infrastructure. The Zoning Map is legally binding for land owners and aims to steer the future development of a municipality. The Local Development Plan is the basis of the Zoning Map. It analyses the current spatial, demographic and socio-economic situation of communities and outlines the targeted development for the next 15 years. The Local Development Plan has to address the topics 'nature and environment', 'settlement area and population', 'economy', and 'technical infrastructure'.

The Hazard Map in Austria is an area-based expert opinion on potential hazards; it serves as a basis for land-use planning as well as for the construction and security sectors (BMLFUW, 2012). The Hazard Map is compiled by the Austrian Service for Torrent and Avalanche Control (WLV) or the Federal Water Engineering Administration (BWV) depending on the type of catchment (mountain channels or lowland rivers). They are made available to all municipalities, provincial and federal authorities in Austria. After detailed examinations in the catchment areas and assessment of earlier events, hazards emanating from torrents and mountain reliefs are identified in the area of relevance to land-use planning. The possible impact of hazards occurrence on individual plots of land is illustrated by means of the distinction between Red and Yellow Hazard Zones, among others. The Red Hazard Zone comprises areas which are so severely endangered by torrents or avalanches, that due to the expected damaging effect of the events considered, their permanent use for settlement and transport is not possible or would only be possible at disproportionately high expenditure. The construction of new buildings is therefore absolutely prohibited in Red Hazard Zones. Already existing buildings in Red Hazard Zones need comprehensive adaptive and protection measures. The Yellow Hazard Zone comprises all other torrent- or avalanche-prone areas whose permanent use for settlement or transport purposes is impaired due to this risk (BMLFUW, 2012). Consequently, the Local Zoning Maps and the Local Development Plan need to refer to the local Hazard Maps.

The case study St. Lorenzen: a debris flow in 2012 in a small village in Austria

St. Lorenzen is part of the municipality of Trieben, in the federal provincial state of Styria and is a small village located on an alluvial fan of the so-called "Lorenzerbach", a mountain channel on the slopes of the Palten Valley (Figure 16.1).

The Palten Valley has a long history of being affected by disasters related to natural hazards. Over the course of the last 300 years the village of St. Lorenzen has been hit several times by floods and debris flows of varying intensities. In the nineteenth and twentieth centuries the community built various mitigation measures, ranging from the reinforcement of the stream's banks, to the construction of a check dam, to name just two (Janu *et al.*, 2012). Despite these measures, after days of intense rain a major debris flow hit the village in the early hours of 21 July 2012. Some 78 buildings (houses, stables, garages) were affected, 75 severely, and 2 buildings were destroyed completely (2012) (Figure 16.2). Shortly after the disaster the construction of new technical mitigation measures started (Figure 16.3).

The event of 2012 in St. Lorenzen has been chosen as a case study for several reasons: first, there exists a precise documentation of the disaster and the emergency response, and second the recently revised Zoning Map and Hazard Map are available. Since the municipalities are in charge of local spatial planning, the municipal council plays an important political and administrative role in the elaboration of Zoning Maps and Local Development Plans. Therefore, the gender-sensitive analysis of the composition of the municipal council gives evidence of the (un)equal opportunities for women and men at the local council level.

An interdisciplinary approach calls for a diversity of methods

Vulnerability to disasters arises due to the combination of physical, social, economic and environmental factors (Cutter *et al.*, 2003; Wisner *et al.*, 2004). With a view to understanding the emergence of social vulnerability, the social context (e.g. ethnicity, socio-economic situation, gender, age) needs to be included in a research design. In addition, spatial and social aspects are also of growing importance in the context of climate change and natural hazards (Damyanovic *et al.*, 2014; Balas *et al.*, 2011). The diversity of methods applied in this case study reflects this concept of vulnerability and the interdisciplinary approach

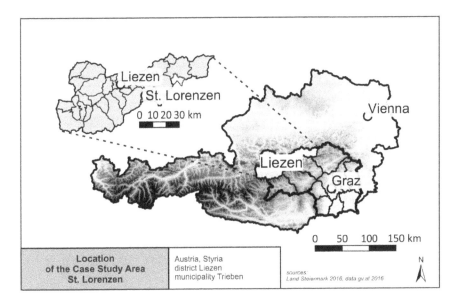

Figure 16.1 The location of the case study area

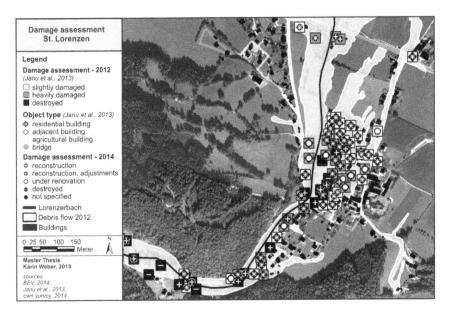

Figure 16.2 Map of damage caused by the debris flow

Figure 16.3 The debris block that was built after the disaster to protect the village from future hazards emanating from the mountain channel

that has become standard in studies on disaster and climate change (e.g. Mercer *et al.*, 2010). Consequently, the research team consisted of landscape planners with gender expertise, experts in mountain risk engineering and media and communication sciences. To identify gender+ aspects of the disaster, qualitative interviews were conducted with experts concerned with the prevention of disasters and risk reduction as well as with members of the emergency response teams and volunteer organisations involved in disaster management. Furthermore, data derived from qualitative interviews with locals – of different gender, age and socio-economic background – affected by the debris flow provided valuable insights into daily life before and after the event, and coping strategies following the event. Document analysis of the Zoning Map, Hazard Maps, Local Development Plan and emergency plans, as well as of the documentation of the emergency response activities, complemented the analysis. In addition, a gender+-sensitive media analysis of local and regional sources was conducted to reveal the perception of gender aspects and the representation of different groups in the management of the disaster. A gender-specific secondary analysis of socio-demographic data of the village revealed the socio-economic status and future social demography for the case study area. Mapping the current land use and building types provided information on the village's spatial structure.

Table 16.1 The 4R-method

Components of 4R-method	Indicators	Aspects
Representation (1 R)	Number of women and men including information on age, socio-economy and social aspects	Participation
Resources (2 R)	Money, time and distribution of space	Factors influencing the access to planning process
Right and Legal Framework (3 R)	Laws and policies	Right to participate
Reality (4 R)	Social norms and values	Socially determined access to the planning process

For the gender+-sensitive analysis of the Zoning Map, Local Development Plan and the Hazard Map and the related planning processes, the 4R-method was applied. The 4R-method uses the number and proportion of women to men involved in decision-making processes as an indicator of how women and men are able to participate in those decision-making processes. In addition, it examines the legal framework, the resources (time, budget) allocated for the process, and the social norms and values underlying the plans and concepts. The investigation of these aspects in the case study aims at revealing how far equal opportunities in the development process of Zoning Map, Local Development Plan and Hazard Map have been implemented.

This 4R-method was originally developed by Gertrud Aström for the Swedish Association of Local Authorities and was translated to spatial and urban planning in the framework of the GenderAlp! project (Damyanovic *et al.*, 2005). The 4R-method, shown in Table 16.1, is structured as follows: the 1st R refers to the representation of men and women in planning projects and processes: Who was involved in the planning process? The 2nd R stands for the equal distribution of resources such as space, time and money: How was the planning process funded? Who invested time or was able to invest time to participate in the process? The legal framework and the assessment of rights are addressed in the 3rd R: How does the legal framework address and/or reflect gender-related differences? The 4th R stands for the realities in terms of social norms, values and existing planning models which make inequalities between men and women visible: What are the underlying planning values? Did the planning process create and/or (re)enforce inequalities? (Pettersson, 2004; Damyanovic *et al.*, 2005; Damyanovic and Roither, 2009; Damyanovic, 2013)

The representation of women and men in the planning processes: 1 R

At the time of the disaster and the elaboration of the Zoning Map and Local Development Plan, the municipal council of Trieben consisted of

16 councillors, 12 men and 4 women, and was headed by a male mayor. In comparison with the demographic structure of the municipality, which was 48.8 per cent men and 51.2 per cent women, it was evident that women were underrepresented in the municipal council. However, with 25 per cent women as members of the council, Trieben was above the average of 23.4 per cent for the federal provincial state (Reinwald *et al.*, 2011: 31). Looking at the administrative component of the planning process at local levels, the participation of women is significantly higher: the council department responsible for spatial planning and building development was headed by a woman and the Zoning Map and Local Development Plan was elaborated by a woman.

By law, the Zoning Map and the Local Development Plan have to be approved by the provincial government. Women are significantly underrepresented at that level of decision-making. For example: the supervisory body for spatial planning in Styria had only one female member out of a total of 19 (Land Steiermark, 2014).

In terms of the involvement of women and men in the process of elaborating the Hazard Map the following can be stated: the Austrian Service for Torrent and Avalanche Control (WLV) is a male domain. In Austria, at the time of the survey, only four women are employed on the academic planning level. In the elaboration of the Hazard Zone Plan and the construction of prevention measures in St. Lorenzen only men were involved, although the on-site construction manager is a woman. No gender-specific approach in the elaboration and implementation of the Hazard Map and the design and construction of mitigation measures could be identified and is, per se, not intended in the legal framework. This can result in overlooking the differences between men and women in terms of their vulnerability and capacity and a male-dominated decision-making process.

The allocation of resources (money and time) to the planning processes: 2 R

An analysis of the resources spent on the elaboration of the Zoning Map and Local Development plan reveals the efforts, options and priorities of the local government. By law the local municipality has to provide the budget for the elaboration of the Zoning Map and Local Development Plan. A participation process accompanying the planning process had not been budgeted for. Apart from the legally obligatory information on the planning process that has to be provided to the citizens, the local council engaged in no other form of participation with the local stakeholders. That means that citizens who do not own property or land in the municipality tend not to get involved in the planning process although the spatial planning law would enable them to do so. Furthermore, because of the lack of an active participation process, citizens with a weak socio-economic background, educational disadvantages or reduced mobility have faced further obstacles to participating in the decision-making process. Women in rural areas such as those in which Trieben is situated are less likely to have educational participation at a higher level and a lower income than women in urban areas, or men in Austria (Bundesministerin für Frauen und Öffentlichen Dienst im Bundeskanzleramt Österreich, 2010: 293).

Besides monetary resources, time resources also play an important role in achieving equal opportunities in the planning processes. Again, these resources differ between men and women and are determined by social norms and values (see section further below on 4 R).

The legal framework of the planning processes related to the Hazard Map: 3 R

The planning process for the Hazard Map in St. Lorenzen involved the WLV, a planner in charge, a committee that assessed the Hazard Map, the mayor and citizens of the municipality who have a 'legitimate interest' e.g. people who own land located in hazard zones. Similar to the Zoning Maps, the Hazard Map indirectly excludes people without property. The main objective of mitigation measures in the context of disasters is the protection of life, health and livelihood (Rudolf-Miklau, 2009), that include, for example, residential buildings, streets and infrastructure, no matter whether in private, public or corporate ownership

Participation is not stipulated in the legal framework of preparing and implementing the hazard map. Therefore, none had to be realised in Trieben. Also, in the 4 week statuary period of public consultation, nobody objected to the Hazard Map. That fact can be interpreted variously as a general agreement with the plan, as inhabitants' lack of interest, or as a lack of information provided to the local community. It can also be seen as an evidence of the unawareness of the local population about the Hazard Map and its implication on the use of a plot, the design of houses and the use of construction material. Interviews with the inhabitants of St. Lorenzen revealed that not everybody was aware of the whole degree of the risks emanating from the Lorenzerbach. Most of the persons interviewed knew of the Hazard Map, but a disaster of the magnitude of 2012 was not at all expected, although the Hazard Map proved to be accurate (Damyanovic *et al.*, 2014; Weber, 2015).

Social norms and values underlying the planning process and the Zoning Map: 4 R

The amount of time an individual can spend on participating in decision-making processes depends not only on socio-economic aspects but also on social norms and values. For example, the low proportion of female members of the local council can be interpreted as evidence for constraints on the time resources available to women. Surveys of women's involvement in local politics show that lack of time is the main reason for their limited involvement in political activities (Reinwald *et al.*, 2011: 51; Oedl-Wieser, 2006: 142). These results support the so-called 'hypothesis of availability' (*Abkömmlichkeitsthese*) which states that due to the existing gender-specific division of labour and the resultant responsibility for the house and care work, women have less time than men for political activities. One can also describe the many tasks and responsibilities (reproductive, productive and community) women have to take over as a 'triple burden' and the resulting lack of time that keeps them out of decision-making processes. Moreover, women tend to trust less in their abilities (e.g. speak in public, and defend their opinions) than men (Reinwald *et al.*, 2011: 51; Oedl-Wieser, 2006: 142, 131).

The lack of time available to many women for getting involved in decision-making processes at the local level might also be related to the relatively high ratio of out-commuters in Trieben: 47 per cent of the inhabitants have an employment commute to a workplace outside of the municipality (ÖEK, 2013: 102). The time spent on the daily commute or the time spent on the care for children and care-dependent relatives because one partner is absent most of the week is missing for the potential engagement in a planning and decision-making process. Furthermore, the out-commuter rate gives an idea of who would be present or absent in the community when a disaster happens.

A relatively high out-commuter rate is one aspect of the socio-demographic structure of the community of Trieben. The community is also characterised by a relatively high proportion of people older than 65 years (26.6 per cent). Especially, the rate of 24 per cent of women older than 65 years in the village of St. Lorenzen is slightly higher than the provincial average of 22.9 per cent. The Zoning Map and Local Development Plan react to the demographic trends in adequate and differentiated measures in terms of spatial development but they do not acknowledge the capacities of the elderly. It rather associates the aging society with negative issues and stresses its vulnerability. Yet, the older generation has local knowledge about and experience with the occurrence of natural hazards. In interviews post the debris flow of 2012, they reported historical events in St. Lorenzen and its surroundings, and mentioned the fact that the residential area is built on an alluvial cone. Hazard events which date further back were mentioned by older interviewees more often: "My father – 94 years old – once told me that this giant rock with a spruce/tree on top of it had been washed away by a giant flood. If I didn't know better, I couldn't believe it" (interview, 2014 – man, 60 year old). The knowledge about historical events could have been made visible and useful in a participation process accompanying the elaboration of the Zoning Map and Local Development Plan and to support the disaster preparedness (Weber, 2015: 132).

Spatial planning instruments and their potential for a gender-sensitive participatory disaster risk reduction

As a technical instrument, the Hazard Map is not legally binding. Only through the integration into spatial planning instruments (the Zoning Maps and Local Development Plans) does it become legally binding and have an impact on the federal subsidies a community can receive. Unlike the Hazard Zone Map, the Zoning Maps and Local Development Plans take socio-economic circumstances and prognoses into account. None of the three instruments, the Hazard Maps, the Zone Maps and Local Development Plans in Austria, apply a gender perspective as a matter of course, neither are they gender-sensitive, as the case study of St. Lorenzen shows.

In terms of potentials and links for participation processes, the planning process for Zoning Maps and Local Development Plans holds more potential than the Hazard Map. Within the latter, no actual citizen participation is intended, because it is understood as a governmental basis of decision-making (Rudolf-Miklau, 2014: 62). But since the Zoning Maps need to refer to the Hazard Maps, DRR could be a topic addressed through a participation process accompanying the elaboration of a Zoning Map and the Local Development Plan for a community. These spatial planning instruments could be used as a way to inform, to raise awareness about disaster risks and to enhance disaster preparedness and therefore to contribute overall to DRR. The discussion about the technical attributes of natural hazards – e.g. location, intensity and dimension – must stay within the experts' community because a significant part of the responsibility for the effect of the Hazard Zone Plan and the accepted risks have to stay with the experts and authorities (2014: 62). Besides, the individual perception of risk which in its nature lacks objectivity, cannot be the basis for such a discussion (2014: 62). However, the public needs to be informed and enabled to add their local knowledge and experience to the expert's opinion.

However, the overall perception of the inhabitants of St. Lorenzen is that the main responsibility for their safety related to disaster lies with the authorities. Personal risk is often not interpreted as personal action but as the failure of federal authorities (Hübl *et al.*, 2009: 68). The perception of risk is also gender-specific. Women and men experience, perceive and identify risks differently. Although the interviewees claimed to feel safe after the construction of preventive measures, especially women expressed their feelings of uncertainties. One woman (84 years old) said that, although "one believes to be capable of solving everything by technical means, again one must realise that this is not the case". Everyone can be equally exposed to a hazard, but women and men have different levels of vulnerability and access to resources, and have therefore developed different coping skills (UNISDR, UNDP, IUCN, 2009: 35). Discussing these differences within a spatial development process that takes into account the daily needs and requirements of different groups could help link individual awareness about the hazard with the requirements of the community and other relevant stakeholders.

Conclusion and outlook

A gender-sensitive participation process integrated into a local spatial development instrument that addresses the occurrence of natural hazards – if relevant to the community – can draw on the different experiences, skills and possible roles of men and women, young and old in DRR combining social and spatial aspects. Such a participation process concerned with gender+ specific aspects of a disaster would need to address topics like: livelihood/ work/activities, socio-demography, mental and physical health, perception of safety and risk, knowledge-building and transfer, communication and information, environment, housing, infrastructure (Mehta, 2007; Damyanovic *et al.*, 2014: 53; Deare, 2004). In general, a gender-sensitive participation process could also be used as knowledge transfer between the generations as well as between women and men. Moreover, it could compensate for the non-stipulated participation in the elaboration of Hazard Maps and foster self-provision and individual and collective responsibility. Such a process would need to address the 4 Rs – Representation, Resources, Right and Reality. The 4th R is particularly crucial because it deals with the issue of gendered power relations and responsibilities at its roots (Damyanovic, 2013). Social norms very much influence the legal framework, policies and consequently the distribution of resources which influences the vulnerability and capacity of men and women and the distribution of and access to resources. They value or hamper participation in the decision-making processes related to the management of natural hazards and climate change adaptation measures.

However, further discussion and gender-sensitive research, particularly on how social norms cause inequality in DRR, needs to be conducted. Also, analysis and discussion on the type, form and intensity of the participation process, as well on the identification of existing participation processes, programmes and instruments that could be used to include crucial gender+ aspects in the management of natural disasters at a local scale, are necessary.

Notes

1 The Manila Declaration resulted from the Third Global Congress of Women in Politics and Governance, on Gender in Climate Change Adaptation and Disaster Risk Reduction, in Manila, Philippines, 19–22 October, 2008
2 The Hyogo Framework for Action (HFA) 2005–2015. Building the Resilience of Nations and Communities to Disaster. The HFA is a 10-year plan to make the world safer from natural hazards. It was endorsed by the UN General Assembly in the Resolution A/RES/60/195 following the 2005 World Disaster Reduction Conference.
3 The 'Sendai Call to Action on Gender and Diversity in DRR. Towards the Post-Hyogo Framework for Action (HFA2). 2015. Available online at http://jwndrr. sakura.ne.jp/en/wp-content/uploads/2014/11/en_Sendai-Call-to-Action.pdf (accessed 14 June, 2014).

Acknowledgements

'GIAClim' has been funded through the Startclim13 research programme by The Austrian Federal Ministry of Agriculture, Forestry, Environment and Water Management (www.austroclim.at/index.php?id=startclim2013).

This chapter would not have been possible without the contributions to the project by all research team members: Prof. Christiane Brandenburg, Brigitte Allex – both from the Institute of Landscape Development, Recreation and Conservation Planning, BOKU Vienna; Eva-Maria Pircher, a media and communication researcher, and Prof. Johannes Hübl, head of the Institute of Mountain Risk Engineering. We would also like to thank the people from St. Lorenzen and Trieben for sharing their experience and for the experts concerned with the management of natural disasters for the insight into their work they gave us during the interviews and many informal conversations.

References

APCC – Austrian Panel on Climate Change. (2014) 'Österreichischer Sachstandsbericht Klimawandel 2014 '(Austrian assessment report 2014 (AAR14)), Austrian Academy of Sciences Press, Vienna.

Baćanović, V. (2015) 'Gender Analysis of the Impact of the 2014 Floods in Serbia', Organization for Security and Co-operation in Europe – OSCE (publisher).

Balas, M., Stickler, T., Lexer, W. and Felderer, A. (2011) 'Ausarbeitung sozialer Aspekte des Klimawandels und von Handlungsempfehlungen für die Raumordnung als Beitrag zum Policy Paper – Auf dem Weg zu einer nationalen Anpassungsstrategie' (Social Aspects of Climate Change and Guidelines for Spatial Planning), Environmental Agency Austria, Vienna.

Bauriedl, S. (2014) 'Geschlechter im Klimawandel: Soziale Differenzierung als Kompetenz sozialwissenschaftlicher Klimaanpassungsforschung' (Gender in Climate Change: Social Differentiation as a Competence for Social Science Research on Climate Change Adaptation), *GAIA* 23(1): 8–10.

Blöschl, G., Viglione, A., Merz, R., Parajka, J., Salinas, J.L. and Schöner, W. (2011) 'Auswirkungen des Klimawandels auf Hochwasser und Niederwasser' (Climate change impacts on floods and low water). *Österreichische Wasser- und Abfallwirtschaft* 63(1–2): 21–30.

BMLFUW. (2012) 'Land-use Planning', information provided by the BMLFUW Öffentlichkeitsarbeit 13 November 2012. Available online at www.bmlfuw.gv.at/en/fields/forestry/Austriasforests/Landplanning.htm (accessed 13 June 2013).

Bradshaw, S. (2004) '*Socio-economic impacts of natural disasters: A gender analysis*', United Nations, Santiago de Chile.

Bradshaw, S. (2015) 'Engendering development and disasters', *Disasters* 39: 54–75.

Bradshaw, S. and Fordham, M. (2013) 'Women, girls and disasters', Report produced for the Department for International Development, Britain.

Bundesministerin für Frauen und Öffentlichen Dienst im Bundeskanzleramt. (2010) 'Frauenbericht 2010 Bericht betreffend die Situation von Frauen in Österreich im Zeitraum von 1998 bis 2008' (Report on the Situation of women in Austria between 1998 and 2008), Bundesministerin für Frauen und Öffentlichen Dienst im Bundeskanzleramt, Vienna.

Chávez-Rodríguez, L. (2013) 'Klimawandel und Gender: Untersuchung der Bedeutung von Geschlecht für die soziale Vulnerabilität in überflutungsgefährdeten Gebieten' (Climate Change and Gender: Analysis of the importance of gender for social vulnerability in flood-prone areas) (Dissertation). University of Bremen, Bremen.

Cutter, S.L., Boruff, B.J. and Shirley, W.L. (2003) 'Social vulnerability to environmental hazards', *Social Science Quarterly*, 84: 242–261.

Damyanovic, D. (2007) 'Landschaftsplanung als Qualitätssicherung zur Umsetzung der Strategie des Gender-Mainstreamings. Bestandteil des örtlichen Entwicklungskonzepts dargestellt an der Fallstudie Tröpolach – Stadtgemeinde Hermagor-Pressegger See (Kärnten)' (Landscape planning as quality assurance for the implementation of the strategy of gender mainstreaming. Theoretical and methodological concepts for a gender-sensitive planning process, as a component of the regional development concept, illustrated by means of the case study of Tröpolach – urban municipality of Hermagor-Pressegger See (Carinthia)), Guthmann-Peterson, Vienna.

Damyanovic, D. (2013) 'Gender mainstreaming as a strategy for sustainable urban planning', in I. Sánchez de Madariaga and M. Roberts, M. (eds), *Fair Shared Cities. The Impact of Gender Planning in Europe*. Farnham: Ashgate, 177–192.

Damyanovic, D. and Roither, A. (2009) 'The method of structuralist planning assessment', in L. Licka and E. Schwab (eds), *Landscape – Great Idea! X-LArch III, Conference Proceedings*. Vienna: Institute of Landscape Architecture BOKU, pp. 36–39.

Damyanovic, D., Reinwald, F. and Schneider, G. (2005) 'Handbuch zur Analyse von Gender Mainstreaming Beispiele für eine Good practise Datenbank. Gender Alp!. Räumliche Entwicklung für Frauen und Männer' (Manual for the input of Gender Mainstreaming examples into the good practice database. Gender Alp!. Spatial development for woman and men) Interreg III B – Alpinspace, Europäische Union.

Damyanovic, D., Fuchs, B., Reinwald, F., Pircher, E., Allex, B., Eisl, J. Brandenburg, C. and Hübl, J. (2014) 'GIAKlim – Gender Impact Assessment im Kontext der Klimawandelanpassung und Naturgefahren. Endbericht von StartClim2013.F in StartClim2013: Anpassung an den Klimawandel in Österreich – Themenfeld Wasser, Auftraggeber: BMLFUW, BMWF, ÖBF, Land Oberösterreich' (GIAClim – Gender Impact Assessment in the context of climate change and natural hazards) Vienna: Institute of Landscape Planning, BOKU Vienna.

David, E. and Enarson, E.P. (2012) *The Women of Katrina: How Gender, Race, and Class Matter in an American Disaster*. Nashville: Vanderbilt University Press.

Deare, F. (2004) 'A methodological approach to gender analysis in natural disaster assessment: a guide for the Caribbean', Naciones Unidas, CEPAL: Sustainable Development and Human Settlements Division, Women and Development Unit, Santiago, Chile.

Enarson, E. (2008) 'Gender Mainstreaming In Emergency Management: Opportunities for Building Community Resilience in Canada', Prepared for the Public Healths Agency of Canada, Centre of Emergency.

Enarson, E. and Scanlon, J. (1999) 'Gender patterns in flood evacuation: A case study in Canada's Red River Valley', *Applied Behavioral Science Review* 7: 103–124.

Frei, C., Schöll, R., Fukutome, S., Schmidli, J. and Vidale, P.L. (2006) 'Future change of precipitation extremes in Europe: Intercomparison of scenarios from regional climate models', *Journal of Geophysical Research* 111: 1–22.

Fuchs, S. (2010) 'Auswirkungen des Klimawandels auf Naturgefahren – Herausforderungen für eine nachhaltige Landnutzung in alpinen Gebieten' (Climate change impacts on natural hazards – Challenges for sustainable landuse in alpine areas). In ÖWAV (ed.), *Auswirkungen des Klimawandels auf Hydrologie und Wasserwirtschaft in Österreich*. Wien: ÖWAV, pp. 169–180.

Gobiet, A., Kotlarski, S., Beniston, M., Heinrich, G., Rajczak, J. and Stoffel, M. (2014) '21st century climate change in the European Alps – A review', *Science of the Total Environment* 493: 1138–1151.

Hübl, J., Keiler, M. and Fuchs, S. (2009) 'Risikomanagement für alpine Naturgefahren' Wildbach- und Lawinenverbau' (Risk management of alpine natural hazards), *Wildbach- und Lawinenverbau* 163: 60–74.

Ibarrarán, M.E., Ruth, M., Ahmad, S. and London, M. (2009) 'Climate change and natural disasters: macroeconomic performance and distributional impacts', *Environment, Development and Sustainability* 11: 549–569.

IPCC. (2014) 'Climate Change 2014: Synthesis Report. Contribution of Working Groups I, II and III to the Fifth Assessment Report of the Intergovernmental Panel on Climate Change' (Core Writing Team, R.K. Pachauri and L.A. Meyer (eds)). Geneva, Switzerland: IPCC.

Janu, S., Mehlhorn, S. and Moser, M. (2012) 'Ereignisdokumentation und Analyse vom 21. Juli 2012 in St. Lorenzen' (Documentation and Analysis of the Prozess in St. Lorenzen on 21 July 2012) Vienna:Institute of Mountain Risk Engineering, Department for Civil Engineering and Natural Hazards, University of Natural Resources and Life Sciences, Vienna.

Köplin, N., Schädler, B., Viviroli, D. and Weingartner, R. (2014) 'Seasonality and magnitude of floods in Switzerland under future climate change: Seasonality and magnitude of floods under climate change', *Hydrological Processes* 28: 2567–2578.

LAND STEIERMARK (2014) Available online at www.verwaltung.steiermark.at/cms/dokumente/11679832_74834953/8a642226/MitgliederErsatzmitglieder%20zur%20Informaton.pdf (accessed 16 January 2014).

Le Masson, V. (2013) *Exploring Disaster Risk Reduction and Climate Change Adaption from a Gender Perspective: Insights from Ladakh, India*. West London: Brunel University Press.

McCallum, S., Dworak, T., Prutsch, A., Kent, N., Mysiak, J., Bosello, F., Klostermann, J., Dlugolecki, A., Williams, E., König, M., Leitner, M., Miller, K., Harley, M., Smithers, R., Berglund, M., Glas, N., Romanovska, L., van de Sandt, K., Bachschmidt, R., Völler, S. and Horrocks, L. (2013) 'Support to the development of the EU Strategy for Adaptation to Climate Change: Background report to the Impact Assessment, Part I –Problem definition, policy context and assessment of policy options'. Vienna: Environment Agency Austria.

Mehta, M. (2007) *Gender Matters: Lessons for Disaster Risk Reduction in South Asia.* Kathmandu: International Centre for Integrated Mountain Development.

Mercer, J., Kelman, I., Taranis, L. and Suchet-Pearson, S. (2010) 'Framework for integrating indigenous and scientific knowledge for disaster risk reduction', *Disasters* 34: 214–239.

Oedl-Wieser, T. (2006) *Frauen und Politik am Land (Women and Politics in Rural Areas)*. Vienna: Bundesanstalt für Bergbauernfragen.

Örtliches Entwicklungskonzept (ÖEK). Local Development Plan of the municipality of Trieben) der Stadtgemeinde Trieben 4.00. (2013) (Local Development Plan of the municipality of Trieben): drafted by DI Martina Kaml. GZ.: 03 / 1108 / RO / 01.2 – ÖEK | 27 March 2013| amended 25 September 2013.

Pettersson, G. (2004) 'Fester Boden – weite Sprünge' (Solid Ground – big Jumps), in K. Lang (ed.), *Die kleine große Revolution (The small big Revolution)*. Hamburg: VSA-Verlag, 44–65.

Prettenthaler, F., Habsburg-Lothringen, C. and Sterner, C. (2008) 'Soziale Aspekte von Climate Change Impacts in Österreich. Erste Beiträge zur Inzidenz der Lasten des Klimawandels' (Social Aspects of Climate Change Impacts in Austria), Global 2000, Vienna.

Pütz, M., Kruse, S. and Butterling, M. (2011) 'Assessing the Climate Change Fitness of Spatial Planning: A Guidance for Planners'. ETC Alpine Space Project CLISP'. Available online at www.clisp.eu/content/sites/default/files/GuidanceForPlanners_E_20110817_0.pdf (accessed 13 June 2014).

Rajczak, J., Pall, P. and Schär, C. (2013) 'Projections of extreme precipitation events in regional climate simulations for Europe and the Alpine Region: Projections of extreme precipitation', *Journal of Geophysical Research: Atmospheres* 118: 3610–3626.

Rathgeber, T. (2009) *'Klimawandel verletzt Menschenrechte über die Voraussetzungen einer gerechten Klimapolitik' (Climate Change harms human rights, when presupposing equitable climate politics).* Berlin: Heinrich-Böll-Stiftung.

Reinwald, F., Damyanovic, D. and Weber, K. (2011) 'Frauen in der burgenländischen Gemeindepolitik', Im Auftrag des Amts der Burgenländischen Landesregierung, LAD-Frauenbüro. (Women in Local Politics), Vienna: Institute of Landscape Planning, University of Natural Resources and Life Sciences.

Rudolf-Miklau, F. (2009) *'Naturgefahren-Management in Österreich: Vorsorge, Bewältigung, Information' (Natural hazards management in Austria: Preparedness, Recovery, Information)* Orac kompakt. Wien: LexisNexis.

Rudolf-Miklau, F. (2014) 'Entscheidungsmechanismen im Hochwasserrisikomanagement' (Decisionmaking processes in flood risk management), *zoll+ Österreichische Schriftenreihe für Landschaft und Freiraum* 24: 59–64.

Schmidli, J., Goodess, C.M., Frei, C., Haylock, M.R., Hundecha, Y., Ribalaygua, J. and Schmith, T. (2007) 'Statistical and dynamical downscaling of precipitation: An evaluation and comparison of scenarios for the European Alps', *Journal of Geophysical Research* 112.

Shaw, C., Van Unen, J. and Lang, V. (2012) 'Women's Voices From The Flood Plains. An Economic Gender Lens on Responses in Disaster affected Areas in Queensland and Victoria', Economic Justice Program of JERA International. Available online at www.security4women.org.au/wp-content/uploads/eS4W_Womens-Voices-from-the-Flood-Plains_Report.pdf (accessed 13 January 2017).

Stoffel, M. (2010) 'Magnitude–frequency relationships of debris flows – A case study based on field surveys and tree-ring records', *Geomorphology* 116: 67–76.

Terry, G. (2009) *Climate Change and Gender Justice: Working in Gender & Development.* Oxfam, Warwickshire; Oxford, UK: Practical Action Publications.

UNDP – United Nations Development Programme. (2009) *Resource Guide on Gender and Climate Change* (1st edn). New York: United Nations Development Programme.

UNISDR, UNDP, IUCN. (2009) 'Making disaster risk reduction gender-sensitive', *Policy and Practical Guidelines.* Geneva, Switzerland: United Nations.

Verloo, M. (2006) 'Multiple inequalities, intersectionality and the European Union', *European Journal of Women's Studies* 13: 211–228.

Verloo, M., Acar, F., Baer, S., Bustelo Ruesta, M., Jalušić, V., Mergaert, L., Pantelidou Maloutas, M., Rönnblom, M., Sauer, B., Vriend, T., Walby, S. and Zentai, V. (2011) 'Final Quing (Quality in Gender+ Equality Policies) Report', Vienna.

Weber, K. (2015) 'Landschaftsplanerische Betrachtung des Umgangs mit Naturgefahren aus genderspezifischer Perspektive – Am Fallbeispiel des Murenabgangs 2012 in St. Lorenzen im Paltental' (A Gender-sensitive Analysis of Natural Disasters – The Case of a debris flow in St. Lorenzen in Austria] Master thesis at University Natural Resources and Life Sciences, Vienna.

Weber, M. (2005) 'Gender, Klimawandel und Klimapolitik, Über Fallstricke bei einer integrativen Betrachtung Diskussionspapier 01/05 des Projektes „Global Governance und Klimawandel' (Gender, Climate Change and Climate Policy). Available online at www.sozial-oekologische-forschung.org/intern/upload/literatur/Weber05_Gender-Klimapolitik.pdf (accessed 14 September 2010).

Wisner, B., Blaikie, P., Cannon, T. and Davis, I. (2004) *At Risk: Natural Hazards, People's Vulnerability, and Disasters* (2nd edn). London; New York: Routledge.

Index